D1188616

Streamlining Digital
Signal Processing

IEEE Press
445 Hoes Lane
Piscataway, NJ 08854

IEEE Press Editorial Board
Lajos Hanzo, *Editor in Chief*

R. Abhari	M. El-Hawary	O. P. Malik
J. Anderson	B-M. Haemmerli	S. Nahavandi
G. W. Arnold	M. Lanzerotti	T. Samad
F. Canavero	D. Jacobson	G. Zobrist

Kenneth Moore, *Director of IEEE Book and Information Services (BIS)*

Streamlining Digital Signal Processing
A Tricks of the Trade Guidebook

Second Edition

Edited by
Richard G. Lyons

IEEE PRESS

A John Wiley & Sons, Inc., Publication

Copyright © 2012 by the Institute of Electrical and Electronics Engineers.

Published by John Wiley & Sons, Inc., Hoboken, New Jersey. All rights reserved.
Published simultaneously in Canada.

No part of this publication may be reproduced, stored in a retrieval system, or transmitted in any form
or by any means, electronic, mechanical, photocopying, recording, scanning, or otherwise, except as
permitted under Section 107 or 108 of the 1976 United States Copyright Act, without either the prior
written permission of the Publisher, or authorization through payment of the appropriate per-copy fee
to the Copyright Clearance Center, Inc., 222 Rosewood Drive, Danvers, MA 01923, (978) 750-8400,
fax (978) 750-4470, or on the web at www.copyright.com. Requests to the Publisher for permission
should be addressed to the Permissions Department, John Wiley & Sons, Inc., 111 River Street,
Hoboken, NJ 07030, (201) 748-6011, fax (201) 748-6008, or online at http://www.wiley.com/go/
permissions.

Limit of Liability/Disclaimer of Warranty: While the publisher and author have used their best efforts
in preparing this book, they make no representations or warranties with respect to the accuracy or
completeness of the contents of this book and specifically disclaim any implied warranties of
merchantability or fitness for a particular purpose. No warranty may be created or extended by sales
representatives or written sales materials. The advice and strategies contained herein may not be
suitable for your situation. You should consult with a professional where appropriate. Neither the
publisher nor author shall be liable for any loss of profit or any other commercial damages, including
but not limited to special, incidental, consequential, or other damages.

For general information on our other products and services or for technical support, please contact our
Customer Care Department within the United States at (800) 762-2974, outside the United States at
(317) 572-3993 or fax (317) 572-4002.

Wiley also publishes its books in a variety of electronic formats. Some content that appears in print
may not be available in electronic formats. For more information about Wiley products, visit our web
site at www.wiley.com.

Library of Congress Cataloging-in-Publication Data:
Streamlining digital signal processing : a tricks of the trade guidebook / edited by Richard G. Lyons.
– 2nd ed.
     p. cm.
  ISBN 978-1-118-27838-3 (pbk.)
  1. Signal processing–Digital techniques. I. Lyons, Richard G., 1948–
TK5102.9.S793 2012
621.382'2–dc23
                                2011047633

Printed in the United States of America.

10 9 8 7 6 5 4 3 2 1

This book is dedicated to all the signal processing engineers who struggle to learn their craft and willingly share that knowledge with their engineering brethren—people of whom the English poet Chaucer would say, "Gladly would he learn and gladly teach."

Contents

Preface

$\mathbf{A}$n updated and expanded version of its first edition, this book presents recent advances in digital signal processing (DSP) to simplify, or increase the computational speed of, common signal processing operations. The topics here describe clever DSP *tricks of the trade* not covered in conventional DSP textbooks. This material is practical, real-world DSP tips and tricks as opposed to the traditional highly specialized, math-intensive research subjects directed at industry researchers and university professors. Here we go beyond the standard *DSP fundamentals* textbook and present new, but tried-n-true, clever implementations of digital filter design, spectrum analysis, signal generation, high-speed function approximation, and various other DSP functions.

Our goal in this book is to create a resource that is relevant to the needs of the working DSP engineer by helping bridge the theory-to-practice gap between introductory DSP textbooks and the esoteric, difficult-to-understand academic journals. We hope the material in this book makes the practicing DSP engineer say, "Wow that's pretty neat—I have to remember this, maybe I can use it sometime." While this book will be useful to experienced DSP engineers, due to its gentle tutorial style it will also be of considerable value to the DSP beginner.

The mathematics used here is simple algebra and the arithmetic of complex numbers, making this material accessible to a wide engineering and scientific audience. In addition, each chapter contains a reference list for those readers wishing to learn more about a given DSP topic. The chapter topics in this book are written in a stand-alone manner, so the subject matter can be read in any desired order.

The contributors to this book make up a *dream team* of experienced DSP engineer-authors. They are not only knowledgeable in signal processing theory, they are "make it work" engineers who build working DSP systems. (They actually know which end of the soldering iron is hot.) Unlike many authors whose writing seems to say, "I understand this topic and I defy you to understand it," our contributors go all-out to convey as much DSP understanding as possible. As such the chapters of this book are postcards from our skilled contributors on their endless quest for signal processing's holy grail: accurate processing results at the price of a bare minimum of computations.

Software simulation code, in the C-language, MATLAB®, and MathCAD® is available for some of the material in this book. The Internet website URL address for such software is http://booksupport.wiley.com.

We welcome you to this DSP tricks of the trade guidebook. I, the IEEE Press, and John Wiley & Sons hope you find it valuable.

RICHARD G. LYONS
E-mail: R.Lyons@ieee.org

"If you really wish to learn then you must mount the machine and become acquainted with its tricks by actual trial."

—Wilbur Wright, co-inventor of the first successful airplane,
1867–1912

Contributors

Mark Allie
University of Wisconsin–Madison
Madison, Wisconsin

François Auger
IREENA, University of Nantes
France

Andor Bariska
Institute of Data Analysis and
Process Design
Zurich University of Applied Sciences
Winterthur, Switzerland

Douglas W. Barker
ITT Corporation
White Plains, New York

Amy Bell
Institute for Defense Analyses
Alexandria, Virginia

Greg Berchin
Consultant in Signal Processing
Colchester, Vermont

Guoan Bi
Nanyang Technological University
Singapore

Mark Borgerding
3dB Labs, Inc.
Cincinnati, Ohio

Joan Carletta
University of Akron
Akron, Ohio

Lionel Cordesses
Technocentre, Renault
Guyancourt, France

Matthew Donadio
Night Kitchen Interactive
Philadelphia, Pennsylvania

Krzysztof Duda
Department of Measurement and
Instrumentation
AGH University of Science and
Technology
Krakow, Poland

Shlomo Engelberg
Jerusalem College of Technology
Jerusalem, Israel

Huei-Wen Ferng
National Taiwan University of Science
and Technology
Taipei, Taiwan, R. O. C.

Bruno Feuvrie
IREENA, University of Nantes
France

Woon-Seng Gan
Nanyang Technological University
Singapore

Maurice Givens
Gas Technology Institute
Des Plaines, Illinois

Hongwei Guo
Shanghai University
Shanghai, R. O. C.

Fred Harris
San Diego State University
San Diego, California

Laszlo Hars
Seagate Research
Pittsburgh, Pennsylvania

Robert Inkol
Defence Research and Development
Ottawa, Canada

Eric Jacobsen
Anchor Hill Communications
Scottsdale, Arizona

Jean Jiang
College of Engineering and Technology
Purdue University North Central
Westville, Indiana

Alain Joyal
Defence Research and Development
Ottawa, Canada

Peter Kootsookos
Emuse Technologies Ltd.
Dublin, Ireland

Kishore Kotteri
Microsoft Corp.
Redmond, Washington

Feng Li
IREENA, University of Nantes
France

Ricardo A. Losada
The MathWorks, Inc.
Natick, Massachusetts

Zhen Luo
Guangdong University of Technology
Guangzhou, R. O. C.

Richard Lyons
Besser Associates
Mountain View, California

James McNames
Portland State University
Portland, Oregon

Sanjit K. Mitra
University of Southern California
Santa Barbara, California

Vincent Pellissier
The MathWorks, Inc.
Natick, Massachusetts

Charles Rader
Retired, formerly with MIT Lincoln
Laboratory
Lexington, Massachusetts

Sreeraman Rajan
Defence Research and Development
Ottawa, Canada

Josep Sala
Technical University of Catalonia
Barcelona, Spain

Abhishek Seth
Synopsys
Karnataka, India

Zhi Shen
Huazhong University of Science and
Technology
Hubei, R.O.C.

David Shiung
MediaTek Inc.
Hsin-chu, Taiwan, R.O.C.

Rainer Storn
Rohde & Schwarz GmbH & Co. KG
Munich, Germany

Li Tan
College of Engineering and Technology
Purdue University North Central
Westville, Indiana

Clay S. Turner
Pace-O-Matic, Inc.
Atlanta, Georgia

Krishnaraj Varma
Hughes Network Systems
Germantown, Maryland

Vladimir Vassilevsky
Abvolt Ltd.
Perry, Oklahoma

Sichun Wang
Defence Research and Development
Ottawa, Canada

Randy Yates
Digital Signal Labs Inc.
Fuquay-Varina, North Carolina

Jyri Ylöstalo
Nokia Siemens Networks
Helsinki, Finland

Part One

Efficient Digital Filters

Chapter 1

Lost Knowledge Refound: Sharpened FIR Filters

Matthew Donadio
Night Kitchen Interactive

What would you do in the following situation? Let's say you are diagnosing a DSP system problem in the field. You have your trusty laptop with your development system and an emulator. You figure out that there was a problem with the system specifications and a symmetric FIR filter in the software won't do the job; it needs reduced passband ripple or, maybe, more stopband attenuation. You then realize you don't have any filter design software on the laptop, and the customer is getting angry. The answer is easy: you can take the existing filter and *sharpen* it. Simply stated, filter sharpening is a technique for creating a new filter from an old one [1]–[3]. While the technique is almost 30 years old, it is not generally known by DSP engineers nor is it mentioned in most DSP textbooks.

1.1 IMPROVING A DIGITAL FILTER

Before we look at filter sharpening, let's consider the first solution that comes to mind, filtering the data twice with the existing filter. If the original filter's transfer function is $H(z)$, then the new transfer function (of the $H(z)$ filter cascaded with itself) is $H(z)^2$. For example, let's assume the original lowpass N-tap FIR filter, designed using the Parks-McClellan algorithm [4], has the following characteristics:

Number of coefficients: $N = 17$

Sample rate: $F_s = 1$

Passband width: $f_{pass} = 0.2$

Passband deviation: $\delta_{pass} = 0.05$ (0.42 dB peak ripple)

Streamlining Digital Signal Processing: A Tricks of the Trade Guidebook, Second Edition. Edited by Richard G. Lyons.
© 2012 the Institute of Electrical and Electronics Engineers. Published 2012 by John Wiley & Sons, Inc.

3

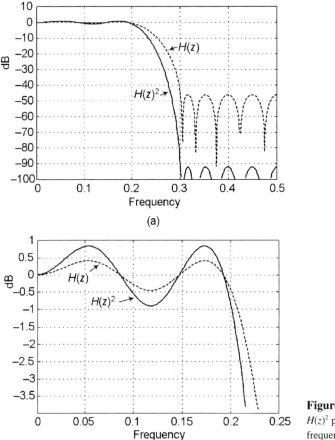

Figure 1–1 $H(z)$ and $H(z)^2$ performance: (a) full frequency response; (b) passband response.

Stopband frequency: $f_{stop} = 0.3$

Stopband deviation: $\delta_{stop} = 0.005$ (−46 dB attenuation)

Figure 1–1(a) shows the performance of the $H(z)$ and cascaded $H(z)^2$ filters. Everything looks okay. The new filter has the same band edges, and the stopband attenuation is increased. But what about the passband? Let's zoom in and take a look at Figure 1–1(b). The squared filter, $H(z)^2$, has larger deviations in the passband than the original filter. In general, the squaring process will:

1. Approximately double the error (response ripple) in the passband.
2. Square the errors in the stopband (i.e., double the attenuation in dB in the stopband).
3. Leave the passband and stopband edges unchanged.

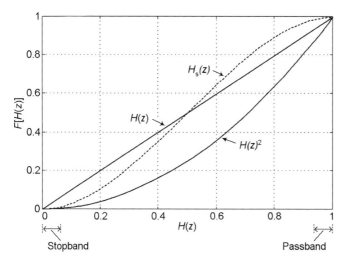

Figure 1–2   Various $F[H(z)]$ functions operating on $H(z)$.

4. Approximately double the impulse response length of the original filter.

5. Maintain filter phase linearity.

It is fairly easy to examine this operation to see the observed behavior if we view the relationship between $H(z)$ and $H(z)^2$ in a slightly unconventional way. We can think of filter squaring as a function $F[H(z)]$ operating on the $H(z)$ transfer function. We can then plot the output amplitude of this function, $H(z)^2$, versus the amplitude of the input $H(z)$ to visualize the amplitude change function.

The plot for $F[H(z)] = H(z)$ is simple; the output is the input, so the result is the straight line as shown in Figure 1–2. The function $F[H(z)] = H(z)^2$ is a quadratic curve. When the $H(z)$ input amplitude is near zero, the $H(z)^2$ output amplitude is closer to zero, which means the stopband attenuation is increased with $H(z)^2$. When the $H(z)$ input amplitude is near one, the $H(z)^2$ output band is approximately twice as far away from one, which means the passband ripple is increased.

The squaring process improved the stopband, but degraded the passband. The improvement was a result of the amplitude change function being horizontal at zero. So to improve $H(z)$ in both the passband and stopband, we want the $F[H(z)]$ amplitude function to be horizontal at both $H(z) = 0$ and $H(z) = 1$ (in other words, have a first derivative of zero at these points). This results in the output amplitude changing slower than the input amplitude as we move away from zero and one, which lowers the ripple in these areas. The simplest function that meets this will be a cubic of the form

$$F(x) = c_0 + c_1 x + c_2 x^2 + c_3 x^3. \tag{1-1}$$

Its derivative (with respect to x) is

$$F'(x) = c_1 + 2c_2 x + 3c_3 x^2. \tag{1-2}$$

Specifying $F(x)$ and $F'(x)$ for the two values of $x = 0$ and $x = 1$ allows us to solve (1–1) and (1–2) for the c_n coefficients as

$$F(x)\big|_{x=0} = 0 \Rightarrow c_0 = 0 \qquad (1\text{–}3)$$

$$F'(x)\big|_{x=0} = 0 \Rightarrow c_1 = 0 \qquad (1\text{–}4)$$

$$F(x)\big|_{x=1} = 1 \Rightarrow c_2 + c_3 = 1 \qquad (1\text{–}5)$$

$$F'(x)\big|_{x=1} = 0 \Rightarrow 2c_2 + 3c_3 = 0. \qquad (1\text{–}6)$$

Solving (1–5) and (1–6) simultaneously yields $c_2 = 3$ and $c_3 = -2$, giving us the function

$$F(x) = 3x^2 - 2x^3 = (3 - 2x)x^2. \qquad (1\text{–}7)$$

Stating this function as the sharpened filter $H_s(z)$ in terms of $H(z)$, we have

$$H_s(z) = 3H(z)^2 - 2H(z)^3 = [3 - 2H(z)]H(z)^2. \qquad (1\text{–}8)$$

The function $H_s(z)$ is the dotted curve in Figure 1–2.

1.2   FIR FILTER SHARPENING

$H_s(z)$ is called the "sharpened" version of $H(z)$. If we have a function whose z-transform is $H(z)$, then we can outline the filter sharpening procedure, with the aid of Figure 1–3, as the following:

1. Filter the input signal, $x(n)$, once with $H(z)$.
2. Double the filter output sequence to obtain $w(n)$.
3. Subtract $w(n)$ from $3x(n)$ to obtain $u(n)$.
4. Filter $u(n)$ twice by $H(z)$ to obtain the output $y(n)$.

Using the sharpening process results in the improved $H_s(z)$ filter performance shown in Figure 1–4, where we see the increased stopband attenuation and reduced passband ripple beyond that afforded by the original $H(z)$ filter.

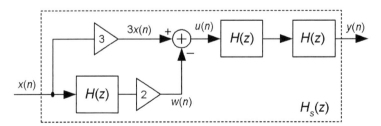

Figure 1–3   Filter sharpening process.

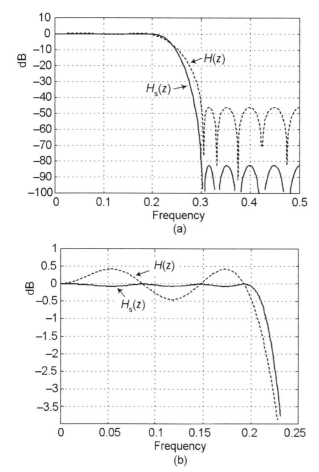

Figure 1–4 $H(z)$ and $H_s(z)$ performance: (a) full frequency response; (b) passband response.

It's interesting to notice that $H_s(z)$ has the same half-amplitude frequency (−6 dB point) as $H(z)$. This condition is not peculiar to the specific filter sharpening example used here—it's true for all $H_s(z)$s implemented as in Figure 1–3. This characteristic, useful if we're sharpening a halfband FIR filter, makes sense if we substitute 0.5 for $H(z)$ in (1–8), yielding $H_s(z) = 0.5$.

1.3 IMPLEMENTATION ISSUES

The filter sharpening procedure is very easy to perform, and is applicable to a broad class of FIR filters; including lowpass, bandpass, and highpass FIR filters having symmetrical coefficients and even-order (an odd number of taps). Even multi-passband FIR filters, under the restriction that all passband gains are equal, can be sharpened.

From an implementation standpoint, to correctly implement the sharpening process in Figure 1–3 we must delay the $3x(n)$ sequence by the group delay, $(N − 1)/2$ samples, inherent in $H(z)$. In other words, we must time-align $3x(n)$ and $w(n)$. This is analogous to the need to delay the real path in a practical Hilbert transformer. Because of this time-alignment constraint, filter sharpening is not applicable to filters having nonconstant group delay, such as minimum phase FIR filters or infinite impulse response (IIR) filters. In addition, filter sharpening is inappropriate for Hilbert transformer, differentiating FIR filters, and filters with shaped bands such as sinc compensated filters and raised cosine filters, because cascading such filters corrupts their fundamental properties.

If the original $H(z)$ FIR filter has a nonunity passband gain, the derivation of (1–8) can be modified to account for a passband gain G, leading to a "sharpening" polynomial of

$$H_{s,\text{gain}>1}(z) = \frac{3H(z)^2}{G} - \frac{2H(z)^3}{G^2} = \left[\frac{3}{G} - \frac{2H(z)}{G^2}\right] H(z)^2. \qquad (1–9)$$

Notice when $G = 1$, $H_{s,\text{gain}>1}(z)$ in (1–9) is equal to our $H_s(z)$ in (1–8).

1.4   CONCLUSIONS

We've presented a simple method for transforming a FIR filter into one with better passband and stopband characteristics, while maintaining phase linearity. While filter sharpening may not be used often, it does have its place in an engineer's toolbox. An optimal (Parks-McClellan-designed) filter will have a shorter impulse response than a sharpened filter with the same passband and stopband ripple, and thus be more computationally efficient. However, filter sharpening can be used whenever a given filter response cannot be modified, such as software code that makes use of an unchangeable filter subroutine. The scenario we described was hypothetical, but all practicing engineers have been in situations in the field where a problem needs to be solved without the full arsenal of normal design tools. Filter sharpening could be used when improved filtering is needed but insufficient ROM space is available to store more filter coefficients, or as a way to reduce ROM requirements. In addition, in some hardware filter applications using *application-specific integrated circuits* (ASICs), it may be easier to add additional chips to a filter design than it is to design a new ASIC.

1.5   REFERENCES

[1] J. KAISER and R. HAMMING, "Sharpening the Response of a Symmetric Nonrecursive Filter by Multiple Use of the Same Filter," *IEEE Trans. Acoustics, Speech, Signal Proc.*, vol. ASSP–25, no. 5, 1977, pp. 415–422.
[2] R. HAMMING, *Digital Filters*, Prentice Hall, Englewood Cliffs, NJ, 1977, pp. 112–117.
[3] R. HAMMING, *Digital Filters*, 3rd ed., Dover, Mineola, NY, 1998, pp. 140–145.

[4] T. PARKS and J. MCCLELLAN, "A Program for the Design of Linear Phase Finite Impulse Response Digital Filters," *IEEE Trans. Audio Electroacoust.*, vol. AU–20, August 1972, pp. 195–199.

EDITOR COMMENTS

When $H(z)$ is a unity-gain filter we can eliminate the multipliers shown in Figure 1–3. The multiply-by-two operation can be implemented with an arithmetic left-shift by one binary bit. The multiply-by-three operation can be implemented by adding a binary signal sample to a shifted-left-by-one-bit version of itself.

To further explain the significance of (1–9), the derivation of (1–8) was based on the assumption that the original $H(z)$ filter to be sharpened had a passband gain of one. If the original filter has a nonunity passband gain of G, then (1–8) will not provide proper sharpening; in that case (1–9) must be used as shown in Figure 1–5. In that figure we've included a *Delay* element, whose length in samples is equal to the group delay of $H(z)$, needed for real-time signal synchronization.

It is important to realize that the $3/G$ and $2/G^2$ scaling factors in Figure 1–5 provide optimum filter sharpening. However, those scaling factors can be modified to some extent if doing so simplifies the filter implementation. For example, if $2/G^2 = 1.8$, for ease of implementation, the practitioner should try using a scaling factor of 2 in place of 1.8 because multiplication by 2 can be implemented by a simple binary left-shift by one bit. Using a scaling factor of 2 will not be optimum but it may well be acceptable, depending on the characteristics of the filter to be sharpened. Software modeling will resolve this issue.

As a historical aside, *filter sharpening* is a process refined and expanded by the accomplished R. Hamming (of Hamming window fame) based on an idea originally proposed by the great American mathematician John Tukey, the inventor of the radix-2 fast Fourier transform (FFT).

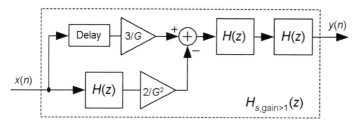

Figure 1–5 Nonunity gain filter sharpening.

Chapter 2

Quantized FIR Filter Design Using Compensating Zeros

Amy Bell
Institute for
Defense Analyses

Joan Carletta
University of Akron

Kishore Kotteri
Microsoft Corp.

This chapter presents a design method for translating a finite impulse response (FIR) floating-point filter design into an FIR fixed-point multiplierless filter design. This method is simple, fast, and provides filters with high performance. Conventional wisdom dictates that finite word-length (i.e., quantization) effects can be minimized by dividing a filter into smaller, cascaded sections. The design method presented here takes this idea a step further by showing how to quantize the cascaded sections so that the finite word-length effects in one section are guaranteed to compensate for the finite word-length effects in the other section. This simple method, called *compensating zeros*, ensures that: (1) the quantized filter's frequency response closely matches the unquantized filter's frequency response (in both magnitude and phase); and (2) the required hardware remains small and fast.

Digital filter design typically begins with a technique to find double-precision, floating-point filter coefficients that meet some given performance specifications— like the magnitude response and phase response of the filter. Two well-known techniques for designing floating-point, FIR filters are the windowing method and the equiripple Parks-McClellan method [1], [2]. If the filter design is for a real-time application, then the filter must be translated to fixed-point, a more restrictive form of mathematics that can be performed much more quickly in hardware. For embedded systems applications, a multiplierless implementation of a filter is advantageous; it replaces multiplications with faster, cheaper shifts and additions. Translation to a fixed-point, multiplierless implementation involves quantizing the original filter coefficients (i.e., approximating them using fixed-point mathematics). The primary difficulty with real-time implementations is that this translation alters the original

Streamlining Digital Signal Processing: A Tricks of the Trade Guidebook, Second Edition. Edited by Richard G. Lyons.
© 2012 the Institute of Electrical and Electronics Engineers. Published 2012 by John Wiley & Sons, Inc.

11

design; consequently, the desired filter's frequency response characteristics are often not preserved.

Multiplierless filter design can be posed as an optimization problem to minimize the degradation in performance; simulated annealing, genetic algorithms, and integer programming are among the many optimization techniques that have been employed [3]. However, in general, optimization techniques are complex, can require long run times, and provide no performance guarantees. The compensating zeros technique is a straightforward, intuitive method that renders optimization unnecessary; instead, the technique involves the solution of a linear system of equations. It is developed and illustrated with two examples involving real-coefficient FIR filters; the examples depict results for the frequency response as well as hardware speed and size.

2.1 QUANTIZED FILTER DESIGN FIGURES OF MERIT

Several important figures of merit are used to evaluate the performance of a filter implemented in hardware. The quantized filter design evaluation process begins with the following two metrics:

1. *Magnitude MSE.* Magnitude mean-squared-error (MSE) represents the average of the squared difference between the magnitude response of the quantized (fixed-point) filter and the unquantized (ideal, floating-point) filter over all frequencies. A linear phase response can easily be maintained after quantization by preserving symmetry in the quantized filter coefficients.

2. *Hardware complexity.* In a multiplierless filter, all mathematical operations are represented by shifts and additions. This requires that each quantized filter coefficient be expressed as sums and differences of powers of two: for each coefficient, a representation called *canonical signed digit* (CSD) is used [3]. CSD format expresses a number as sums and differences of powers of two using a minimum number of terms. Before a quantized filter design is implemented in actual hardware, the hardware complexity is estimated in terms of T, the total number of non-zero terms used when writing all filter coefficients in CSD format. In general, the smaller T is, the smaller and faster will be the hardware implementation. For application-specific integrated circuit and field programmable gate array filter implementations, a fully parallel hardware implementation requires $T - 1$ adders; an embedded processor implementation requires $T - 1$ addition operations.

Once the filter has been implemented in hardware, it can be evaluated more directly. Important metrics from a hardware perspective include: hardware size; throughput (filter outputs per second); and latency (time from filter input to corresponding filter output). The relative importance of these metrics depends on the application.

The goal of the quantized filter design is to achieve a small magnitude MSE while keeping the hardware costs low. In general, the higher the value of T, the closer the quantized filter coefficients are to the unquantized coefficients and the smaller the magnitude MSE. Conversely, smaller T implies worse-magnitude MSE. Hence, there is a trade-off between performance and hardware cost; T can be thought of as the parameter that controls this trade-off.

2.2 FILTER STRUCTURES

Filter designs can be implemented in hardware using various structures. The three most common structures are *direct*, *cascade*, and *lattice*. In general, pole-zero, infinite impulse response (IIR) filters are more robust to quantization effects when the cascade and lattice structures are employed; performance degrades quickly when the direct structure is used [1], [2].

For all-zero, FIR filters, the direct structure usually performs well (if the zeros are not very clustered, but are moderately uniformly distributed) [1], [2]. Moreover, since most FIR filters have linear phase (the filter coefficients are symmetric), the lattice structure cannot be used because at least one reflection coefficient equals ± 1. Although the direct structure is a good choice for many FIR filter implementations, the cascade structure offers at least one advantage. When an FIR filter is quantized using a direct structure, the quantization of one coefficient affects all of the filter's zeros. In contrast, if an FIR filter is quantized with a cascade structure, the quantization of coefficients in one of the cascaded sections affects only those zeros in its section—the zeros in the other cascaded sections are isolated and unaffected. Depending on the application, it may be important to more closely approximate the unquantized locations of some zeros than others.

The compensating zeros method uses a cascade structure. However, it goes beyond a "simple quantization" technique that uniformly divvies up the given T non-zero terms in CSD format across the coefficients in the cascaded sections. The next section first illustrates a simple quantization approach for an FIR filter design using a cascade structure; then the compensating zeros method is developed and used to redesign the same FIR filter. The result is an improvement in the magnitude MSE for the same T.

2.3 EXAMPLE 1: A WINDOWED FIR FILTER

Consider a lowpass, symmetric, length-19 FIR filter designed using a rectangular window. The filter has a normalized (such that a frequency of one corresponds to the sampling rate) passband edge frequency of 0.25 and exhibits linear phase. The floating-point filter coefficients are listed in Table 2–1, and Figure 2–1 shows the unquantized magnitude response of this filter.

Figure 2–2 illustrates the pole-zero plot for $h(n)$. To implement this filter in the cascade form, $h(n)$ is split into two cascaded sections whose coefficients are $c_1(n)$ and $c_2(n)$. This is accomplished by distributing the zeros of $h(n)$ between $c_1(n)$ and $c_2(n)$.

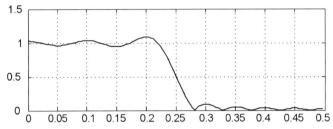

Figure 2–1   Frequency magnitude response of $h(n)$ for the windowed FIR filter.

Table 2–1   Unquantized Windowed FIR Filter Coefficients, $h(n)$

n	$h(n)$
0, 18	0.03536776513153
1, 17	−1.94908591626e–017
2, 16	−0.04547284088340
3, 15	1.94908591626e–017
4, 14	0.06366197723676
5, 13	−1.9490859163e–017
6, 12	−0.10610329539460
7, 11	1.94908591626e–017
8, 10	0.31830988618379
9	0.50000000000000

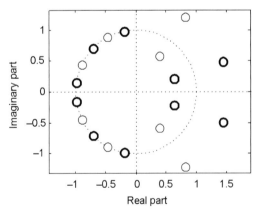

Figure 2–2   Pole-zero plot for $h(n)$. The zeros are divided into two cascaded sections by placing the thin zeros in the first section, $c_1(n)$, and the bold zeros in the second section, $c_2(n)$.

To separate the zeros of $h(n)$ into the two cascaded sections, the z-plane is scanned from $\omega = 0$ to $\omega = \pi$. As they are encountered, the zeros are placed alternately in the two sections. The first zero encountered is at $z = 0.66e^{j0.324}$. This zero, its conjugate, and the two reciprocals are put in one section. The next zero at $z = 0.69e^{j0.978}$, its conjugate, and the reciprocal pair are placed in the other section. This proceeds until all of the zeros of the unquantized filter are divided among the two cascade sections.

The steps in this zero-allocation process are as follows: compute the roots of $h(n)$; partition those roots into two sets of roots as described above; and determine the two sets of coefficients, $c_1(n)$ and $c_2(n)$, for the two polynomials associated with the two sets of roots. The section with fewer zeros becomes the first section in the cascade, $c_1(n)$, and the section with more zeros becomes the second section, $c_2(n)$ (this approach provides more degrees of freedom in our design method—see design rule-of-thumb number 6 in Section 2.7).

For the example, the zero allocation is illustrated in Figure 2–2 where the 8 thin zeros go to $c_1(n)$, which has length 9, and the 10 bold zeros go to $c_2(n)$, which has length 11. This method of splitting up the zeros has the advantage of keeping the zeros relatively spread out within each section, thereby minimizing the quantization effects within each section. Because complex conjugate pairs and reciprocal pairs of zeros are kept together in the same cascade section, the two resulting sections have symmetric, real-valued coefficients. The resulting floating-point cascade $c_1(n)$ and $c_2(n)$ coefficients are shown in Table 2–2.

Figure 2–3 depicts the block diagram corresponding to $h(n)$ and the equivalent cascaded form. Coefficients $c_1(n)$ and $c_2(n)$ are normalized so that the first

Table 2–2   Unquantized Cascaded Coefficients for $h(n)$

$c_1(n)$		$c_2(n)$	
n_1		n_2	
0, 8	1.0000000	0, 10	1.0000000
1, 7	0.3373269	1, 9	−0.3373269
2, 6	0.9886239	2, 8	−2.1605488
3, 5	1.9572410	3, 7	−0.8949404
4	3.0152448	4, 6	1.8828427
		5	3.5382228
		$k = 0.0353678$	

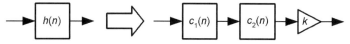

Figure 2–3   Direct form of $h(n)$ and the equivalent cascade form using $c_1(n)$, $c_2(n)$, and k.

and last coefficients in each section are 1; this ensures that at least two of the coefficients in each cascade section are efficiently represented in CSD format. Consequently, it is necessary to include a gain factor, k in Table 2–2, following the cascade sections. The magnitude response of the cascaded filter is identical to Figure 2–1.

Now consider a fixed-point, quantized, multiplierless design of this cascade structure so that it can be implemented in fast hardware. Assume that there are a fixed total number of CSD terms, T, for representing the two unquantized cascaded sections and the unquantized gain factor. Two different techniques are considered for quantizing the filter: a *simple quantization* method that treats each filter section independently, and our proposed *compensating zeros* method in which the quantization errors in one section are compensated for in the next section.

2.4   SIMPLE QUANTIZATION

For the simple quantization method, in the process of distributing a fixed number of CSD terms T to a single cascade section with n coefficients, all reasonable distributions are examined. These "reasonable distributions" consider all of the "mostly uniform" T allocation schemes to n coefficients: all coefficients receive at least one CSD term and the remaining CSD terms are allocated to those coefficients that are most different (in terms of percent different) from their unquantized values. Extremely nonuniform allocation schemes (e.g., one coefficient receives all of the T and the remaining coefficients are set to zero) are not considered.

Of all the distribution schemes examined, the distribution that gives the best result (i.e., the smallest magnitude MSE) is chosen. (*Note*: This does not require an optimization technique; for reasonably small values of T, it is simple to organize a search that looks in the area around the floating-point coefficients, which is the only area where high-quality solutions lie). This process ensures that there is no better simple quantization scheme for the given cascaded filter.

In applying the simple quantization method to the windowed FIR filter example, the unquantized cascade coefficients, $c_1(n)$ and $c_2(n)$, are independently quantized to the simple quantized cascade coefficients, $c_1'(n)$ and $c_2'(n)$. In this example, a total of $T = 25$ CSD terms was chosen; this choice results in small hardware while still providing a reasonable approximation to the desired filter. Based on the relative lengths of the sections, 9 CSD terms are used for $c_1'(n)$, 14 terms are used for $c_2'(n)$, and 2 terms are used for the quantized gain factor k'. The resulting simple quantized coefficients are listed in Table 2–3 (in the CSD format, an underline indicates that the power of 2 is to be subtracted instead of added).

The frequency response of the simple quantized filter, $c_1'(n)$ and $c_2'(n)$, is compared with the unquantized filter, $h(n)$, in Figure 2–4. Although a linear phase response is retained after simple quantization (i.e., the simple quantized coefficients are symmetric), the magnitude response is significantly different from the original unquantized case.

Table 2–3   Simple-Quantized Cascaded Coefficients for $h(n)$, $(T = 25)$

$c_1'(n)$ $(T = 9)$			$c_2'(n)$ $(T = 14)$		
n_1	Decimal	CSD	n_2	Decimal	CSD
0, 8	1.00	001.00	0, 10	1.000	001.000
1, 7	0.25	000.01	1, 9	−0.250	000.0$\underline{1}$0
2, 6	1.00	001.00	2, 8	−2.000	0$\underline{1}$0.000
3, 5	2.00	010.00	3, 7	−1.000	00$\underline{1}$.000
4	4.00	100.00	4, 6	1.875	010.00$\underline{1}$
			5	3.500	100.$\underline{1}$00
	$k' = 0.0351563$			0.00001001 $(T = 2)$	

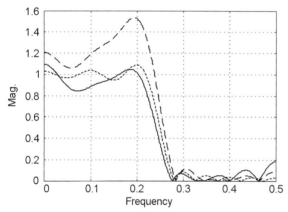

Figure 2–4   Frequency responses of the unquantized windowed filter $h(n)$ (dotted), simple quantization (dashed), and compensating zeros quantization (solid).

2.5 COMPENSATING ZEROS QUANTIZATION

The proposed compensating zeros quantization method takes the quantization error of the first cascaded section into account when quantizing the second section. The key to this method is the desire that the frequency response of the quantized cascade structure match the frequency response of the original, unquantized direct structure.

Quantization using the compensating zeros method begins with the quantization of the first cascade section $c_1(n)$ to $c_1'(n)$ and the gain factor k to k' (using the simple quantization method described in the previous section). Next, instead of quantizing $c_2(n)$ to $c_2'(n)$, $c_{comp}(n)$ is computed such that $c_1'(n)$ cascaded with $c_{comp}(n)$ is as close as possible to the original filter $h(n)$. Coefficients $c_{comp}(n)$ are called the compensating section, since their aim is to compensate for the performance degradation resulting from the quantization of $c_1'(n)$; the computation of $c_{comp}(n)$ is developed below.

If $C_1(z)$, $C_2(z)$, $C_1'(z)$, $C_2'(z)$, and $C_{comp}(z)$ are the transfer functions of $c_1(n)$, $c_2(n)$, $c_1'(n)$, $c_2'(n)$, and $c_{comp}(n)$, respectively, then the transfer function of the unquantized cascaded filter $H(z)$ can be written as

$$H(z) = kC_1(z)C_2(z), \tag{2-1}$$

where k is the gain factor. The transfer function of the semiquantized filter using the compensating zeros method is given by

$$H_{comp}'(z) = k'C_1'(z)C_{comp}(z). \tag{2-2}$$

$H_{comp}'(z)$ is called the semiquantized filter because $c_{comp}(n)$ has floating-point coefficients. The goal is for the semiquantized and unquantized transfer functions to be equal, that is, $H_{comp}'(z) = H(z)$, or

$$k'C_1'(z)C_{comp}(z) = kC_1(z)C_2(z). \tag{2-3}$$

In (2–3), $C_1(z)C_2(z)$ and k on the right-hand side are the known, unquantized cascade filters. After $c_1(n)$ and k are quantized, $C_1'(z)$ and k' on the left-hand side are known. Thus, (2–3) can be solved for $c_{comp}(n)$.

Since the 11-tap $c_{comp}(n)$ is symmetric (with the first and last coefficients normalized to 1), it can be expressed in terms of only five unknowns. In general, for a length-N symmetric filter with normalized leading and ending coefficients, there are M unique coefficients where $M = \lceil (N–2)/2 \rceil$. (The $\lceil x \rceil$ notation means: the next integer larger than x; or if x is an integer, $\lceil x \rceil = x$.)

Equation (2–3) can be evaluated at $M = 5$ values of z to solve for the five unknowns in $c_{comp}(n)$. Since the frequency response is the primary concern, these values of z are on the unit circle. For the example, (2–3) is solved at the frequencies $f = 0, 0.125, 0.2, 0.3,$ and 0.45 (i.e., $z = 1, e^{j0.25\pi}, e^{j0.4\pi}, e^{j0.6\pi}, e^{j0.9\pi}$).

Table 2–4 lists the computed floating-point coefficients of $c_{comp}(n)$. Now that $c_{comp}(n)$ has been obtained, it is quantized (using simple quantization and the remaining T) to arrive at $c_2'(n)$. The final, quantized $c_2'(n)$ filter coefficients using this compensating zeros method are also given in Table 2–4. Thus the compensating zeros quantized filter coefficients are the $c_1'(n)$ and k' from Table 2–2 in cascade with the $c_2'(n)$ in Table 2–4.

The frequency response of the compensating zeros quantized filter is compared with the unquantized filter and the simple quantized filter in Figure 2–4; the overall frequency response of the compensating zeros quantized implementation is closer to the unquantized filter than the simple quantized implementation. The small value of $T = 25$ employed in this example hampers the ability of even the compensating zeros quantized magnitude response to closely approximate the unquantized magnitude response. The magnitude MSE for compensating zeros quantization (7.347e–3) is an order of magnitude better than the magnitude MSE for simple quantization (4.459e–2).

For a fixed value of T, there are two ways to quantize the cascaded coefficients: simple and compensating zeros. If T is large enough, then simple quantization can

Table 2–4 Unquantized $c_{comp}(n)$ and Compensating Zeros-Quantized $c_2'(n)$ for $h(n)$ ($T = 25$)

n	$c_{comp}(n)$	$c_2'(n)$ ($T = 14$) Decimal	$c_2'(n)$ ($T = 14$) CSD
0, 10	1.0000000	1.00	001.00
1, 9	−0.7266865	−0.75	00$\underline{1}$.01
2, 8	−1.1627044	−1.00	00$\underline{1}$.00
3, 7	−0.6238289	−0.50	000.$\underline{1}$0
4, 6	1.2408329	1.00	001.00
5	2.8920707	3.00	10$\underline{1}$.00

achieve the desired frequency response. However, when T is restricted, compensating zeros quantization provides an alternative that outperforms simple quantization. In the example, it turns out that for $T = 26$, the simple quantization method can achieve the same magnitude MSE as the $T = 25$ compensating zeros quantization method. This improvement is achieved when the one extra T is assigned to the first cascaded section; no improvement is realized when it is assigned to the second cascaded section. There is an *art* to how to assign the extra T: These terms should be allocated to the coefficients—in either cascaded section—that are most different (in terms of percent different) from their unquantized values.

The compensating zeros quantization procedure is outlined as follows.

Step 1: Derive the unquantized cascade coefficients, $c_1(n)$, $c_2(n)$, and k, from the given, direct unquantized coefficients, $h(n)$.

Step 2: Quantize $c_1(n)$ to $c_1'(n)$ and k to k' using the simple quantization method.

Step 3: Select a set of M unique positive frequencies. For symmetric filters, M is given by $M = \lceil (N - 2)/2 \rceil$; for non-symmetric filters, M is given by $M = \lceil (N - 1)/2 \rceil$, where N is the length of the second cascaded section's $c_2(n)$.

Step 4: Solve for the M unknown $c_{comp}(n)$ coefficients by solving (2–3). For symmetric filters, (2–3) is evaluated at the M positive frequencies selected in step 3. For nonsymmetric filters, (2–3) is solved at M positive frequencies and the corresponding M negative frequencies.

Step 5: Using the simple quantization method, quantize the floating-point $c_{comp}(n)$ coefficients—with the remaining T from step 2—to arrive at $c_2'(n)$.

Step 6: Compute the direct form coefficients of the compensating zeros quantized filter using $H'(z) = k'C_1'(z)C_2'(z)$ (if desired).

Although the quantized $c_2'(n)$ values obtained using the compensating zeros method (in Table 2–4) are not very different from the values obtained using the simple method (in Table 2–3), it is important to note that simple quantization could not possibly find these improved coefficients. In the simple quantization case, the

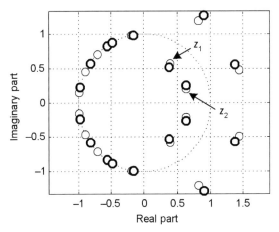

Figure 2–5   Zero plot comparing the unquantized zeros of $h(n)$ (thin circles) and the compensating zeros quantized zeros of $h'(n)$ (bold circles).

Table 2–5   Distance of the Zeros in Figure 2–5 from the Origin

	$|z_2|$	$|z_1|$
Unquantized	0.66061185	0.69197298
Comp. quantized	0.67774476	0.644315248

goal is to closely approximate the floating-point $c_1(n)$ and $c_2(n)$ coefficients in Table 2–2. However, in the compensating zeros quantization case, the goal is to closely approximate a different set of floating-point coefficients—$c_1(n)$ and $c_{comp}(n)$. Figure 2–5 compares the zero location plots of the compensating zeros quantized filter (bold circles) and the unquantized filter (thin circles).

Figure 2–5 also shows the zeros z_1 and z_2 in the first quadrant; recall that z_1 is in $c_1(n)$ and z_2 is in $c_2(n)$. The quantization of $c_1(n)$ to $c_1'(n)$ moves z_1 away from the unit circle. Consequently, the compensating zeros quantization method moves z_2 toward the unit circle (Table 2–5 shows the distances of z_1 and z_2 from the origin before and after quantization). This compensating movement of the quantized zeros is the reason the quantized magnitude response more closely resembles the unquantized magnitude response; it also motivates the name of the quantization method.

The hardware performance of the simple quantized and compensating zeros quantized filters was evaluated by implementing the filters on an Altera field-programmable gate array (FPGA). For each filter, given the desired CSD coefficients, filter synthesis software, written in C, was used to generate a synthesizable

Table 2–6 Hardware Metrics for the Windowed FIR Example

	Simple quantized	Compensating zeros quantized
Hardware complexity (logic elements)	1042	996
Throughput (Mresults/second)	82.41	83.93
Latency (clock cycles)	15	15
Input data format	(8, 0)	(8, 0)
Output data format	(23, −13)	(21, −13)

description of the filter in VHDL, a hardware description language. The software automatically chooses appropriate bit widths for internal signals such that errors due to truncation and overflow are completely avoided. In this way, an implemented fixed-point filter provides exactly the same outputs as the same filter would in infinite precision, given the quantized, finite precision filter coefficients. Fixed-point, two's complement adders are used, but the number of integer and fraction bits for each hardware adder are chosen so that no round-off or overflow can occur.

The VHDL description generated is a structural one. A high-performance, multiplierless implementation is achieved. For each section in a filter, a chain of registers is used to shift the data in, and the data is shifted in accordance with the filter coefficients before being summed. For example, if one of the filter coefficients were $18 = 2^4 + 2^1$, the corresponding data word would go to two shifters and be shifted four and one places, respectively, before being summed. A pipelined tree of *carry save adders* (CSAs) is used for the summation. The CSA tree produces two outputs that must be summed, or *vector-merged*, to produce the final filter output. For the results presented here, we use a ripple carry adder for the vector merge, taking care to exploit special-purpose routing (*carry chains*) provided on Altera FPGAs to make ripple carry addition fast.

Table 2–6 summarizes the hardware performance of the filters. Data formats for all signals are shown as (n,l), where n is the total number of bits, including the sign bit, and 2^l is the weight of the least significant bit. Both filters take in inputs of data format (8, 0) (i.e., eight-bit two's complement integers). They vary in terms of their output data formats, depending on the precision of the coefficients used. Throughput is the most important performance metric; it measures how many inputs can be processed per second. Latency also bears on performance, but is less critical; it measures how many clock cycles a particular set of data takes to pass through the system, from input to corresponding output.

The results of Table 2–6 show that the compensating zeros quantized filter is slightly smaller and faster than the simple quantized filter. This is because the coefficients for the compensating zeros quantized case turn out to be slightly less wide (in terms of bits) than those for the simple quantized case, so that the adders also turn out to be slightly less wide.

2.6 EXAMPLE 2: A BIORTHOGONAL FIR FILTER

Now consider another example that illustrates the cascade structure's advantage of isolating some of the zeros in an FIR filter. The biorthogonal FIR wavelet filters are best known for their inclusion in the most recent version of the international image compression standard, JPEG2000 [4]; in this example, the 20-tap, biorthogonal, FIR, lowpass analysis wavelet filter is examined [5]. This filter has a passband edge frequency of 0.31 and exhibits linear phase. This filter has 9 zeros clustered at $z = -1$, and 10 zeros on the right-hand side of the z-plane.

For biorthogonal filters, the stopband magnitude response is characterized by the cluster of zeros at $z = -1$; it is advantageous to employ the cascade structure to place all of the zeros at $z = -1$ into an isolated cascade section. This cascade section will keep the zeros exactly at $z = -1$, have only integer coefficients, and require no quantization. Furthermore, this one large cascade section can be split up into smaller T-friendly sections having 4, 2, or 1 zeros to reduce the total T required. The remaining zeros on the right-hand side of the z-plane determine the passband characteristics and are separated into two cascade sections, $c_1(n)$ and $c_2(n)$, as before. Here the nine zeros at $z = -1$ are divided into two sections of four zeros, $c_3(n)$ and $c_4(n)$, and one section of one zero, $c_5(n)$.

Using the simple quantization method, the unquantized cascade sections $c_1(n)$ and $c_2(n)$ are quantized, but $c_3(n)$, $c_4(n)$, and $c_5(n)$ are captured exactly; a total of $T = 38$ CSD is chosen. The compensating zeros quantization method follows the procedure outlined above for $c_1(n)$ and $c_2(n)$; the remaining three cascaded sections are again captured exactly; a total of $T = 38$ CSD was used.

Figure 2–6 illustrates that the magnitude response of the compensating zeros quantized implementation is closer to the unquantized filter than the simple quantized implementation. The magnitude MSE for compensating zeros quantization (9.231e–3) is an order of magnitude better than the magnitude MSE for simple quantization (8.449e–2). For the biorthogonal FIR filter, it turns out that four extra

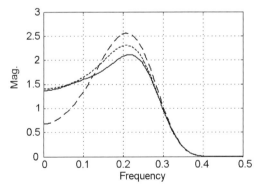

Figure 2–6 Frequency response of the unquantized biorthogonal filter (dotted) compared with the two quantized filters: simple quantization (dashed) and compensating zeros quantization (solid).

T, ($T = 42$), are required for the simple quantization method to achieve the same magnitude MSE as the $T = 38$ compensating zeros quantization method.

2.7 DESIGN RULES-OF-THUMB

The following guidelines recommend how to derive the maximum benefit from the compensating zeros quantization technique.

1. As T increases for a given filter design, the performance of simple quantization approaches compensating zeros quantization. The performance advantage of the compensating zeros method is only realized when T presents a real constraint on the fixed-point filter design.

2. In the compensating zeros technique, the first cascaded section must be quantized so that it is different from the unquantized filter (i.e., $C_1'(z)$ must be different from $C_1(z)$). This affords the second cascaded section the opportunity to compensate for the quantization effects in the first section.

3. For nonlinear phase FIR filters the filter coefficients are no longer symmetric; consequently, (2–3) must be solved for both positive and negative frequencies to ensure that real coefficients are maintained.

4. The set of frequencies at which (2–3) is evaluated determines how well the compensated magnitude response matches the unquantized response. There is no optimal set of frequencies; instead, several frequency sets may need to be evaluated in order to identify the set that yields the closest match.

5. The original filter must be long enough so that when it is divided into the cascade structure, the sections have sufficient length so that compensation can occur.

6. It is desirable to quantize the shorter cascaded section first and then compensate with the longer cascaded section. The longer the compensating section, the larger the set of frequencies at which (2–3) is evaluated, and the better the match between the quantized and unquantized frequency responses.

7. The compensating zeros method can be employed for multiple cascaded sections. If a filter is divided into N cascaded sections, c_1 through c_N, then the design begins as described in the example for the first two cascaded sections, c_1 and c_2. Next, c_1 and c_2 are combined into one quantized filter and c_3 is designed to compensate for the quantization effects in both c_1 and c_2 (by solving (2–3)). This process continues until the last cascaded section, c_N, is designed. Furthermore, not all cascaded sections have to be used to perform the quantization compensation; the choice is up to the designer.

2.8 CONCLUSIONS

For hardware filter implementations that use the cascade multiplierless structure, the compensating zeros quantization method outperforms simple quantization given a small fixed-value of T. The compensating zeros technique quantizes the cascaded sections so that the finite word-length effects in one section are guaranteed to compensate for the finite word-length effects in the other section. The algorithm involves no optimization—just the solution of a linear system of equations. Moreover, compensating zeros quantization ensures that: (1) the quantized filter's frequency response closely matches the unquantized filter's frequency response (magnitude and phase), and (2) the required hardware remains small and fast. This technique can be applied to any unquantized FIR filter and it can exploit the cascade structure's ability to isolate some of the zeros in the filter.

2.9 REFERENCES

[1] J. PROAKIS and D. MANOLAKIS, *Digital Signal Processing Principles, Algorithms, and Applications*, Prentice Hall, 3rd ed., 1995, pp. 500–652.
[2] A. OPPENHEIM, R. SCHAFER, and J. BUCK, *Discrete-Time Signal Processing*, Prentice Hall, 2nd ed., 1999, pp. 366–510.
[3] C. LIM, R. YANG, D. LI, and J. SONG, "Signed Power-of-Two Term Allocation Scheme for the Design of Digital Filters," *IEEE Transactions on Circuits and Systems – II: Analog and Digital Signal Proc.*, vol. 46, no. 5, May 1999, pp. 577–584.
[4] "T.800: Information Technology – JPEG2000 Image Coding System," ISO/IEC 15444-1:2002. [Online: http://www.itu.int.]
[5] I. DAUBECHIES, "Ten Lectures on Wavelets," *SIAM*, 1992, pp. 277.

Chapter 3

Designing Nonstandard Filters with Differential Evolution

Rainer Storn
Rohde & Schwarz GmbH & Co. KG

Many filter design tasks deal with standard filter types such as lowpass, highpass, bandpass, or bandstop filters and can be solved with off-the-shelf filter design tools [1–3]. In some cases, however, the requirements for the digital filters deviate from the standard ones. This chapter describes a powerful technique for designing nonstandard filters such as the following:

- *Minimum phase filters.* These often appear in voice applications where minimum signal delay is an issue. The constraint here is that the zeros of the transfer function must remain inside the unit circle.

- *Recursive filters with linearized phase.* If high selectivity and low hardware expense are requirements in data applications, then IIR-filters with linearized phase are often the best choice. Phase linearization is traditionally done with all-pass filters, but direct design in the z-domain generally yields a lower filter order.

- *Recursive filters with pole radius restrictions.* Such restrictions may apply to reduce the sensitivity to coefficient quantization or to get a short impulse response length. The latter can be important in communication applications like modems.

- *Constraints in the frequency and time domains.* A typical application is filters for ISDN (Integrated Services Digital Network) equipment where constraint masks are defined for the time and frequency domains.

- *Magnitude constraints that deviate from the standard LP, HP, BP, BS type.* Filters with "odd-looking" magnitude requirements occur fairly often in the DSP engineer's world. Sinc-compensated filters, filters that are

Streamlining Digital Signal Processing: A Tricks of the Trade Guidebook, Second Edition. Edited by Richard G. Lyons.
© 2012 the Institute of Electrical and Electronics Engineers. Published 2012 by John Wiley & Sons, Inc.

supposed to remove specific unwanted tones in addition to another filter task (e.g., a lowpass with a notch), and differentiators are common examples.

- **Postfilters.** Some developments have to work in conjunction with existing hardware, be it analog or digital. Additional filtering (e.g., to improve certain behavior) must take the already existing filter into account in order to meet an overall filter specification.

Such designs are often resolved by resorting to filter design experts and/or expensive filter design software, both of which can incur substantial costs. There is a simpler and less expensive way, however, that offers a solution to many unusual filter design tasks. It uses the power of a genetic algorithm called *differential evolution* (DE), which is still not widely known in the signal processing community [4]. The approach is to recast the filter design problem as a minimization problem and use the poles and zeros as parameters. In this approach the ensuing error function must be minimized or ideally be driven to zero.

Filter design problems that are recast as minimization problems generally yield multimodal results that are very difficult to minimize. Yet DE is powerful enough to successfully attack even these types of error functions. The approach that will be described in the following has even found its way into a commercial product [5].

3.1 RECASTING FILTER DESIGN AS A MINIMIZATION PROBLEM

The most general form of a digital filter is

$$H(z) = \frac{U(z)}{D(z)} = \frac{\displaystyle\sum_{n=0}^{N_z} a(n) \cdot z^{-n}}{1 + \displaystyle\sum_{m=1}^{M_p} b(m) \cdot z^{-m}} = A_0 \frac{\displaystyle\prod_{n=0}^{N_z - 1} (z - z_0(n))}{\displaystyle\prod_{m=0}^{M_p - 1} (z - z_p(m))} \tag{3-1}$$

The degree of $H(z)$ is defined as the maximum of N_z and M_p. The parameters $a(n)$ and $b(m)$ are called the *coefficients* of the filter while the $z_0(n)$ and $z_p(m)$ denote the zeros and poles of the filter, respectively.

The idea of treating filter design as a minimization problem is not new [6]. How we approach the minimization here, however, is radically different from the classical methods. While classical methods resort to calculus-based minimization algorithms like Newton and Quasi-Newton methods rendering a complicated system of equations to solve, a very simple yet powerful genetic algorithm called differential evolution (DE) will be applied. Before explaining how DE works, let us formulate our problem. In digital filter design the frequency magnitude response $A(\Omega)$, the phase angle $\varphi(\Omega)$, and the group delay $G(\Omega)$ are important quantities. Some or all of the quantities $A(\Omega)$, $\varphi(\Omega)$, and $G(\Omega)$ may be subject to certain constraints in a filter design problem. $A(\Omega)$ is used here as an example to explain the principles of a constraint-based design.

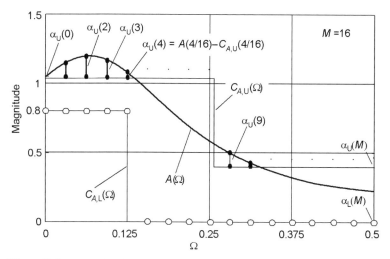

Figure 3–1   Example tolerance scheme for the magnitude $A(\Omega)$.

Upper and lower constraints $C_{A,U}(\Omega)$ and $C_{A,L}(\Omega)$ define a tolerance scheme as shown in Figure 3–1.

What is of interest is how much $A(\Omega)$ violates the tolerance scheme in the frequency domain. The task is to define an error function e_A that becomes larger the more $A(\Omega)$ violates the tolerance scheme, and that becomes zero when $A(\Omega)$ satisfies the constraints. It is straightforward to define the area outside the tolerance scheme enclosed by $A(\Omega)$ as the error. However, it is computationally simpler and more efficient to use the sum of squared error samples $\alpha_U(m)$, $\alpha_L(m)$ as the error function e_A.

If e_A is zero then all constraints are met, so the *stopping criterion* for the minimization is clearly defined as $e_A = 0$. Optimization problems that are defined by constraints only and are solved once all constraints are met are called *constraint satisfaction problems*. For other quantities like, for example, the group delay, the same principle can applied to build e_G. If, for example, constraints exist for both magnitude and group delay the stopping criterion may be defined as $e_{total} = e_A + e_G = 0$.

The principle of just adding error terms that define the constraints to be met can be applied for all constraints levied on the filter design. Among these may be phase constraints, constraints to keep the poles inside the unit circle (stability), keeping the zeros inside the unit circle (minimum phase), and so on. Note that the partial error terms need not be weighted in order to end up with a successful design. DE has enough optimization quality to not require weights.

The natural parameters to vary in order to drive the total error to zero are either the coefficients $a(n)$ and $b(m)$ or the radii and angles of the zeros $z_0(n)$ and the poles $z_p(m)$, along with the gain factor A_0. While varying the coefficients may seem attractive at first, because they represent what is finally needed to implement a filter, the trick here is that working with the poles and zeros is much more convenient. The primary reason is that it is much simpler to ensure stability this way.

For each iteration of the minimization a stability criterion has to be checked—at least for IIR filters. If the parameters are zeros and poles, one has just to prevent the poles from moving outside the unit circle. If the coefficients are used as parameters, the stability criterion is more computationally intensive because it requires some knowledge about the roots of $H(z)$. Using poles and zeros as parameters also simplifies the design of minimum phase filters, for example, to ensure that the zeros stay inside the unit circle.

3.2 MINIMIZATION WITH DIFFERENTIAL EVOLUTION

In order to adjust the parameters $z_0(n)$ and $z_p(m)$ such that the error e_{total} becomes zero we use the differential evolution (DE) method. DE belongs to the class of direct search minimization methods [4]. These methods are in contrast to gradient-based methods, which rely heavily on calculus and usually lead to rather complicated system equations [6]. Gradient-based methods also require that the function to be minimized can always be differentiated. This requirement cannot be fulfilled, for example, when coefficient quantization is incorporated into the minimization.

Direct search minimization methods act a bit like a person in a boat on a lake who wants to find the deepest spot of the lake. The person cannot see to the bottom of the lake so they have to use a plumbline to probe the deepness. If the bottom of the lake has several valleys of varying depth, it becomes difficult to find the deepest spot.

DE operates according to the following scenario. N_P base points are spread over the the *error function surface* to initialize the minimization. This is shown by an example in Figure 3–2(a) that depicts the contour lines of a multimodal function (called *peaks function*) and the initialization points chosen for DE. Each dot represents a candidate solution for the values of parameters $p1$ and $p2$ (the axes in Figure 3–2) and we seek to find the values of those parameters that minimize the error function. After initialization each starting point is perturbed by a random weighted difference vector V and its error value is compared with the error value of a randomly selected other point of the population. The perturbed point with the lower error value wins in this pairwise competition. This scheme is repeated from generation to generation.

The difference vector V is built from two other randomly chosen points of the population itself. It is then weighted with some constant factor F before it is used for perturbation. For each point to be perturbed a new difference vector defined by two randomly selected points of the population is chosen. As a rule of thumb F is chosen around 0.85. Figure 3–2(a) illustrates this idea. Once each point of the population is perturbed, the perturbed points compete against the original points. Each new point has to compete against a distinct old point. The winner of each match is the point that has the lower error function value. Eventually a new population of points has emerged that is ready for the next iteration. Because the competition is a

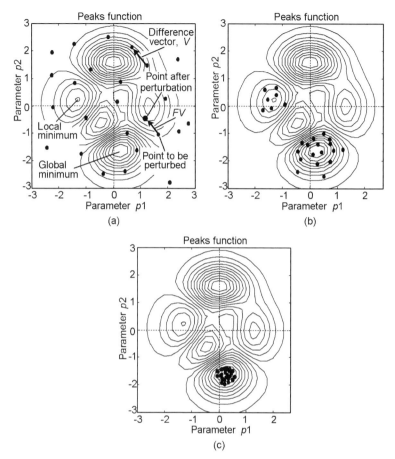

Figure 3–2   The convergence behavior of DE.

1-to-1 competition, the number of population points (population vectors) is the same in each iteration.

Figures 3–2(b) and (c) show how the point population converges toward the global minimum. One can also see that some points stay at a local minimum for a while, but finally they have to vanish because they lose in the competition against points that are in the vicinity of the true global minimum. Now it becomes evident why the use of difference vectors for perturbation is advantageous: While the population is moving toward the global minimum the difference vectors become shorter, which is exactly what should happen, and their directions adapt to the error function surface at hand. This is the great advantage of basing perturbation on the population itself rather than applying a predefined probability density function (e.g., gaussian), the parameters of which (e.g., standard deviation) are not straightforward to choose. All mathematical details of differential evolution, along with computer source code in C, MATLAB®, and other programming languages can be found in [4].

3.3   A DIFFERENTIAL EVOLUTION DESIGN EXAMPLE

Now it is time to look at an example to show that the DE design method really works. The following example has been computed by the filter design program FIWIZ, which is a commercial program that, among others, employs the above method for its filter design tasks [5]. Figure 3–3 shows a lowpass tolerance scheme, the dotted curves, needed for a graphics codec that has tight constraints in both the magnitude response (in dB) and group delay [8]. The y-axis of the group delay plot is scaled in units of the sampling time T_s. The tolerance constraint for the group delay may be shifted along the y-axis. In addition, the coefficient wordlength shall be constrained to 24 bits and biquad stages shall be used for filter realization. Figure 3–3 shows a fifth-order design that uses an elliptical filter approach being able to satisfy the magnitude constraints. The severe group delay constraints, however, are not met.

However, DE offers much more flexibility so that group delay constraints can be considered during the filter design. The result, a twelfth-order IIR filter, can be seen as the performance shown in Figure 3–4. Due to forcing the zeros outside the unit circle this design was able to be finished in 130,500 iterations (roughly 18 minutes on a Pentium III 650 MHz PC). Despite the large number of 37 parameters

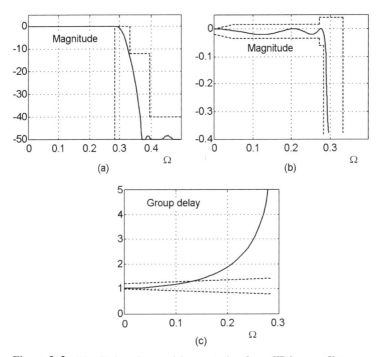

Figure 3–3   Magnitude and group delay constraints for an IIR lowpass filter.

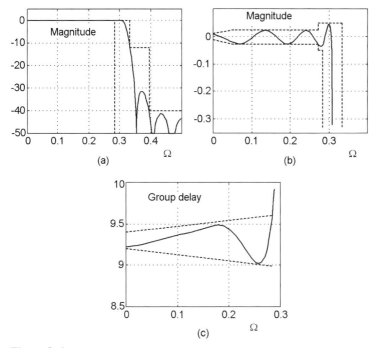

Figure 3–4   DE-designed twelfth-order IIR filter performance.

a value of $N_P = 30$ was sufficient. We see that the DE-designed IIR filter satisfies the stringent group delay constraints.

3.4 CONCLUSIONS

In this chapter an alternative method for nonstandard filter design has been described. This method recasts the filter design problem as a minimization problem and solves the minimization via the DE minimizer, for which public domain software has been made available in [7]. The advantages of this method are its simplicity as well as the capability to design unconventional filter types. A great asset of this approach is that it can be applied with a minimal knowledge of digital filter design theory.

3.5 REFERENCES

[1] "Digital Filter Design Software." [Online: http://www.dspguru.com/sw/tools/filtdsn2.htm.]
[2] "GUI: Filter Design and Analysis Tool (Signal Processing Toolbox)." [Online: http://www. mathworks.com/access/helpdesk/help/toolbox/signal/fdtoo11a.html.]
[3] "Filter Design Software." [Online: http://www.poynton.com/Poynton-dsp.html.]
[4] K. PRICE, R. STORN, and J. LAMPINEN, *Differential Evolution—A Practical Approach to Global Optimization*, Springer, 2005.

[5] "Digital Filter Design Software Fiwiz." [Online: http://www.icsi.berkeley.edu/~storn/fiwiz.html.]

[6] A. ANTONIOU, *Digital Filters—Analysis, Design, and Applications*, McGraw-Hill, 1993.

[7] "Differential Evolution Homepage." [Online: http://www.icsi.berkeley.edu/~storn/code.html.]

[8] R. STORN, "Differential Evolution Design of an IIR-Filter with Requirements for Magnitude and Group Delay," *IEEE International Conference on Evolutionary Computation ICEC 96*, pp. 268–273.

Chapter 4

Designing IIR Filters with a Given 3 dB Point

Ricardo A. Losada and Vincent Pellissier
The MathWorks, Inc.

Often in IIR filter design our critical design parameter is the cutoff frequency at which the filter's power decays to half (−3 dB) the nominal passband value. This chapter presents techniques that enable designs of discrete-time Chebyshev and elliptic filters given a 3 dB attenuation frequency point. These techniques place Chebyshev and elliptic filters on the same footing as Butterworth filters, which traditionally have been designed for a given 3 dB point. The result is that it is easy to replace a Butterworth design with either a Chebyshev or an elliptic filter of the same order and obtain a steeper rolloff at the expense of some ripple in the passband and/or stopband of the filter.

We start by presenting a technique that solves the problem of designing discrete-time Chebyshev type I and II IIR filters given a 3 dB attenuation frequency point. Traditionally, to design a lowpass Chebyshev (type I) IIR filter we start with the following set of desired specifications: $\{N, \omega_p, A_p\}$. N is the filter order, ω_p the passband-edge frequency, and A_p is the desired attenuation at ω_p (see Figure 4–1). The problem is that it's impractical to set $A_p = 3$ dB and design for the specification set $\{N, \omega_p, 3\}$; due to the filter's equiripple behavior, all the ripples in the passband would reach the −3 dB point, yielding intolerable passband ripple.

To solve this problem, our designs are based on analytic relations that can be found in the analog domain between the passband-edge frequency and the 3 dB cutoff frequency in the case of type I Chebyshev filters and between the stopband-edge frequency and the 3 dB cutoff frequency in the case of type II Chebyshev filters. We use the inverse bilinear transformation in order to map the specifications given in the digital domain into analog specifications. We then use the analytic relation between the frequencies we have mentioned above to translate a set of

Streamlining Digital Signal Processing: A Tricks of the Trade Guidebook, Second Edition. Edited by Richard G. Lyons.
© 2012 the Institute of Electrical and Electronics Engineers. Published 2012 by John Wiley & Sons, Inc.

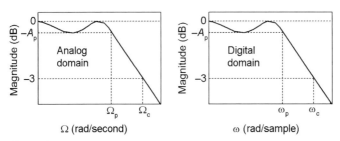

Figure 4–1   Definition of analog/digital frequency parameters.

specifications into another one we know how to handle, and finally we use the bilinear transformation to map the new set of specifications we can handle back to the digital domain.

In the case of highpass, bandpass, and bandstop filters, we show a trick where we use an arbitrary *prototype* lowpass filter, along with the Constantinides spectral transformations, in order to build upon what was done in the lowpass case and allow for the design of these other types of filters with given 3 dB points.

We then turn our attention to elliptic filters. For this case, we present a simple technique using the Constantinides spectral lowpass to lowpass transformation on a half-band elliptic filter to obtain the desired filter.

We will use Ω to indicate analog frequency (radians per second), and a set such as $\{N, \Omega_c, A_p\}$ to indicate analog design specifications. Normalized (or "digital") frequency is denoted by ω (radians per sample or simply radians), and we will use a set such as $\{N, \omega_c, A_p\}$ to indicate discrete-time or digital design specifications. The various analog and digital frequency-domain variables used in this chapter are illustrated in Figure 4–1. Example design specifications are described below. The lowpass case is described first and then the highpass, bandpass, and bandstop cases are developed.

4.1   LOWPASS CHEBYSHEV FILTERS

For the lowpass Chebyshev filter case, to solve the 3 dB passband ripple problem cited earlier, we can translate a given set of specifications that include the cutoff frequency ω_c to the usual set $\{N, \omega_p, A_p\}$ in order to use the existing algorithms to design the filter. This amounts to exchanging one parameter in the specifications set with ω_c. So the given specification set for the design would be $\{N, \omega_c, A_p\}$ or $\{N, \omega_p, \omega_c\}$. (We concentrate here on designs with a fixed filter order. It would also be possible to translate the set $\{\omega_p, \omega_c, A_p\}$ to $\{N, \omega_p, A_p\}$ but it would require rounding the required order to the next integer, thereby exceeding one of the specifications.) We will show how to do this, but first we show how to find the 3 dB point in an analog Chebyshev type I lowpass filter given $\{N, \Omega_p, A_p\}$.

4.2 DETERMINING THE 3 DB POINT OF AN ANALOG CHEBYSHEV TYPE I FILTER

Analog Chebyshev type I lowpass filters have a magnitude squared response given by

$$|H(j\Omega)|^2 = \frac{1}{1 + \varepsilon_p^2 C_N^2\left(\dfrac{\Omega}{\Omega_p}\right)}$$

where the quantity ε_p controlling the passband ripples is related to the passband attenuation A_p by

$$\varepsilon_p = \sqrt{10^{A_p/10} - 1} \qquad (4\text{--}1)$$

and the Chebyshev polynomial of degree N is defined by

$$C_N(x) = \begin{cases} \cos\left(N\cos^{-1}(x)\right) & \text{if } |x| \le 1 \\ \cosh\left(N\cosh^{-1}(x)\right) & \text{if } |x| > 1 \end{cases}$$

For lowpass filters it is reasonable to assume $\Omega_c/\Omega_p > 1$, therefore we can determine the 3 dB frequency by determining ε_p from (4–1) and finding the point at which the magnitude squared is equal to 0.5 by solving for Ω_c in

$$\frac{1}{2} = \frac{1}{1 + \varepsilon_p^2 C_N^2\left(\dfrac{\Omega_c}{\Omega_p}\right)}$$

Solving for Ω_c, it has been shown in [1] that we get

$$\Omega_c = \Omega_p \cosh\left(\frac{1}{N}\cosh^{-1}\left(\frac{1}{\varepsilon_p}\right)\right) \qquad (4\text{--}2)$$

4.3 DESIGNING THE LOWPASS FILTER

Now that we have derived the relation between Ω_c and Ω_p, we outline the design of the digital Chebyshev filter. The design process consists of three steps:

1. Given $\{N, \omega_c, A_p\}$ or $\{N, \omega_p, \omega_c\}$ determine $\{N, \Omega_c, A_p\}$ or $\{N, \Omega_p, \Omega_c\}$ as appropriate by prewarping the digital frequencies using $\Omega = \tan(\omega/2)$ (inverse bilinear transformation).

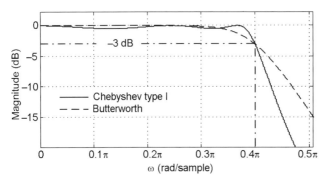

Figure 4–2   A Chebyshev filter with $\{N = 5, \omega_c = 0.4\pi, A_p = 0.5 \text{ dB}\}$. For comparison purposes, a Butterworth filter with $\{N = 5, \omega_c = 0.4\pi\}$ is shown.

2. Using (4–1) and (4–2) find the missing parameter, either Ω_p or A_p.
3. If the missing parameter was Ω_p, determine ω_p from $\omega_p = 2\tan^{-1}(\Omega_p)$ (bilinear transformation).

At this point we should have the specification set $\{N, \omega_p, A_p\}$ so we can use existing design algorithms readily available in filter design software packages (such as the cheby1 command in MATLAB) to design the Chebyshev type I lowpass filter.

EXAMPLE 1

Suppose we want to design a filter with the following specification set: $\{N = 5, \omega_c = 0.4\pi, A_p = 0.5 \text{ dB}\}$. We can use the technique outlined above to translate the cutoff frequency $\omega_c = 0.4\pi$ to the passband-edge frequency $\omega_p = 0.3827\pi$. The design is shown in Figure 4–2. A Butterworth filter of the same order and same 3 dB point is shown for reference. This example illustrates the application of our technique. It allows us to easily keep the same cutoff frequency as a Butterworth design, yet allowing some passband ripple as a tradeoff for a steeper rolloff.

4.4 CHEBYSHEV TYPE II FILTERS

A similar approach can be taken for Chebyshev type II (inverse Chebyshev) filters. However, in this case, it is a matter of translating the specification sets $\{N, \omega_c, A_s\}$ or $\{N, \omega_c, \omega_s\}$ (ω_s is the stopband-edge frequency and A_s is the stopband attenuation) to $\{N, \omega_s, A_s\}$ and then using filter design software (such as the cheby2 command in MATLAB) to obtain the filter coefficients. To do this, we need to use the expression for the magnitude squared of the frequency response of Chebyshev type II analog frequencies, described in [1], in order to determine Ω_c in terms of Ω_s:

$$|H(j\Omega)|^2 = \frac{C_N^2\left(\dfrac{\Omega_s}{\Omega}\right)}{\varepsilon_s^2 + C_N^2\left(\dfrac{\Omega_s}{\Omega}\right)}$$

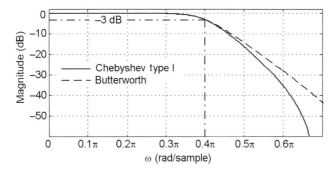

Figure 4–3   A Chebyshev type II filter with $\{N = 5,\ \omega_c = 0.4\pi,\ A_s = 60\ \text{dB}\}$. For comparison purposes, a Butterworth filter with $\{N = 5,\ \omega_c = 0.4\pi\}$ is shown.

Once again, setting the magnitude squared to 0.5 and solving for Ω_c we obtain:

$$\Omega_c = \frac{\Omega_s}{\cosh\left(\dfrac{1}{N}\cosh^{-1}(\varepsilon_s)\right)}$$

where the quantity ε_s controlling the depth of the stopband is related to the stopband attenuation A_s by

$$\varepsilon_s = \sqrt{10^{A_s/10} - 1}$$

EXAMPLE 2

Chebyshev type II filters provide a good way of attaining a very similar response to Butterworth in the passband, but with a sharper rolloff for a fixed filter order. Using the technique we have described we can design a Chebyshev type II filter with the following specifications: $\{N = 5,\ \omega_c = 0.4\pi,\ A_s = 60\ \text{dB}\}$. The result is shown in Figure 4–3. Again, we include the same Butterworth filter response from the previous example.

4.5 HIGHPASS, BANDPASS, AND BANDSTOP CHEBYSHEV FILTERS

The relation between the analog cutoff frequency and the analog passband-edge or stopband-edge frequency we used in the previous section is valid only for lowpass filters. In order to design highpass and other types of filters, we can build on what we have done for lowpass filters and use the Constantinides frequency transformations in [2], [3]. For instance, given the highpass specifications $\{N,\ \omega_c',\ \omega_p'\}$ or $\{N,\ \omega_c',\ A_p\}$, we need to translate the desired cutoff frequency ω_c' to either A_p or ω_p' in order to use filter design software to design for the usual set $\{N,\ \omega_p',\ A_p\}$.

The trick is to take advantage of interesting properties of the Constantinides frequency transformations. The passband ripples are conserved and the transformation projecting the lowpass cutoff frequency ω_c to the highpass cutoff frequency ω_c'

will also project the lowpass passband-edge frequency ω_p to the highpass passband-edge frequency ω_p'. The problem comes down to determining the lowpass specification set $\{N, \omega_c, \omega_p\}$ that designs a filter that when transformed meets the given highpass specifications $\{N, \omega_c', \omega_p'\}$. We never actually design the lowpass filter, however.

The full procedure is as follows:

1. Given the highpass specifications $\{N, \omega_c', \omega_p'\}$, choose *any* value for the cutoff frequency ω_c of a digital lowpass filter.

2. Knowing ω_c and ω_c' find the parameter α used in the lowpass to highpass Constantinides transformation:

$$\alpha = -\frac{\cos\left(\dfrac{\omega_c + \omega_c'}{2}\right)}{\cos\left(\dfrac{\omega_c - \omega_c'}{2}\right)} \tag{4-3}$$

3. Given the desired passband-edge frequency ω_p' of the highpass filter and α, determine the passband-edge frequency ω_p of the digital lowpass filter from

$$z^{-1} = -\frac{z'^{-1} + \alpha}{1 + \alpha z'^{-1}} \tag{4-4}$$

evaluated at $z = e^{j\omega_p}$ and $z' = e^{j\omega_p'}$.

4. At this point, we have the lowpass specifications $\{N, \omega_c, \omega_p\}$. With this, we can determine A_p as we have seen in the previous section to translate the specifications set to $\{N, \omega_c, A_p\}$.

5. A_p being conserved by the frequency transformation, we can now substitute A_p for ω_c' in the highpass specification set and use the new set $\{N, \omega_p', A_p\}$ to design highpass Chebyshev type I filters with filter design software.

Notice that the procedure is similar if the highpass specification set is $\{N, \omega_c', A_p\}$. Once again we choose a value for ω_c. We now have the lowpass specification set $\{N, \omega_c, A_p\}$; we can now translate to the lowpass specification set $\{N, \omega_p, A_p\}$ as seen previously and, similar to what is done above, determine ω_p through the Constantinides frequency transformations to end up with the highpass specification set $\{N, \omega_p', A_p\}$.

For bandpass and bandstop filters, we need to halve the order when we convert from the original specifications to the lowpass specifications since this is implicit in the Constantinides transformation, but other than that the procedure is basically the same.

For instance, in the bandpass case, say we start with the specification set $\{N', \omega_{c1}', \omega_{c2}', A_p\}$ where $\omega_{c1}', \omega_{c2}'$ are the lower and upper 3 dB frequencies, respectively. We once again choose any value for ω_c, the cutoff frequency of the lowpass filter. Now we have the lowpass specification set $\{N, \omega_c, A_p\}$ where $N = N'/2$. We translate that set to the lowpass specification set $\{N, \omega_p, A_p\}$. We can then use the lowpass-to-bandpass transformation (see [2] or [3]) and find the two solutions (since the transformation is a quadratic) for ω' when $\omega = \omega_p$. These two solutions are $\omega'_{p1}, \omega'_{p2}$, the lower and higher passband-edge frequencies. We end up with the specification set $\{N', \omega'_{p1}, \omega'_{p2}, A_p\}$, which our design algorithm can accommodate to obtain the bandpass filter that has an attenuation of A_p at the passband-edges $\omega'_{p1}, \omega'_{p2}$ and has an attenuation of 3 dB at $\omega'_{c1}, \omega'_{c2}$.

If the original bandpass specification set we have is $\{N', \omega'_{c1}, \omega'_{c2}, \omega'_{p1}, \omega'_{p2}\}$, we can again choose ω_c and using the lowpass-to-bandpass transformation and $\omega'_{p1}, \omega'_{p2}$ compute ω_p and then determine A_p. Note that in this case we don't really need ω'_{p1}, ω'_{p2} but only the difference between them, $\omega'_{p2} - \omega'_{p1}$, since this is all that is needed to compute the lowpass-to-bandpass transformation.

EXAMPLE 3

We design a Chebyshev type I bandpass filter with the following set of specifications: $\{N' = 8$, $\omega'_{c1} = 0.2\pi$, $\omega'_{c2} = 0.7\pi$, $\omega'_{p2} - \omega'_{p1} = 0.48\pi\}$. The results are shown in Figure 4–4.

4.6 ELLIPTIC FILTERS

In the elliptic filter case, a simple way to obtain a desired 3 dB point is to design a (lowpass) half-band elliptic filter and use the Constantinides spectral transformations to obtain a lowpass, highpass, bandpass, or bandstop filter with its 3 dB point located at a desired frequency.

Half-band IIR filters have the property that their 3 dB point is located exactly at the half-band frequency, 0.5π. We can use this frequency as the originating point in the Constantinides spectral transformations. The destination point will be the final frequency where a 3 dB attenuation is desired.

It is worth noting that passband and stopband ripple of a half-band IIR filter are related. Therefore, it is not possible to have independent control over each when we design such a filter. However, it turns out that for even relatively modest stopband attenuations, the corresponding passband ripple of a half-band IIR filter is extremely

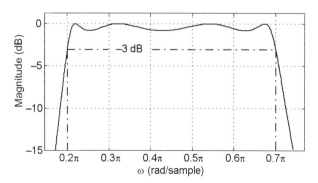

Figure 4–4   A Chebyshev type I bandpass filter with $\{N' = 8, \omega'_{c1} = 0.2\pi,$ $\omega'_{c2} = 0.7\pi,$ $\omega'_{p2} - \omega'_{p1} = 0.48\pi\}$.

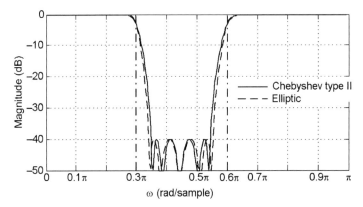

Figure 4–5   A Chebyshev type II and an elliptical bandstop filter with $\{N = 10, \omega'_{c1} = 0.3\pi,$ $\omega'_{c2} = 0.6\pi\}$.

small. As an example, for a stopband attenuation of only 40 dB, the passband ripple is about 4×10^{-4} dB. With such a small passband ripple, elliptical filters designed using the technique outlined here are an attractive alternative to Chebyshev type II designs. The small passband ripple is of little consequence in many applications; however, it allows for a steeper transition between passband and stopband than a comparable Chebyshev type II filter.

EXAMPLE 4

We design a fifth-order elliptic half-band filter with a stopband attenuation of 40 dB. We then transform this filter to a tenth-order bandstop filter with lower and upper 3 dB points given by 0.3π and 0.6π. The results are shown in Figure 4–5. For comparison purposes, a tenth-order Chebyshev type II bandstop filter with the same 40 dB of attenuation is shown. Notice the steeper transitions from passband to stopband in the elliptic filter.

If full control of both passband and stopband ripples is desired for elliptical designs, a solution that allows for such control along with control over the location of the 3 dB point has been given in [4].

4.7 CONCLUSIONS

The design of IIR filters with a given 3 dB frequency point is common in practical applications. We presented ways of designing lowpass, highpass, bandpass, and bandstop Chebyshev type I and II filters that meet a 3 dB constraint. The methodology presented uses bilinear transformations, analytic expressions for the magnitude squared response of lowpass analog Chebyshev type I and II filters, and—in the case of highpass, bandpass, and bandstop filters—the Constantinides spectral transformations.

In the case of elliptical filters, we presented an approach based on the design of half-band elliptical filters and the use of the Constantinides spectral transformations that is a viable alternative to Chebyshev type II designs. We also included a reference to a more general solution that can be used if the half-band approach is inadequate.

The designs attainable with the techniques presented here can be used to substitute practical Butterworth designs by trading off some passband/stopband ripple for better transition performance while meeting the same 3 dB specification. The differences in the phase behavior between Butterworth, Chebyshev, and elliptic filters should be brought into account when phase is an issue. The techniques presented here are robust and applicable to a wide range of IIR filter design specifications; they have been included in the Filter Design Toolbox for MATLAB.

4.8 REFERENCES

[1] S. ORFANIDIS, *Introduction to Signal Processing*. Prentice Hall, Upper Saddle River, NJ, 1996.

[2] A. CONSTANTINIDES, "Spectral Transformations for Digital Filters," *Proc. IEE*, 117, August 1970, pp. 1585–1590.

[3] S. MITRA, *Digital Signal Processing: A Computer-Based Approach*, 2nd ed. McGraw-Hill, New York, 2001.

[4] S. ORFANIDIS, "High-Order Digital Parametric Equalizer Design," *J. Audio Eng. Soc.*, vol. 53, November 2005, pp. 1026–1046.

Chapter 5

Filtering Tricks for FSK Demodulation

David Shiung
MediaTek Inc.

Huei-Wen Ferng
National Taiwan
Univ. of Science and
Technology

Richard Lyons
Besser Associates

In the past decades, economical implementations of digital systems have always been appealing research topics. In this chapter we present a helpful trick used to make the implementation of a digital noncoherent frequency shift keying (FSK) demodulator more economical from a hardware complexity standpoint, with the goal of minimizing its computational workload.

5.1 A SIMPLE WIRELESS RECEIVER

The RF front-end of a wireless receiver, shown in Figure 5–1, is mainly designed using analog components. The information of a noncoherent FSK signal is conveyed at zero-crossing points during each bit period. One often uses a limiter rather than an analog-to-digital converter (ADC) for the sake of hardware simplicity because the output of a limiter contains only two states and functionally behaves as a one-bit ADC. Thus the signal at node B is a sequence of ones and zeros [1]. In the following, we shall call the signal at node A the intermediate-frequency (IF) signal.

Although a variety of receiver architectures (e.g., low-IF and zero IF receivers) exist, we use the superheterodyne system as an example [2]. Owing to fabrication variation, the duty cycle of the IF signal is frequently not equal to 50% [2], [3]. That is, the time duration of pulse "1" is not equal to that of pulse "0". The curve in Figure 5–2(a) shows our analog IF signal and the discrete samples in the figure show the binary sequence at the output of the limiter (node B in Figure 5–1).

Streamlining Digital Signal Processing: A Tricks of the Trade Guidebook, Second Edition. Edited by Richard G. Lyons.
© 2012 the Institute of Electrical and Electronics Engineers. Published 2012 by John Wiley & Sons, Inc.

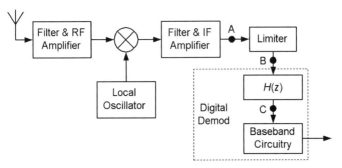

Figure 5–1   A noncoherent FSK demodulator preceded by a limiter and RF front-end circuitry.

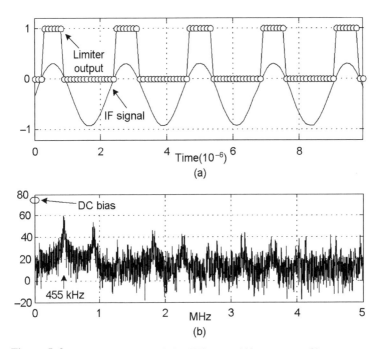

Figure 5–2   Limiter signals: (a) limiter IF input and binary output; (b) output spectrum.

In our application, the f_s sampling frequency of the digital demodulator is 10 MHz; the IF signal is located at frequency $f_I = 455$ kHz, the frequency deviation f_D is set to ± 4 kHz; the data rate R is set to 4 kbps; and a 30% duty cycle is assumed. The signal assumed in this example has a modulation index of 2 (i.e., $|2f_D/R| = 2$) so that the best bit error rate (BER) performance may be achieved for many popular FSK demodulators [4].

Examining the spectra of limiter output sequence in Figure 5–2(b) shows that there are harmonics located at integral multiples of $f_I = 455$ kHz. In addition to those harmonics, there is a very large DC bias (zero-Hz spectral component), on the limiter output signal, that must be eliminated. The primary DSP trick in this discussion is using a comb filter to solve this DC bias (DC offset) problem, which, in addition, will also help us filter the limiter output signal's harmonics.

5.2 USING A COMB FILTER

The comb filter we use to eliminate the DC bias and minimize follow-on filter complexity is the standard N-delay comb filter shown in Figure 5–3(a) [5]. Its periodic passbands and nulls are shown in Figure 5–3(b) for $N = 8$.

The z-domain transfer function of K cascaded comb filters is

$$H(z) = (1 - z^{-N})^K \qquad (5\text{--}1)$$

where in our application $K = 1$.

The following two equations allow us to locate the local minimum (L_{min}) and local maximum (L_{max}) in the positive-frequency range of $|H(\omega)|$.

$$L_{min}: \omega = \frac{2\pi}{N} \cdot i, \quad i = 0, 1, \ldots, \left\lfloor \frac{N}{2} \right\rfloor \qquad (5\text{--}2)$$

$$L_{max}: \omega = \frac{\pi}{N} + \frac{2\pi}{N} \cdot i, \quad i = 0, 1, \ldots, \left\lfloor \frac{N-1}{2} \right\rfloor, \qquad (5\text{--}3)$$

where $\lfloor q \rfloor$ means the integer part of q. Notice that these results are valid only for $N \geq 2$. If we regard the locations of L_{max} as passbands and treat L_{min} as stopbands of $H(z)$, the comb filter can be viewed as a multiband filter.

In general, the coefficient for each tap of an FIR filter is in floating-point format. Thus, multipliers and adders are generally required to perform signal filtering. However, $H(z)$ needs no multiplier and requires only an adder and a few registers. If the input sequence to such a filter is binary, as in our application, the adder of the comb filter becomes a simple XOR logic operation!

If there exists unwanted interference near DC, we may use multiple comb filters in cascade to further suppress these interfering signals. This may sometimes occur when low-frequency noise is present. However, the price paid in this case is

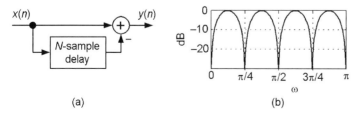

Figure 5–3 An $N = 8$ comb filter: (a) structure; (b) response $|H(\omega)|$.

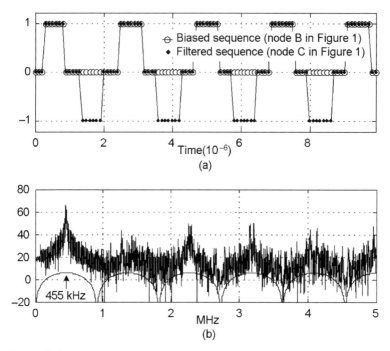

Figure 5–4 DC component removal for a binary IF signal using $H(z)$ with $N = 11$. (Multi-bit comb filter used. See Figure 5–6(b).)

nontrivial adders and multipliers. Some approaches have been proposed to simplify the hardware [6].

If the parameter N of $H(z)$ is carefully chosen such that $2\pi f_1/f_s$ is located at the first L_{max} of the comb filter's magnitude response, all harmonics located at even multiples of f_1 and the zero Hz (DC) component of the hard-limited signal applied to the comb filter can be suppressed. This operation is very helpful in reducing the hardware complexity of the following circuitry and enhancing the system performance of the noncoherent FSK receiver.

The signals into and out of the $H(z)$ comb filter, when $N = 11$, are shown in Figure 5–4(a). Examining the spectra of the filtered output signal, in Figure 5–4(b), we find that only our signal of interest at 455 kHz and odd harmonics of the fundamental frequency f_1 exist. For reference, the comb filter's $|H(\omega)|$ is shown in Figure 5–4(b). Since no low-frequency interference is assumed, we use $H(z)$, whose frequency response is superimposed in Figure 5–4(b), to suppress the DC offset of the IF signal. Comparing Figure 5–2(b) and Figure 5–4(b), it is found that even harmonics and the DC component of the input signal are all diminished to an unnoticeable degree. The odd harmonics are attenuated by follow-on filters where their computational workload is reduced because of the comb filter. Note that this efficient comb filtering also works for an M-ary FSK signaling by properly choosing the system parameters (e.g., f_1, f_s, f_D, R, and N).

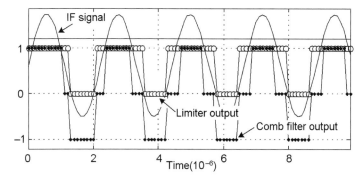

Figure 5–5   Limiter output signal at 60% duty cycle and $N = 11$ comb filter output.

5.3 BENEFICIAL COMB FILTER BEHAVIOR

The locations of the comb filter output harmonics remain unchanged even if the duty cycle of the blue curve in Figure 5–2(a) changes. The length of the comb filter (N) is determined only by the sampling frequency f_s of the digital system and the IF frequency f_I. This fact, which is not at all obvious at first glance, means that our comb filter trick is inherently immune to varying peak-to-peak amplitudes, or changing DC bias levels, of the analog IF signal that occurs in many practical applications. To demonstrate this favorable comb filter behavior, let us assume the bias level (the average) of the analog IF signal in Figure 5–2(a) increased to a positive value such that the limiter's binary output sequence had a duty cycle of 60% as shown in Figure 5–5. There we see how our $N = 11$ comb filter's output sequence is bipolar (DC bias removal has occurred), maintains a 50% duty cycle, and has a fundamental frequency equal to that of the IF signal.

A comb filter may also be used at the baseband to solve the same DC bias removal problem at demodulator output. Unlike that in the IF signal, the DC offset at demodulator output is a consequence of uncorrected frequency drift between the transmitter and the receiver [7]. Thus, we can effectively combat variation in the bit duration of the demodulated data caused by the DC offset occurred in the baseband circuitry using the same methodology.

5.4 CONCLUSIONS

Here we show how an efficient comb filter can be used to both remove the DC offset of our signals and reduce the complexity of follow-on filtering. The advantages of the comb filter are its implementation simplicity (a single exclusive-or circuit) and its ability to combat the problem of DC offset for both IF and baseband signals. Through examples, we show that the duty cycle for signals at the IF can be easily corrected to a 50% duty cycle. This corrected signal is amenable to additional processing at its IF frequency or may be down-converted to a baseband frequency for further processing.

5.5 REFERENCES

[1] M. SIMON and J. SPRINGETT, "The Performance of a Noncoherent FSK Receiver Preceded by a Bandpass Limiter," *IEEE Trans. Commun.*, vol. COM-20, December 1972, pp. 1128–1136.

[2] B. RAZAVI, *RF Microelectronics*, 1st ed. Prentice Hall, Englewood Cliffs, NJ, 1998.

[3] P. HUANG, et al., "A 2-V CMOS 455-kHz FM/FSK Demodulator Using Feedforward Offset Cancellation Limiting Amplifier," *IEEE J. Solid-State Circuits*, vol. 36, January 2001, pp. 135–138.

[4] J. PROAKIS, *Digital Communications*, 3rd ed. McGraw-Hill, New York, 1995.

[5] E. HOGENAUER, "An Economical Class of Digital Filters for Decimation and Interpolation," *IEEE Trans. Acoust. Speech, Signal Processing*, vol. ASSP-29, no. 2, April 1981, pp. 155–162.

[6] K. PARHI, *VLSI Digital Signal Processing Systems*, 1st ed. New York: John Wiley & Sons, 1999.

[7] F. WESTMAN, et al., "A Robust CMOS Bluetooth Radio/Modem System-on-Chip," *IEEE Circuits & Device Mag.*, November 2002, pp. 7–16.

EDITOR COMMENTS

To further describe the comb filter used in this chapter, the details of that filter are shown in Figure 5–6(a), where the BSR means a binary (single-bit) shift register. This filter, whose input is bipolar binary bits $(+1, -1)$, is a specific form of the more general comb filter shown in Figure 5–6(b) where the $w(n)$ input is a multi-bit binary word. In the literature of DSP, such a comb filter is typically depicted as that shown in Figure 5–6(c) where, in this case, $N = 11$.

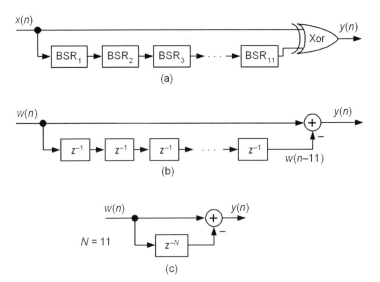

Figure 5–6   Comb filter: (a) single-bit $x(n)$; (b) multi-bit $x(n)$; (c) typical comb filter depiction.

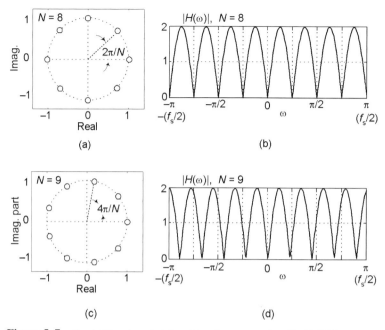

Figure 5–7   Comb filter characteristics: (a) z-plane zeros when $N = 8$; (b) frequency response when $N = 8$; (c) z-plane zeros when $N = 9$; (d) frequency response when $N = 9$.

The Figure 5–6(c) comb filter's z-domain transfer function is

$$H(z) = (1 - z^{-N})   \qquad (5\text{--}4)$$

yielding z-plane zeros located at multiples of the Nth root of one as shown in Figure 5–7(a) when, for example, $N = 8$. That filter's frequency magnitude response is plotted in Figure 5–7(b). Comb filter behavior when $N = 9$ is shown in Figure 5–7(c) and (d).

The frequency magnitude response of this comb filter is

$$|H(\omega)| = 2|\sin(\omega N/2)| \qquad (5\text{--}5)$$

having maximum magnitude of 2 as plotted in Figure 5–7. When N is odd, for example, $N = 9$, a comb filter's z-plane zeros are those shown in Figure 5–7(c) with the filter's associated frequency magnitude response depicted in Figure 5–7(d).

As an aside, we mention here that an *alternate* comb filter can be built using the network in Figure 5–8(a) where addition is performed as opposed to the subtraction in Figure 5–6(c). This alternate comb filter (not applicable to DC bias removal) gives us a bit of design flexibility in using comb filters because it passes low frequency signals due to its frequency magnitude peak at zero Hz. This filter's z-domain transfer function is

$$H_{\text{alt}}(z) = (1 + z^{-N})   \qquad (5\text{--}6)$$

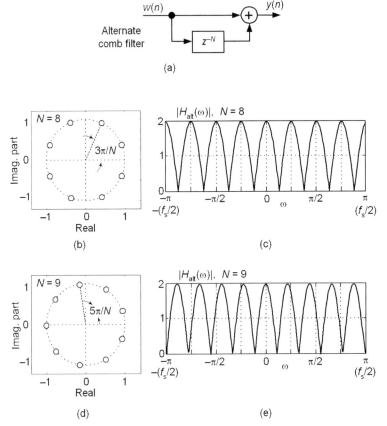

Figure 5–8   Alternate comb filter: (a) structure; (b) z-plane zeros when $N = 8$; (c) frequency response when $N = 8$; (d) z-plane zeros when $N = 9$; (e) frequency magnitude response when $N = 9$.

with its z-plane zeros are located as shown in Figure 5–8(b) when, for example, $N = 8$. That filter's frequency magnitude response is plotted in Figure 5–8(c). When N is odd, for example $N = 9$, the alternate comb filter's z-plane zeros are those shown in Figure 5–8(d) with the $N = 9$ filter's associated frequency magnitude response depicted in Figure 5–8(e).

Following the notation in (5–2) and (5–3), the alternate comb filter's local minimum (L_{min}) and local maximum (L_{max}) in the positive-frequency range of $|H_{alt}(\omega)|$ are

$$L_{min}: \omega = \frac{\pi}{N} + \frac{2\pi}{N} \cdot i, \quad i = 0, 1, \ldots, \left\lfloor \frac{N-1}{2} \right\rfloor \tag{5-7}$$

$$L_{max}: \omega = \frac{2\pi}{N} \cdot i, \quad i = 0, 1, \ldots, \left\lfloor \frac{N}{2} \right\rfloor. \tag{5-8}$$

Chapter 6

Reducing CIC Filter Complexity

Ricardo A. Losada   **Richard Lyons**
The MathWorks, Inc.   **Besser Associates**

Cascaded integrator-comb (CIC) filters are used in high-speed interpolation and decimation applications. This chapter provides tricks to reduce the complexity and enhance the usefulness of CIC filters. The first trick shows a way to reduce the number of adders and delay elements in a multistage CIC interpolation filter. The result is a multiplierless scheme that performs high-order linear interpolation using CIC filters. The second trick shows a way to eliminate the integrators from CIC decimation filters. The benefit is the elimination of unpleasant data word growth problems.

6.1 REDUCING INTERPOLATION FILTER COMPLEXITY

CIC filters are widely used for efficient multiplierless interpolation. Typically, such filters are not used stand-alone; instead, they are usually used as part of a multisection interpolation scheme, generally as the last section where the data have already been interpolated to a relatively high data rate. The fact that the CIC filters need to operate at such high rates makes their multiplierless nature attractive for hardware implementation.

Typical CIC interpolator filters usually consist of cascaded single stages reordered in such a way that all the comb filters are grouped together as are all the integrator filters. By looking closely at a single-stage CIC interpolator, we will show a simple trick to reduce the complexity of a multistage implementation. Because a multistage CIC interpolator has a single-stage CIC interpolator at its core, this trick will simplify the complexity of any CIC interpolator.

Streamlining Digital Signal Processing: A Tricks of the Trade Guidebook, Second Edition. Edited by Richard G. Lyons.
© 2012 the Institute of Electrical and Electronics Engineers. Published 2012 by John Wiley & Sons, Inc.

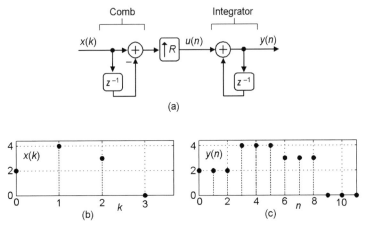

Figure 6–1 CIC interpolation filter: (a) structure; (b) input sequence; (c) output sequence for $R = 3$.

Consider the single-stage CIC interpolator in Figure 6–1(a). The "$\uparrow R$" symbol means insert $R - 1$ zero-valued samples in between each sample of the output of the first adder comb. For illustration purposes, assume $R = 3$. Now imagine an arbitrary $x(k)$ input sequence and assume the delays' initial conditions are equal to zero. When the first $x(0)$ sample is presented at the CIC input, $u(n) = \{x(0), 0, 0\}$. The first $y(n)$ output will be $x(0)$, then this output is fed back and added to zero. So the second $y(n)$ output will be $x(0)$ as well; same for the third $y(n)$. Overall, the first $x(0)$ filter input sample produces the output sequence $y(n) = \{x(0), x(0), x(0)\}$. The next sample input to the comb is $x(1)$ making $u(n) = \{x(1) - x(0), 0, 0\}$. The integrator delay has the value $x(0)$ stored. We add it to $x(1) - x(0)$ to get the next output $y(n) = x(1)$. The value $x(1)$ is stored in the integrator delay and is then added to zero to produce the next output $y(n) = x(1)$. Continuing in this manner, the second input sample to the CIC filter, $x(1)$, produces the output sequence $y(n) = \{x(1), x(1), x(1)\}$. This behavior repeats so that for a given CIC input sequence $x(k)$, the output $y(n)$ is a sequence where each input sample is repeated R times. This is shown in Figures 6–1(b) and 6–1(c) for $R = 3$.

Naturally, when implementing a single-stage CIC filter in real hardware, it is not necessary to use the adders and delays (or the "zero-stuffer") shown in Figure 6–1(a). It is simply a matter of repeating each input sample $R - 1$ times, imposing no hardware cost.

Let us next consider a multistage CIC filter as the one shown in Figure 6–2(a) having three stages. At its core, there is a single-stage CIC interpolator. Our first trick, then, is to replace the innermost single-stage interpolator with a black box, which we call a *hold interpolator*, whose job is to repeat each input sample $R - 1$ times as explained above. Such a reduced-complexity CIC scheme is shown in Figure 6–2(b).

Note that in the comb sections, the number of bits required for each adder tends to increase as we move from left to right. Therefore, the adder and delay that can

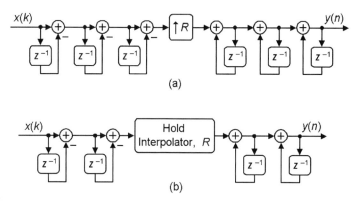

Figure 6–2   Three-stage CIC interpolation filter: (a) standard structure; (b) reduced-complexity structure.

be removed from the comb section will typically be the ones that require the most number of bits in the entire comb section for a standard implementation of a CIC interpolator. So this trick enables us to remove the adder and delay in that section that will save us the most number of bits. However, this is not the case in the integrator section, where we remove the adder and delay that would require the least number of bits of the entire section (but still as many or more than any adder or delay from the comb section).

6.2   AN EFFICIENT LINEAR INTERPOLATOR

Linear interpolators, as their name implies, interpolate samples between two adjacent samples (of the original signal to be interpolated) by placing them in an equidistant manner on the straight line that joins said two adjacent samples [1]. The behavior is illustrated in Figure 6–3 for the case when $R = 3$. Using an example, we now present a very efficient scheme to compute those interpolated samples in a way that requires no multiplies.

As with CIC filters, the performance of linear interpolators is not that great when used on their own. However, linear interpolators are usually not used that way. The reason is that if the interpolation factor R is high, the error introduced by assuming a straight line between two adjacent samples can be large. On the other hand, if interpolation is done in multiple sections, linear interpolation at the end when the signal samples are already very close together will introduce only a small error. Linear interpolation requires a relatively small amount of computation, which is why linear interpolators are used at very high sample rates.

To compute the interpolated $y(n)$ samples using a digital filter, we can use the simple structure shown in Figure 6–4. Let's not concern ourselves with startup transients and assume the filter is in steady state and we have already computed $y(n)$ for $n = 1, 2, 3$ in Figure 6–3(b). We now show how $y(n)$ is computed for $n = 4, 5,$

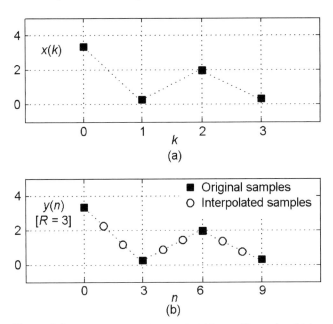

(a)

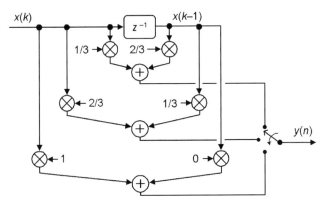

(b)

Figure 6–3   Linear interpolation, $R = 3$: (a) input $x(k)$ samples; (b) interpolated $y(n)$ samples.

Figure 6–4   Interpolation filter, polyphase implementation.

6. The procedure repeats itself from there on. The input to the filter is the signal $x(2)$, while the delay register holds the value of the previous input sample $x(1)$. To compute $y(4)$, we form a weighted average of the $x(1)$ and $x(2)$ samples. Because $y(4)$ is twice as close to $x(1)$ as it is to $x(2)$, we multiply $x(1)$ by 2/3 and $x(2)$ by 1/3. Next, the input signal and the content in the delay element remain the same since the input is operating at the slow rate. In order to compute $y(5)$, we change the weights to be 1/3 and 2/3. Finally, we "compute" $y(6)$ using the weights 0 and 1.

The procedure we have described is in fact a polyphase implementation of a linear interpolator with $R = 3$, as shown in Figure 6–4. There are three polyphase

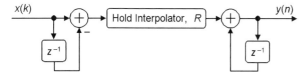

Figure 6–5   Multiplierless linear interpolator.

branches since $R = 3$. Because every input sample needs to be multiplied by each one of the polyphase coefficients, the polyphase implementation of the linear interpolator requires four multiplications for each input sample (we don't count the multiplication by 1 or by 0). In general, for an interpolation factor of R, the number of multiplications required per input sample is $2R - 2$.

By grouping these polyphase coefficients, we can form the transfer function of the $R = 3$ linear interpolator as:

$$H_{\text{linear}}(z) = \frac{1}{3} + \frac{2z^{-1}}{3} + z^{-2} + \frac{2z^{-3}}{3} + \frac{z^{-4}}{3}. \tag{6-1}$$

Now we notice that this transfer function is nothing but a scaled version of a two-stage CIC interpolator. Indeed, for a two-stage CIC we have

$$H_{\text{cic}}(z) = \left(\frac{1-z^{-3}}{1-z^{-1}}\right)^2 = (1 + z^{-1} + z^{-2})^2 = 3H_{\text{linear}}(z). \tag{6-2}$$

Equation (6–2) shows that by using two-stage CIC interpolators, we can implement linear interpolation by $R = 3$ without the need for the $2R - 2$ multipliers. Next, we can use the hold interpolator trick, presented earlier, to simplify the linear interpolator even further.

Using a hold interpolator that inserts $R - 1 = 2$ repeated values for each input sample we can perform efficient linear interpolation as in Figure 6–5. This implementation requires only two adders and two delays no matter what the value of R. The order of this efficient linear interpolator can, of course, be increased by merely increasing the sample repetition factor R.

6.3 NONRECURSIVE CIC DECIMATION FILTERS

CIC filters are computationally efficient and simple to implement. However, there's trouble in paradise. One of the difficulties in using CIC filters is accommodating large data word growth, particularly when implementing integrators in multistage CIC filters. Here's a clever trick that eases the word-width growth problem using nonrecursive CIC decimation filter structures, obtained by means of *polynomial factoring*. These nonrecursive structures achieve computational simplicity through *polyphase decomposition* if the sample rate reduction factor R is an integer power of two.

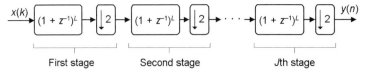

First stage   Second stage   Jth stage

Figure 6–6   Multistage Lth-order nonrecursive CIC structure.

Recall that the transfer function of an Lth-order decimation CIC filter can be expressed in either a recursive form or a nonrecursive form as given by:

$$H_{cic}(z) = \left(\frac{1-z^{-R}}{1-z^{-1}}\right)^L = \left[\sum_{n=0}^{R-1} z^{-n}\right]^L$$

$$= (1+z^{-1}+z^{-2}+\ldots+z^{-R+1})^L. \tag{6–3}$$

Now if the sample rate change factor R is an integer power of two, then $R = 2^J$ where J is a positive integer, and the Lth-order nonrecursive polynomial form of $H_{cic}(z)$ in (6–3) can be factored as

$$H_{cic}(z) = (1+z^{-1})^L(1+z^{-2})^L(1+z^{-4})^L \ldots \left(1+z^{-2^{J-1}}\right)^L. \tag{6–4}$$

The benefit of the factoring given in (6–4) is that the CIC decimation filter can then be implemented with J nonrecursive stages as shown for the multistage CIC filter in Figure 6–6 . This implementation trick eliminates filter feedback loops with their unpleasant binary word-width growth. The data word-widths increase by L bits per stage, while the sampling rate is reduced by a factor of two for each stage.

   This nonrecursive structure has been shown to consume less power than the Figure 6–2(a) recursive implementation for filter orders greater than three and decimation factors larger than eight. Thus the power savings from sample rate reduction is greater than the power consumption increase due to data word-width growth. By the way, the cascade of nonrecursive subfilters in Figure 6–6 are still called CIC filters even though they have no integrators!

   Lucky for us, further improvements are possible with each stage of this nonrecursive structure [2]–[4]. For example, assume $L = 5$ for the first stage in Figure 6–6. In that case the first stage's transfer function is

$$H(z) = (1+z^{-1})^5 = 1+5z^{-1}+10z^{-2}+10z^{-3}+5z^{-4}+z^{-5}$$

$$= 1+10z^{-2}+5z^{-4}+(5+10z^{-2}+z^{-4})z^{-1} = H_1(z)+H_2(z)z^{-1}. \tag{6–5}$$

The last step in (6–5), known as polyphase decomposition, enables a polyphase implementation having two parallel paths as shown in Figure 6–7(a). Because we implement decimation by 2 before the filtering, the new polyphase components are $H_{1'}(z) = 1 + 10z^{-1} + 5z^{-2}$, and $H_{2'}(z) = 5 + 10z^{-1} + z^{-2}$ implemented at half the data rate into the stage. (Reducing data rates as early as possible is a key design goal in

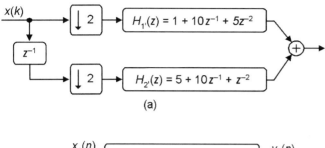

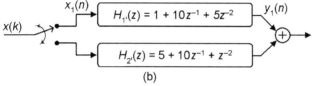

Figure 6–7   Polyphase structure of a single nonrecursive fifth-order CIC stage.

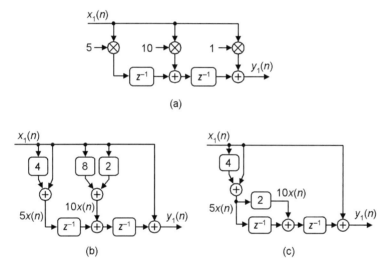

Figure 6–8   Filter component $H_1(z)$: (a) delay line structure; (b) transposed structure; (c) simplified multiplication; (d) substructure sharing.

the implementation of CIC decimation filters.) The initial delay element and the dual decimation by two operations can be implemented by routing the odd-index input samples to $H_{1'}(z)$, and the even-index samples to $H_{2'}(z)$ as shown in Figure 6–7(b). Of course the $H_{1'}(z)$ and $H_{2'}(z)$ polyphase components are implemented in a tapped-delay line fashion.

Fortunately, we can further simplify the $H_{1'}(z)$ and $H_{2'}(z)$ polyphase components. Let's consider the $H_{1'}(z)$ polyphase filter component (implemented in a tapped-delay line configuration) shown in Figure 6–7(b). The transposed version of this filter is presented in Figure 6–8(a) with its flipped coefficient sequence. The adder used in the standard tapped-delay implementation to implement $H_{1'}(z)$ in Figure 6–7(b) must perform two additions per output data sample, while in the transposed structure no adder need perform more than one addition per output sample. So the transposed structure can operate at a higher speed.

The next improvement, proposed by Gao *et al.* [4], uses simplified multiplication, as shown in Figure 6–8(b), by means of arithmetic left shifts and adds. Thus a factor of 5 is implemented as $2^2 + 1$, eliminating all multiplications. Finally, because of the transposed structure, we can use the technique of *substructure sharing* in Figure 6–8(c) to reduce the hardware component count.

The nonrecursive CIC decimation filters described above have the restriction that the R decimation factor must be an integer power of two. That constraint is loosened due to a clever scheme assuming R can be factored into the product of prime numbers. Details of that process, called *prime factorization*, are available in [2] and [5].

6.4 CONCLUSIONS

Here we showed CIC filter tricks to: (1) eliminate one stage in a multistage CIC interpolation filter, (2) perform computationally efficient linear interpolation using a CIC interpolation filter, (3) use nonrecursive structures to eliminate the integrators (along with their unpleasant data word-width growth problems) from CIC decimation filters, and (4) use *polyphase decomposition* and *substructure sharing* to eliminate multipliers in the nonrecursive CIC decimation filters.

6.5 REFERENCES

[1] S. ORFANIDIS, *Introduction to Signal Processing*. Prentice Hall, Upper Saddle River, NJ, 1996.
[2] R. LYONS, Understanding Digital SIGNAL *Processing*, 2nd ed. Prentice Hall, Upper Saddle River, NJ, 2004.
[3] L. ASCARI, et al., "Low Power Implementation of a Sigma Delta Decimation Filter for Cardiac Applications," *IEEE Instrumentation and Measurement Technology Conference*, Budapest, Hungary, May 21–23, 2001, pp. 750–755.
[4] Y. GAO, et al., "Low-Power Implementation of a Fifth-Order Comb Decimation Filter for Multi-Standard Transceiver Applications," *Int. Conf. on Signal Proc. Applications and Technology (ICSPAT)*, Orlando, FL, October 1999. [Online: http://www.pcc.lth.se/events/workshops/1999/posters/Gao.pdf.]
[5] Y. JANG and S. YANG, "Non-recursive Cascaded Integrator-Comb Decimation Filters with Integer Multiple Factors," *44th IEEE Midwest Symposium on Circuits and Systems (MWSCAS)*, Dayton, OH, August 2001, pp. 130–133.

Chapter 7

Precise Filter Design

Greg Berchin
Consultant in Signal Processing

You have just been assigned to a new project at work, in which the objective is to replace an existing analog system with a functionally equivalent digital system. Your job is to design a digital filter that matches the magnitude and phase response of the existing system's analog filter over a broad frequency range.

You are running out of ideas. The bilinear transform and impulse invariance methods provide poor matches to the analog filter response, particularly at high frequencies. Fast convolution requires more computational resources than you have and creates more input/output latency than you can tolerate. What will you do?

This chapter describes an obscure but simple and powerful method for designing a digital filter that approximates an arbitrary magnitude and phase response. If applied to the problem above, it can create a filter roughly comparable in computational burden and latency to the bilinear transform method, with fidelity approaching that of fast convolution. In addition, the method presented here can also be applied to a wide variety of other system identification tasks.

7.1 PROBLEM BACKGROUND

Filter specifications are commonly expressed in terms of passband width and flatness, transition bandwidth, and stopband attenuation. There may also be some general specifications about phase response or time-domain performance, but the exact magnitude and phase responses are usually left to the designer's discretion.

An important exception occurs, however, when a digital filter is to be used to emulate an analog filter. This is traditionally a very difficult problem, because analog systems are described by Laplace transforms, using integration and differentiation, whereas digital systems are described by Z-transforms, using delay. Since the conversion between them is nonlinear, the response of an analog system can only be approximated by a digital system, and vice versa.

Streamlining Digital Signal Processing: A Tricks of the Trade Guidebook, Second Edition. Edited by Richard G. Lyons.
© 2012 the Institute of Electrical and Electronics Engineers. Published 2012 by John Wiley & Sons, Inc.

7.2 TEXTBOOK APPROXIMATIONS

A common method used to create digital filters from analog prototypes is the bilinear transform. This technique can be very effective when specifications are given as passband, transition band, and stopband parameters, as described earlier. And implementation can be very efficient, because the number of coefficients in the digital filter is comparable to the number in its analog prototype. But the bilinear transform suffers from two problems that make a close match to an analog frequency response impossible:

- It squeezes the entire frequency range from zero to infinity in the analog system into the range from DC to half the sampling frequency, inducing *frequency warping*, in the digital system.
- It can only match a prototype frequency response at three frequencies. At all other frequencies the response falls where it may, though its behavior is predictable.

Another design method, called impulse invariance, matches the impulse response of the digital filter to the sampled impulse response of the prototype analog filter. Since there is a one-to-one mapping between impulse response and frequency response in the continuous-time case, one might assume that matching the digital impulse response to the analog impulse response will cause the digital frequency response to match the analog frequency response. Unfortunately, aliasing causes large frequency response errors unless the analog filter rolls off steeply at high frequencies.

A popular method for realizing arbitrary-response filters, called fast convolution, is implemented in the frequency domain by multiplying the FFT of the signal by samples of the analog filter's frequency response, computing the inverse FFT, and so on. While it can be very effective, it is computationally intensive and suffers from high input-output latency.

7.3 AN APPROXIMATION THAT YOU WON'T FIND IN ANY TEXTBOOK

The filter approximation method we present here is called *frequency-domain least-squares* (FDLS). I developed FDLS while I was a graduate student [1] and described it in some conference papers [2], [3], but the technique was all but forgotten after I left school. The FDLS algorithm produces a transfer function that approximates an arbitrary frequency response. The input to the algorithm is a set of magnitude and phase values at a large number (typically thousands) of arbitrary frequencies between 0 Hz and half the sampling rate. The algorithm's output is a set of transfer function coefficients. The technique is quite flexible in that it can create transfer functions containing poles and zeros (IIR), only zeros (FIR), or only poles (autoregressive). Before we can see how the technique works, we need to review some linear algebra and matrix concepts. The FDLS algorithm uses nothing more esoteric than basic linear algebra.

7.4 LEAST SQUARES SOLUTION

Hopefully the reader remembers that in order to uniquely solve a system of equations we need as many equations as unknowns. For example, the single equation with one unknown, $5x = 7$, has the unique solution $x = 7/5$. But the single equation with two unknowns, $5x + 2y = 7$, has multiple solutions for x that depend upon the unspecified value of y: $x = (7 - 2y)/5$.

If another equation is added, such as

$$5x + 2y = 7$$

$$-6x + 4y = 9,$$

then there are unique solutions for both x and y that can be found algebraically or by matrix inversion (denoted in the following by a "-1" superscript):

$$\begin{bmatrix} 5 & 2 \\ -6 & 4 \end{bmatrix} \begin{bmatrix} x \\ y \end{bmatrix} = \begin{bmatrix} 7 \\ 9 \end{bmatrix}.$$

$$\begin{bmatrix} x \\ y \end{bmatrix} = \begin{bmatrix} 5 & 2 \\ -6 & 4 \end{bmatrix}^{-1} \begin{bmatrix} 7 \\ 9 \end{bmatrix} = \begin{bmatrix} \dfrac{1}{8} & -\dfrac{1}{16} \\ \dfrac{3}{16} & \dfrac{5}{32} \end{bmatrix} \begin{bmatrix} 7 \\ 9 \end{bmatrix} = \begin{bmatrix} \left(\dfrac{7}{8} - \dfrac{9}{16} \right) \\ \left(\dfrac{21}{16} + \dfrac{45}{32} \right) \end{bmatrix}.$$

Now let us consider what happens if we add another equation to the pair that we already have (we will see later why we might want to do this), such as

$$5x + 2y = 7$$

$$-6x + 4y = 9$$

$$x + y = 5.$$

There are no values of x and y that satisfy all three equations simultaneously. Well, matrix algebra provides something called the *pseudoinverse* to deal with this situation. It determines the values of x and y that come *as close as possible*, in the least-squares sense, to satisfying all three equations.

Without going into the derivation of the pseudoinverse or the definition of least-squares, let us simply jump to the solution to this new problem:

$$\begin{bmatrix} 5 & 2 \\ -6 & 4 \\ 1 & 1 \end{bmatrix} \begin{bmatrix} x \\ y \end{bmatrix} = \begin{bmatrix} 7 \\ 9 \\ 5 \end{bmatrix} \text{ or }$$

$$\begin{bmatrix} x \\ y \end{bmatrix} \approx \left(\begin{bmatrix} 5 & 2 \\ -6 & 4 \\ 1 & 1 \end{bmatrix}^{\mathrm{T}} \begin{bmatrix} 5 & 2 \\ -6 & 4 \\ 1 & 1 \end{bmatrix} \right)^{-1} \begin{bmatrix} 5 & 2 \\ -6 & 4 \\ 1 & 1 \end{bmatrix}^{\mathrm{T}} \begin{bmatrix} 7 \\ 9 \\ 5 \end{bmatrix} \approx \begin{bmatrix} 0.3716 \\ 2.8491 \end{bmatrix}.$$

(The "T" superscript denotes the matrix transpose.) The pseudoinverse always computes the set of values that comes as close as possible to solving *all* of the equations, when there are more equations than unknowns.

As promised, everything that we have discussed is plain vanilla linear algebra. The mathematical derivation of the matrix inverse and pseudoinverse, and the definition of least-squares, can be found in any basic linear algebra text. And our more mathematically inclined readers will point out that there are better ways than this to compute the pseudoinverse, but this method is adequate for our example.

Now that we remember how to solve simultaneous equations, we need to figure out how to get the equations in the first place. To do that, we first need to review a little DSP.

We will start with the transfer function, which is a mathematical description of the relationship between a system's input and its output. We will assume that the transfer function of our digital filter is in a standard textbook form:

$$\frac{Y(z)}{U(z)} = \frac{b_0 + b_1 z^{-1} + \ldots + b_N z^{-N}}{1 + a_1 z^{-1} + \ldots + a_D z^{-D}}$$

where $U(z)$ is the z-transform of the input signal and $Y(z)$ is the z-transform of the output signal. Furthermore, we assume that the filter is causal, meaning that its response to an input does not begin until after the input is applied. (The filter cannot see into the future.) Under these circumstances the time-domain difference equation that implements our filter is:

$$y(k) = -a_1 y(k-1) - \ldots - a_D y(k-D) + b_0 u(k) + \ldots + b_N u(k-N)$$

where the a and b coefficients are exactly the same as in the transfer function above, k is the time index, $u(k)$ and $y(k)$ are the current values of the input and output, respectively, $u(k-N)$ was the input value N samples in the past, and $y(k-D)$ was the output value D samples in the past. We can write the equation above in matrix form as

$$y(k) = \begin{bmatrix} -y(k-1) & \cdots & -y(k-D) & u(k) & \cdots & u(k-N) \end{bmatrix} \begin{bmatrix} a_1 \\ \vdots \\ a_D \\ b_0 \\ \vdots \\ b_N \end{bmatrix}.$$

We conclude our review of DSP with a consideration of exactly what a frequency response value means. If, for example, the frequency response of a system at a frequency ω_1 is given in magnitude/phase form to be $A_1 \angle \phi_1$, it means that the output amplitude will be A_1 times the input amplitude, and the output phase will be shifted an angle ϕ_1 relative to the input phase, when a steady-state sine wave of frequency ω_1 is applied to the system.

Let's look at an example. If the input to the system described above, at time k, is

$$u_1(k) = \cos(k\omega_1 t_s)$$

where t_s is the sampling period (equal to one over the sampling frequency), then the output will be

$$y_1(k) = A_1 \cos(k\omega_1 t_s + \phi_1).$$

The input and output values at *any* sample time can be determined in a similar manner. For example, the input sample value N samples in the past was

$$u_1(k - N) = \cos((k - N)\omega_1 t_s)$$

and the output sample value D samples in the past was

$$y_1(k - D) = A_1 \cos((k - D)\omega_1 t_s + \phi_1).$$

That's all there is to it. For our purposes, since k represents the current sample time its value can conveniently be set to zero.

That is the end of our review of frequency response, transfer function, and pseudoinverse. Now we will put it all together into a filter design technique. We have just demonstrated that the relationship between input u and output y at any sample time can be inferred from the frequency response value $A \angle \phi$ at frequency ω. We know from our discussion of transfer function that the output is a combination of present and past input and output values, each scaled by a set of b or a coefficients, respectively, the values of which are not yet known. Combining the two, our frequency response value $A_1 \angle \phi_1$ at ω_1, in the example above, provides us with one equation in $D + N + 1$ unknowns:

$$y_1(0) = \begin{bmatrix} -y_1(-1) & \cdots & -y_1(-D) & u_1(0) & \cdots & u_1(-N) \end{bmatrix} \begin{bmatrix} a_1 \\ \vdots \\ a_D \\ b_0 \\ \vdots \\ b_N \end{bmatrix}$$

(Note that k, the current-sample index, has been set to zero.) If we do exactly the same thing again, except this time using frequency response $A_2 \angle \phi_2$ at a *different* frequency ω_2, we obtain a second equation in $D + N + 1$ unknowns:

$$
\begin{bmatrix} y_1(0) \\ y_2(0) \end{bmatrix} = \begin{bmatrix} -y_1(-1) & \cdots & -y_1(-D) & u_1(0) & \cdots & u_1(-N) \\ -y_2(-1) & \cdots & -y_2(-D) & u_2(0) & \cdots & u_2(-N) \end{bmatrix} \begin{bmatrix} a_1 \\ \vdots \\ a_D \\ b_0 \\ \vdots \\ b_N \end{bmatrix}
$$

And if we keep doing this at many more *different* frequencies M than we have unknowns $D + N + 1$, we know from our review of linear algebra that the pseudo-inverse will compute values for the set of coefficients $a_1 \ldots a_D$ and $b_0 \ldots b_N$ that come as close as possible to solving *all* of the equations, *which is exactly what we need to design our filter.* So now we can write:

$$
\begin{bmatrix} y_1(0) \\ y_2(0) \\ \vdots \\ y_M(0) \end{bmatrix} = \begin{bmatrix} -y_1(-1) & \cdots & -y_1(-D) & u_1(0) & \cdots & u_1(-N) \\ -y_2(-1) & \cdots & -y_2(-D) & u_2(0) & \cdots & u_2(-N) \\ \vdots & & \vdots & \vdots & & \vdots \\ -y_M(-1) & \cdots & -y_M(-D) & u_M(0) & \cdots & u_M(-N) \end{bmatrix} \begin{bmatrix} a_1 \\ \vdots \\ a_D \\ b_0 \\ \vdots \\ b_N \end{bmatrix}.
$$

We can employ shortcut matrix notation by defining the following vectors and matrix:

$$
Y = \begin{bmatrix} y_1(0) \\ y_2(0) \\ \vdots \\ y_M(0) \end{bmatrix}, \quad \Theta = \begin{bmatrix} a_1 \\ \vdots \\ a_D \\ b_0 \\ \vdots \\ b_N \end{bmatrix}, \text{ and}
$$

$$
X = \begin{bmatrix} -y_1(-1) & \cdots & -y_1(-D) & u_1(0) & \cdots & u_1(-N) \\ -y_2(-1) & \cdots & -y_2(-D) & u_2(0) & \cdots & u_2(-N) \\ \vdots & & \vdots & \vdots & & \vdots \\ -y_M(-1) & \cdots & -y_M(-D) & u_M(0) & \cdots & u_M(-N) \end{bmatrix};
$$

then

$$Y = X\Theta,$$

and the pseudoinverse is

$$(X^T X)^{-1} X^T Y \approx \Theta,$$

where vector Θ contains the desired filter coefficients.

That's it. That is the entire algorithm. We can now describe our filter design trick as follows:

1. Choose the numerator order N and the denominator order D, where N and D do not have to be equal and either one (but not both) may be zero.

2. Define the M separate input u_m cosine sequences, each of length $(N + 1)$.

3. Compute the M separate output y_m cosine sequences, each of length D (based on $A_m \angle \phi_m$).

4. Fill the X matrix with the input u_m and output y_m cosine sequences.

5. Fill the Y vector with the M output cosine values, $y_m(0) = A_m \cos(\phi_m)$.

6. Compute the pseudoinverse and the Θ vector contains the filter coefficients.

Figures 7–1 and 7–2 show the magnitude and phase, respectively, of a real-world example analog system (dotted), and of the associated bilinear transform (dot-dash), impulse invariance (dashed), and FDLS (solid) approximations. The sampling rate is 240 Hz, and $D = N = 12$. The dotted analog system curves are almost completely obscured by the solid FDLS curves. In this example, the FDLS errors are often 3 to 4 orders of magnitude smaller than those of the other methods.

(In Figure 7–2, the bilinear transform curve is obscured by the FDLS curve at low frequencies and by the impulse invariance curve at high frequencies.)

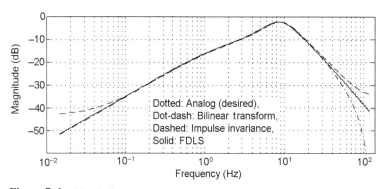

Figure 7–1 Magnitude responses.

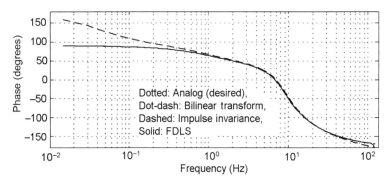

Figure 7–2 Phase responses.

7.5 IMPLEMENTATION NOTES

- I have found no rule of thumb for defining N and D. They are best determined experimentally.
- The selection of a cosine input, rather than sine, was *not* arbitrary. A sine formulation suffers from *zero-crossing* problems at frequencies near half the sampling frequency [1].
- For similar reasons, the algorithm can have some difficulties modeling a system whose phase response approaches odd multiples of 90° near half the sampling frequency. Adding a few samples of artificial delay to the frequency response data, in the form of a linear phase shift, solves this problem. "δ" samples of delay result in a phase shift of "$-\delta\omega_m t_s$", so each equation for $y_m(k)$,

$$y_m(k) = A_m \cos(k\omega_m t_s + \phi_m),$$

becomes:

$$y_m(k) = A_m \cos(k\omega_m t_s + \phi_m - \delta\omega_m t_s).$$

- The algorithm can be used to design 1-, 2-, or even 3-dimensional beamformers [2]. (*Hint*: Making the delay value p, in equations of the general form $u(k - p) = \cos((k - p)\omega t_s)$, an integer in the range $(0, 1, \ldots, N)$ is overly restrictive. Variable p does not have to be an integer and there are some very interesting designs that can be achieved if the integer restriction upon p is removed.)
- A complex-number form of the algorithm exists, in which the inputs and outputs are complex sinusoids e^{jkwt_s}, the filter coefficients can be complex, and the frequency response can be asymmetrical about 0 Hz (or the beam pattern can be asymmetrical about array-normal).
- A *total-least-squares* formulation exists that concentrates all estimator errors into a single diagonal submatrix, for convenient analysis [3].

In terms of the computational complexity of an FDLS-designed filter, the number of feedback and feedforward coefficients are determined by the variables D and N, respectively. As such, an FDLS-designed filter requires $(N + D + 1)$ multiplies and $(N + D)$ additions per filter output sample.

7.6 CONCLUSIONS

FDLS is a powerful method for designing digital filters. As is the case with all approximation techniques, there are circumstances in which the FDLS method works well, and others in which it does not. It does not replace other filter design methods; it provides one more method from which to choose. It is up to the designer to determine whether to use it in any given situation.

7.7 REFERENCES

[1] G. BERCHIN, "A New Algorithm for System Identification from Frequency Response Information," Master's Thesis, University of California–Davis, 1988.
[2] G. BERCHIN and M. SODERSTRAND, "A Transform-Domain Least-Squares Beamforming Technique," *Proceedings of the IEEE Oceans '90 Conference*, Arlington VA, September 1990.
[3] G. BERCHIN and M. SODERSTRAND, "A Total Least Squares Approach to Frequency Domain System Identification," *Proceedings of the 32nd Midwest Symposium on Circuits and Systems*, Urbana IL, August 1989.

EDITOR COMMENTS

Here we present additional examples to illustrate the use of the FDLS algorithm.

Algebraic Example

Recall the FDLS matrix expression

$$\begin{bmatrix} y_1(0) \\ y_2(0) \\ \vdots \\ y_M(0) \end{bmatrix} = \begin{bmatrix} -y_1(-1) & \ldots & -y_1(-D) & u_1(0) & \ldots & u_1(-N) \\ -y_2(-1) & \ldots & -y_2(-D) & u_2(0) & \ldots & u_2(-N) \\ \vdots & & \vdots & \vdots & & \vdots \\ -y_M(-1) & \ldots & -y_M(-D) & u_M(0) & \ldots & u_M(-N) \end{bmatrix} \begin{bmatrix} a_1 \\ \vdots \\ a_D \\ b_0 \\ \vdots \\ b_N \end{bmatrix}$$

which we wrote as $Y = X\Theta$.

Each individual element in the Y column vector is of the form $A_m\cos(\phi_m)$, and each element in the X matrix is of the form $A_1\cos(k\omega_1 t_s + \phi_1)$ or $\cos(k\omega_1 t_s)$. Because all of these elements are of the form of a product (Amplitude)[cos(angle)], each element in Y and X is equal to a constant.

Now if, say, $D = 10$ and $N = 9$, then

$$y_1(0) = A_1 \cos(\phi_1)$$

is the first element of the Y column vector and

$$[-y_1(-1)\ldots -y_1(-10)u_1(0)\ldots u_1(-9)]$$

is the top row of the X matrix expression where

$$-y_1(-1) = -A_1 \cos[(-1)\omega_1 t_s + \phi_1] = -A_1 \cos(-\omega_1 t_s + \phi_1)$$

$$-y_1(-2) = -A_1 \cos[(-2)\omega_1 t_s + \phi_1] = -A_1 \cos(-2\omega_1 t_s + \phi_1)$$

...

$$-y_1(-10) = -A_1 \cos[(-10)\omega_1 t_s + \phi_1] = -A_1 \cos(-10\omega_1 t_s + \phi_1)$$

and

$$u_1(0) = \cos[(0)\omega_1 t_s] = 1$$

$$u_1(-1) = \cos[(-1)\omega_1 t_s] = \cos(-\omega_1 t_s)$$

$$u_1(-2) = \cos[(-2)\omega_1 t_s] = \cos(-2\omega_1 t_s)$$

...

$$u_1(-9) = \cos[(-9)\omega_1 t_s] = \cos(-9\omega_1 t_s).$$

So the top row of the X matrix looks like:

$$[-A_1 \cos(-\omega_1 t_s + \phi_1) \quad -A_1 \cos(-2\omega_1 t_s + \phi_1) \quad \ldots$$
$$-A_1 \cos(-10\omega_1 t_s + \phi_1) \quad 1 \quad \cos(-\omega_1 t_s) \quad \cos(-2\omega_1 t_s) \quad \ldots \quad \cos(-9\omega_1 t_s)].$$

The second row of the X matrix looks like:

$$[-A_2 \cos(-\omega_2 t_s + \phi_2) \quad -A_2 \cos(-2\omega_2 t_s + \phi_2) \quad \ldots$$
$$-A_2 \cos(-10\omega_2 t_s + \phi_2) \quad 1 \quad \cos(-\omega_2 t_s) \quad \cos(-2\omega_2 t_s) \quad \ldots \quad \cos(-9\omega_2 t_s)].$$

And so on.

Numerical Example

Here is an example of the above expressions using actual numbers. Suppose we need to approximate the transfer function coefficients for the system whose frequency magnitude and phase response is that shown in Figure 7–3. Assume that our discrete-system sample rate is 1000 Hz, thus $t_s = 10^{-3}$ seconds, and $N = D = 2$. (The $N = D = 2$

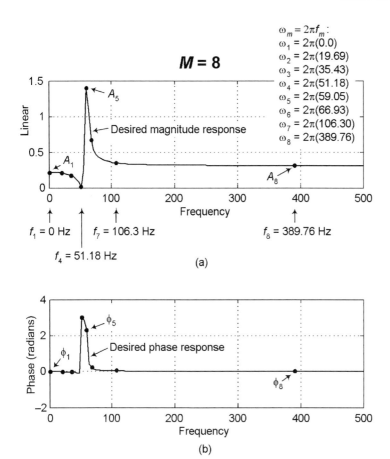

Figure 7–3   Desired magnitude and phase responses.

values means that our final filter will be a second-order filter.) Also assume $M = 8$ and we have the eight A_1-to-A_8 magnitude sample values and the eight ϕ_1-to-ϕ_8 phase samples, shown as dots in Figure 7–3, available to us as input values to the FDLS algorithm.

In matrix form, the target analog system parameters are

$$f_m = \begin{bmatrix} 0.0 \\ 19.6850 \\ 35.4331 \\ 51.1811 \\ 59.0551 \\ 66.9291 \\ 106.299 \\ 389.764 \end{bmatrix} \quad \omega_m t_s = \begin{bmatrix} 0.0 \\ 0.1237 \\ 0.2226 \\ 0.3216 \\ 0.3711 \\ 0.4205 \\ 0.6679 \\ 2.449 \end{bmatrix} \quad A_m = \begin{bmatrix} 0.2172 \\ 0.2065 \\ 0.1696 \\ 0.0164 \\ 1.3959 \\ 0.6734 \\ 0.3490 \\ 0.3095 \end{bmatrix} \quad \phi_m = \begin{bmatrix} 0.0 \\ -0.0156 \\ -0.0383 \\ 3.0125 \\ 2.3087 \\ 0.955 \\ 0.0343 \\ 0.0031 \end{bmatrix}$$

where the f_m vector is in Hz, the $\omega_m t_s$ vector is in radians, and $1 \le m \le 8$. The first two elements of the Y vector are:

$$y_1(0) = A_1 \cos(\phi_1) = 0.2172 \cos(0) = 0.2172.$$

$$y_2(0) = A_2 \cos(\phi_2) = 0.2065 \cos(-0.0156) = 0.2065.$$

The complete Y vector is:

$$Y = \begin{bmatrix} A_1 \cos(\phi_1) \\ A_2 \cos(\phi_2) \\ A_3 \cos(\phi_3) \\ A_4 \cos(\phi_4) \\ A_5 \cos(\phi_5) \\ A_6 \cos(\phi_6) \\ A_7 \cos(\phi_7) \\ A_8 \cos(\phi_8) \end{bmatrix} = \begin{bmatrix} 0.2172 \\ 0.2065 \\ 0.1695 \\ -0.0162 \\ -0.9390 \\ 0.6605 \\ 0.3488 \\ 0.3095 \end{bmatrix}$$

The two elements of the "y_1" part of the first row of the X vector are:

$$-y_1(-1) = -A_1 \cos(-\omega_1 t_s + \phi_1)$$
$$= -0.2172 \cos(-0 + 0) = -0.2172.$$
$$-y_1(-2) = -A_1 \cos(-2\omega_1 t_s + \phi_1)$$
$$= -0.2172 \cos(-0 + 0) = -0.2172.$$

The two elements of the "y_8" part of the eighth row of the X vector are:

$$-y_8(-1) = -A_8 \cos(-\omega_8 t_s + \phi_8)$$
$$= -0.3095 \cos(-2.449 + 0.0031) = 0.2376.$$
$$-y_8(-2) = -A_8 \cos(-2\omega_8 t_s + \phi_8)$$
$$= -0.3095 \cos(-4.898 + 0.0031) = -0.562.$$

The three elements of the "u_1" part of the first row of the X matrix are:

$$u_1(0) = \cos(0) = 1$$
$$u_1(-1) = \cos(-\omega_1 t_s) = \cos(-0) = 1$$
$$u_1(-2) = \cos(-2\omega_1 t_s) = \cos(-0) = 1.$$

The three elements of the "u_8" part of the eighth row of the X matrix are:

$$u_8(0) = \cos(0) = 1$$
$$u_8(-1) = \cos(-\omega_8 t_s) = \cos(-2.449) = -0.7696$$

$$u_8(-2) = \cos(-2\omega_8 t_s) = \cos(-4.898) = 0.1845.$$

The complete X matrix is:

$$X = \begin{bmatrix} -A_1\cos(-1\alpha_1+\phi_1) & -A_1\cos(-2\alpha_1+\phi_1) & \cos(0) & \cos(-1\alpha_1) & \cos(-2\alpha_1) \\ -A_2\cos(-1\alpha_2+\phi_2) & -A_2\cos(-2\alpha_2+\phi_2) & \cos(0) & \cos(-1\alpha_2) & \cos(-2\alpha_2) \\ -A_3\cos(-1\alpha_3+\phi_3) & -A_3\cos(-2\alpha_3+\phi_3) & \cos(0) & \cos(-1\alpha_3) & \cos(-2\alpha_3) \\ -A_4\cos(-1\alpha_4+\phi_4) & -A_4\cos(-2\alpha_4+\phi_4) & \cos(0) & \cos(-1\alpha_4) & \cos(-2\alpha_4) \\ -A_5\cos(-1\alpha_5+\phi_5) & -A_5\cos(-2\alpha_5+\phi_5) & \cos(0) & \cos(-1\alpha_5) & \cos(-2\alpha_5) \\ -A_6\cos(-1\alpha_6+\phi_6) & -A_6\cos(-2\alpha_6+\phi_6) & \cos(0) & \cos(-1\alpha_6) & \cos(-2\alpha_6) \\ -A_7\cos(-1\alpha_7+\phi_7) & -A_7\cos(-2\alpha_7+\phi_7) & \cos(0) & \cos(-1\alpha_7) & \cos(-2\alpha_7) \\ -A_8\cos(-1\alpha_8+\phi_8) & -A_8\cos(-2\alpha_8+\phi_8) & \cos(0) & \cos(-1\alpha_8) & \cos(-2\alpha_8) \end{bmatrix}$$

$$= \begin{bmatrix} -0.2172 & -0.2172 & 1.0 & 1.0 & 1.0 \\ -0.2045 & -0.994 & 1.0 & 0.9924 & 0.9696 \\ -0.1639 & -0.1502 & 1.0 & 0.9753 & 0.9025 \\ 0.0147 & 0.0117 & 1.0 & 0.9487 & 0.8002 \\ 0.5007 & -0.0059 & 1.0 & 0.939 & 0.7370 \\ -0.6564 & -0.5378 & 1.0 & 0.9129 & 0.6667 \\ -0.2812 & -0.0928 & 1.0 & 0.7851 & 0.2328 \\ 0.2376 & -0.0562 & 1.0 & -0.7696 & 0.1845 \end{bmatrix}$$

where $\alpha_1 = \omega_1 t_s$, $\alpha_2 = \omega_2 t_s$, ... $\alpha_8 = \omega_8 t_s$. Given the above Y vector and the X matrix, the FDLS algorithm computes the 2nd-order ($N = D = 2$) transfer function coefficients vector $\theta_{M=8}$ as

$$\theta_{M=8} = \begin{bmatrix} -1.8439 \\ 0.9842 \\ 0.3033 \\ -0.5762 \\ 0.3034 \end{bmatrix}.$$

Treated as filter coefficients, we can write vector $\theta_{M=8}$ as:

$$a_0 = 1$$
$$a_1 = -1.8439$$
$$a_2 = 0.9842$$
$$b_0 = 0.3033$$
$$b_1 = -0.5762$$

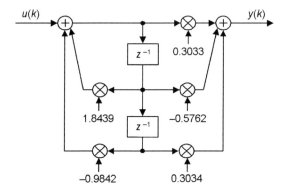

Figure 7–4 2nd-order filter.

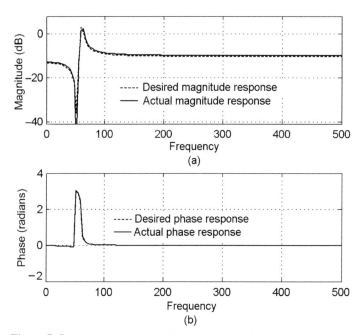

Figure 7–5 Actual versus desired filter magnitude and phase responses.

$$b_2 = 0.3034$$

implemented as the recursive filter network shown in Figure 7–4.

The frequency-domain performance of the filter are the solid curves shown in Figure 7–5. There we see that the $\theta_{M=8}$ coefficients provide an accurate approximation to the desired frequency response in Figure 7–3.

Chapter 8

Turbocharging Interpolated FIR Filters

Richard Lyons
Besser Associates

Interpolated finite impulse response (IFIR) filters—a significant innovation in the field of digital filtering—drastically improve the computational efficiency of traditional Parks-McClellan–designed lowpass FIR filters. IFIR filters is indeed a topic with which every digital filter designer needs to be familiar and, fortunately, tutorial IFIR filter material is available [1]–[5]. This chapter presents two techniques that improve the computational efficiency of IFIR filters.

Before we present the IFIR filter enhancement schemes, let's review the behavior and implementation of standard vanilla-flavored IFIR filters.

8.1 TRADITIONAL IFIR FILTERS

Traditional IFIR filters comprise a *band-edge shaping* subfilter in cascade with a lowpass *masking* subfilter as shown in Figure 8–1, where both subfilters are traditionally implemented as linear-phase tapped-delay FIR filters. The band-edge shaping subfilter has a sparse $h_{be}(k)$ impulse response, with all but every Lth sample being zero, that shapes the final IFIR filter's passband, transition band, and stopband responses. (Integer L is called the *expansion* or *stretch* factor of the band-edge shaping subfilter.) Because the band-edge shaping subfilter's frequency response contains $L - 1$ unwanted periodic passband images, the masking subfilter is used to attenuate those images and can be implemented with few arithmetic operations.

To further describe IFIR filter behavior, consider the $h_{pr}(k)$ impulse response (the coefficients) of a tapped-delay line lowpass FIR *prototype* filter shown in Figure 8–2(a). That prototype filter has the frequency magnitude response shown in Figure 8–2(b). (The f_{pass} and f_{stop} frequency values are normalized to the filter input sample

Streamlining Digital Signal Processing: A Tricks of the Trade Guidebook, Second Edition. Edited by Richard G. Lyons.
© 2012 the Institute of Electrical and Electronics Engineers. Published 2012 by John Wiley & Sons, Inc.

Figure 8–1 Traditional IFIR filter structure.

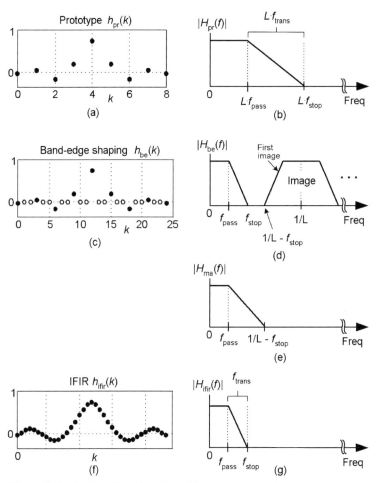

Figure 8–2 IFIR subfilters' impulse and frequency magnitude responses.

rate, f_s, in hertz. For example, $f_{pass} = 0.1$ is equivalent to a cyclic frequency of $f_{pass} = f_s/10$ Hz.) To create a band-edge shaping subfilter, each of the prototype filter's unit delays are replaced with L-unit delays, with the expansion factor L being an integer. If the $h_{pr}(k)$ impulse response of a 9-tap FIR prototype filter is that shown in Figure 8–2(a), the impulse response of the band-edge shaping subfilter, where, for example, $L = 3$, is the $h_{be}(k)$ in Figure 8–2(c). The variable k is merely an integer time-domain index where $0 \le k \le N - 1$.

As we should expect, an L-fold expansion of a time-domain filter impulse response causes an L-fold compression (and repetition) of the frequency-domain $|H_{pr}(f)|$ magnitude response as in Figure 8–2(d). Those repetitive passbands in $|H_{be}(f)|$ centered at integer multiples of $1/L$ (f_s/L Hz)—for simplicity only one passband is shown in Figure 8–2(d)—are called *images*, and those images must be eliminated.

If we follow the band-edge shaping subfilter with a lowpass *masking* subfilter, having the frequency response shown in Figure 8–2(e), whose task is to attenuate the image passbands, we can realize a multistage filter whose $|H_{ifir}(f)|$ frequency magnitude response is shown in Figure 8–2(g). The cascaded $|H_{ifir}(f)|$ frequency magnitude response that we originally set out to achieve is

$$|H_{ifir}(f)| = |H_{be}(f)| \cdot |H_{ma}(f)|. \tag{8–1}$$

The cascade of the two subfilters is the so-called IFIR filter shown in Figure 8–1, with its cascaded impulse response given in Figure 8–2(f). Keep in mind, now, the $h_{ifir}(k)$ sequence in Figure 8–2(f) does not represent the coefficients used in an FIR filter. Sequence $h_{ifir}(k)$ is the convolution of the $h_{be}(k)$ and $h_{ma}(k)$ impulse responses (coefficients).

The goal in IFIR filter design is to find the optimum value for L, denoted as L_{opt}, that minimizes the total number of non-zero coefficients in the band-edge shaping and masking subfilters. Although design curves are available for estimating expansion factor L_{opt} to minimize IFIR filter computational workload [1], [3], the following expression due to Mehrnia and Willson [6] enables L_{opt} to be computed directly using:

$$L_{opt} = \frac{1}{f_{pass} + f_{stop} + \sqrt{f_{stop} - f_{pass}}}. \tag{8–2}$$

Note that once L_{opt} is computed using (8–2) its value is then rounded to the nearest integer.

8.2   TWO-STAGE MASKING IFIR FILTERS

Of this chapter's two techniques for improving the computational efficiency of IFIR filters, the first technique is a scheme called *two-stage-masking*, where the masking subfilter in Figure 8–1 is itself implemented as an IFIR filter using the cascaded arrangement shown in Figure 8–3. This two-stage masking method reduces the computational workload of traditional IFIR filters by roughly 20–30%.

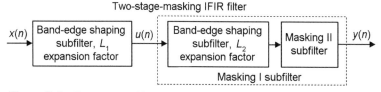

Figure 8–3   Two-stage-masking IFIR filter implementation.

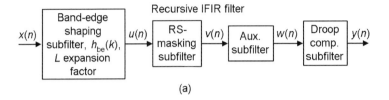

(a)

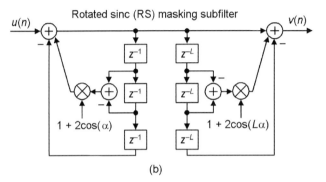

(b)

Figure 8–4   Recursive IFIR filter: (a) subfilters; (b) RS-masking subfilter structure.

The major design issue with the two-stage masking method is that the expansion factors L_1 and L_2 are not directly related to a traditional IFIR filter's optimum L_{opt} value. Fortunately, reference [6] also provides a technique to determine the optimum values for L_1 and L_2 for the two-stage-masking design method.

8.3 RECURSIVE IFIR FILTERS

Our second technique to improve the computational efficiency of IFIR filters replaces Figure 8–1's masking subfilter with a cascade of subfilters as shown in Figure 8–4(a). The detailed structure of the rotated sinc (RS)-masking subfilter is shown in Figure 8–4(b), where L is the band-edge shaping subfilter's expansion factor and a z^{-L} block represents a cascade of L unit-delay elements.

The RS-masking subfilter was originally proposed for use with sigma-delta A/D converters [7]. Factor α is the angular positions (in radians) of the subfilter's poles/zeros on the z-plane near $z = 1$ as shown in Figure 8–5(a). The RS-masking subfilter's z-domain transfer function is

$$H_{RS}(z) = \frac{1 - Az^{-L} + Az^{-2L} - z^{-3L}}{1 - Bz^{-1} + Bz^{-2} - z^{-3}} \tag{8–3}$$

where $A = 1 + 2\cos(L\alpha)$ and $B = 1 + 2\cos(\alpha)$. There are multiple ways to implement (8–3); however, the structure in Figure 8–4(b) is the most computationally efficient.

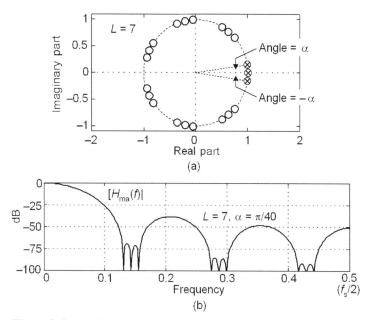

Figure 8–5   Masking subfilter: (a) pole/zero locations; (b) frequency magnitude response.

An RS-masking subfilter with $L = 7$, for example, has transfer function zeros and poles as shown in Figure 8–5(a) with a total of L triplets of zeros evenly spaced around the unit circle. The triplet of zeros near $z = 1$ are overlayed with three poles where pole-zero cancellation makes our masking subfilter a lowpass filter, as illustrated by the $|H_{ma}|$ magnitude response curve in Figure 8–5(b) when $\alpha = \pi/40$, for example.

The expansion factor L in (8–1) determines how many RS-masking subfilter response notches (triplets of zeros) are distributed around the unit circle, and angle α determines the width of those notches. For proper operation, angle α must be less than π/L. As it turns out, thankfully, the RS-masking subfilter's impulse response is both finite in duration and symmetrical so the subfilter exhibits linear phase.

8.4 MASKING ATTENUATION IMPROVEMENT

Because the RS-masking subfilter often does not provide sufficient notch attenuation we can improve that attenuation, at a minimal computational cost, by applying the masking subfilter's $v(n)$ output to an *auxiliary* subfilter as shown in Figure 8–4(a). Using the auxiliary subfilter I in Figure 8–6(a) we can achieve an additional 15–25 dB of notch attenuation. That auxiliary subfilter, when two's complement fixed-point math is used, is a guaranteed-stable cascaded integrator-comb (CIC) filter that places an additional z-domain transfer function zero at the center of each of the band-edge shaping subfilter's passband image frequencies.

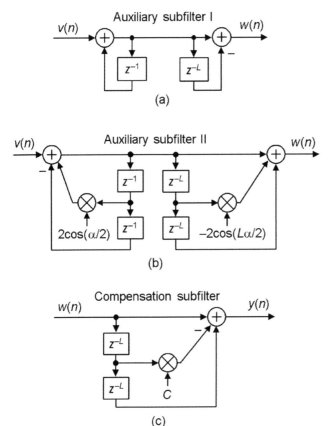

Figure 8–6   Recursive IFIR subfilters: (a) auxiliary subfilter I; (b) auxiliary subfilter II; (c) compensation subfilter.

To achieve greater than 90 dB of masking notch attenuation the auxiliary subfilter II in Figure 8–6(b) may be used. That subfilter places a pair of z-domain zeros within each triplet of zeros of the RS-masking subfilter (except near $z = 1$).

The RS-masking and auxiliary subfilters have some *droop* in their passband magnitude responses. If their cascaded passband droop is intolerable in your application, then some sort of passband droop-compensation must be employed as indicated in Figure 8–4(a). We could incorporate droop compensation in the design of the band-edge shaping subfilter but doing so drastically increases the order (the computational workload) of that subfilter. We could employ the *filter sharpening* technique, described in Chapter 1, to flatten the recursive IFIR filter's drooping passband response. However, that droop-compensation scheme significantly increases the computational workload and the number of delay elements needed in the recursive IFIR filter, as well as restricting the value of L that can be used because we want integer-only delay line lengths. Rather, we suggest using the compensation subfilter shown in Figure 8–6(c) that was originally proposed for use with high-order

CIC lowpass filters [8]. That compensation subfilter has a monotonically rising magnitude response beginning at 0 Hz—just what we need for passband droop compensation. The coefficient C, typically in the range of 4 to 10, is determined empirically.

8.5   RECURSIVE IFIR FILTER DESIGN EXAMPLE

We can further understand the behavior of a recursive IFIR filter by considering the design scenario where we desire a narrowband lowpass IFIR filter with $f_{pass} = 0.01 \ f_s$, a peak–peak passband ripple of 0.2 dB, a transition region bandwidth of $f_{trans} = 0.005 \ f_s$, and 90 dB of stopband attenuation. (A traditional Parks-McClellan–designed FIR lowpass filter satisfying these very demanding design specifications would require a 739-tap filter.) Designing a recursive IFIR filter to meet our design requirements yields an $L = 14$ band-edge shaping subfilter having 52-taps, and an RS-masking subfilter with $\alpha = \pi/42$ radians.

The dashed curve in Figure 8–7(a) shows the design example band-edge shaping subfilter's $|H_{be}(f)|$ frequency magnitude response with its periodically spaced image passbands that must be removed by the RS-masking subfilter. Selecting $\alpha = \pi/42$ for the RS-masking subfilter yields the $|H_{ma}(f)|$ magnitude response shown by the solid curve. The magnitude response of the cascade of just the band-edge shaping and the RS-masking subfilters is shown in Figure 8–7(b). The magnitude response of our final recursive IFIR filter, using an auxiliary subfilter II and a $C = 6$ compensation subfilter, is shown in Figure 8–7(c). The inset in Figure 8–7(c) shows the final recursive IFIR filter's passband flatness measured in dB.

For comparison purposes, Table 8–1 lists the design example computational workload, per filter output sample, for the various filter implementation options discussed above. The "PM FIR" table entry means a single Parks-McClellan–designed, tapped-delay line FIR filter. The "Efficiency gain" column indicates the percent reduction in additions plus multiplications with respect to a standard IFIR filter. (Again, reference [6] proposed the smart idea of implementing a traditional IFIR filter's masking subfilter as, itself, a separate IFIR filter in order to reduce overall computational workload. The results of applying that *two-stage masking* scheme to our IFIR filter design example are also shown in Table 8–1.)

As Table 8–1 shows, for cutting the computational workload of traditional IFIR filters the recursive IFIR filter is indeed a sharp knife.

8.6   RECURSIVE IFIR FILTER IMPLEMENTATION ISSUES

The practical issues to keep in mind when using a recursive IFIR filter are:

- Due to its coefficient symmetry, the band-edge shaping subfilter can be implemented with a "folded delay line" structure to reduce its number of multipliers by a factor of two.

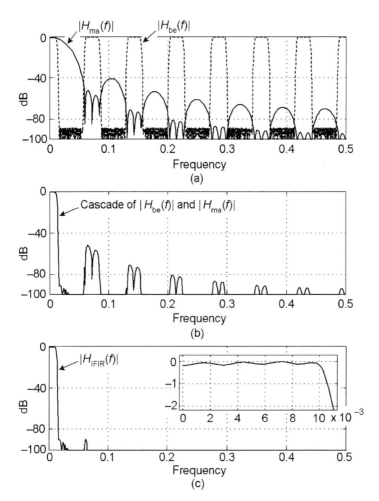

Figure 8–7   Recursive IFIR filter performance: (a) band-edge shaping and RS-masking subfilters' responses; (b) response with no auxiliary filter; (c) response using cascaded auxiliary subfilter II and a compensation subfilter.

Table 8–1   Recursive IFIR Filter Design Example Computational Workload

Lowpass filter:	Adds	Mults	Efficiency gain
PM FIR [Order = 738]	738	739	–
Standard IFIR [$L = 10$]	129	130	–
Two-stage-masking IFIR [$L_1 = 18$, $L_2 = 4$]	95	96	26%
Recursive, RS-masking, IFIR [$L = 14$]	63	57	54%

- While the above two-stage-masking IFIR filter method can be applied to any desired linear-phase lowpass filter, the recursive IFIR filter scheme is more applicable to lowpass filters whose passbands are very narrow with respect to the filter's input sample rate. To optimize an IFIR filter, it's prudent to implement both the two-stage-masking and recursive IFIR filter schemes to evaluate their relative effectiveness.

- The gain of a recursive IFIR filter can be very large (in the hundreds or thousands), particularly for large L and small α, depending on which auxiliary subfilter is used. As such, the recursive IFIR filter scheme is best suited for floating-point numerical implementations. Three options exist that may enable a fixed-point recursive IFIR filter implementation: (1) when filter gain scaling methods are employed; (2) swapping the internal feedback and feed forward sections of a subfilter to minimize data word growth; and (3) reduce the gains of the auxiliary subfilter II and the compensation subfilter by a factor of Q by changing their coefficients from [1, $-2\cos(L\alpha/2)$, 1] and [1, $-C$, 1] to [$1/Q$, $-2\cos(L\alpha/2)/Q$, $1/Q$] and [$1/Q$, $-C/Q$, $1/Q$]. If Q is an integer power of two, then the subfilters' multiply by $1/Q$ can be implemented with binary arithmetic right shifts.

- Whenever we see filter poles lying on the z-domain's unit circle, as in Figure 8–5(a), we should follow Veronica's warning in the movie *The Fly*, and "Be afraid. Be very afraid." Such filters run the risk of being unstable should our finite-precision filter coefficients cause a pole to lie just slightly outside the unit circle. If we find that our quantized-coefficient filter implementation is unstable, at the expense of a few additional multiplications per filter output sample we can use the subfilters shown in Figure 8–8 to guarantee stability while maintaining linear phase. The stability factor r is a constant just as close to, but less than, one as our number format allows.

8.7 RECURSIVE IFIR FILTER DESIGN

Designing a recursive IFIR filter comprises the following steps:

1. Based on the desired lowpass filter's f_{pass} and f_{stop} frequencies, use (8–2) to determine a preliminary value for the band-edge shaping subfilter's integer expansion factor L.

2. Choose an initial value for the RS-masking subfilter's α using $\alpha = 2\pi f_{pass}$. Adjust α to maximize the attenuation of the band-edge shaping subfilter's passband images.

3. If the RS-masking subfilter does not provide sufficient passband image attenuation, employ one of the auxiliary filters in Figure 8–6.

4. Choose an initial value for C (starting with $4 < C < 10$) for the compensation subfilter. Adjust C to provide the desired passband flatness.

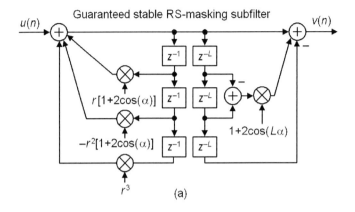

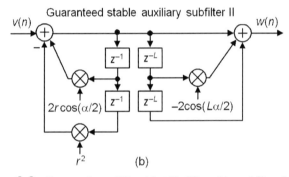

Figure 8–8 Guaranteeing stability: (a) stable RS-masking subfilter; (b) stable auxiliary subfilter II.

5. Continue by increasing L by one (larger values of L yield lower-order band-edge shaping subfilters) and repeat steps 2 through 4 until the either the RS-masking/auxiliary subfilter combination no longer supplies sufficient passband image attenuation or the compensation subfilter no longer can achieve acceptable passband flatness.

6. Sit back and enjoy a job well done.

8.8 RECURSIVE IFIR FILTERS FOR DECIMATION

If our lowpass filtering application requires the $y(n)$ output to be decimated, fortunately the RS-masking subfilter lends itself well to such a sample rate change process. To decimate $y(n)$ by L, we merely rearrange the order of the subfilters' elements so that all feedforward paths and the band-edge shaping subfilter follow the downsample-by-L process as shown in Figure 8–9. Doing so has two advantages: First, the zero-valued coefficients in the band-edge shaping subfilter are eliminated, reducing that subfilter's order by a factor of L. (This converts the band-edge shaping

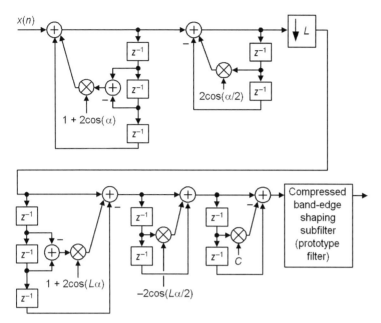

Figure 8–9   RS-masking subfilter cascaded with the auxiliary II and compensation subfilters for decimation by L.

subfilter back to the original prototype filter.) Second, the z^{-L} delay lines in the other subfilters become z^{-1} unit-delays, which reduces signal data storage requirements.

In this decimation scenario the feedback multiplies by the $2\cos(\alpha)$ and $2\cos(\alpha/2)$ coefficients must be performed at the high input sample rate. Reference [7] discusses the possibility of replacing those coefficients with $2 - 2^k$ in order to implement the multiplies with simple high-speed binary shifts and a subtraction.

8.9 AN ALTERNATE MASKING SUBFILTER

For completeness we introduce an alternate structure, shown in Figure 8–10, that can be used in place of the RS-masking subfilter in Figure 8–2(a). This *modified CIC* (MCIC) masking subfilter was inspired by, but is an optimized version of, a cascaded integrator-comb (CIC) filter proposed in reference [9].

The MCIC-masking subfilter, when two's complement fixed-point math is used, is a guaranteed-stable linear phase filter whose frequency response has an infinite-attenuation notch at the center of each of the band-edge shaping subfilter's passband image frequencies. Coefficient K controls the width of the notches, and this subfilter's z-domain transfer function is

$$H_{\text{MCIC}}(z) = \left[\frac{1 - z^{-L}}{1 - z^{-1}}\right]^2 - Kz^{-L+1} \tag{8–4}$$

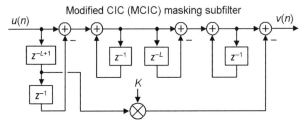

Figure 8–10   MCIC-masking subfilter.

When used in place of the RS-masking subfilter, a $K = 6$ MCIC-masking sub-filter also meets the stringent requirements of the above design example. While the MCIC-masking subfilter requires one fewer multiplier and fewer delay elements than the RS-masking subfilter, sadly the number of delay elements of an MCIC-masking subfilter is not reduced in decimation applications as is an RS-masking subfilter.

8.10 REFERENCES

[1] R. Lyons, "Interpolated Narrowband Lowpass FIR Filters," *IEEE Signal Processing Magazine*, DSP Tips and Tricks Column, vol. 20, no. 1, January 2003, pp. 50–57.

[2] R. Lyons, "Interpolated FIR Filters," *GlobalDSP On-line Magazine*, June 2003. [Online: http://www.globaldsp.com/index.asp?ID=8.]

[3] R. Lyons, *Understanding Digital Signal Processing*, 2nd ed. Prentice Hall, Upper Saddle River, NJ, 2004, pp. 319–331.

[4] F. Harris, *Multirate Signal Processing for Communication Systems*. Prentice Hall, Upper Saddle River, NJ, 2004, pp. 370–375.

[5] P. Vaidyanathan, *Multirate Systems and Filter Banks*. Prentice Hall PTR, Upper Saddle River, NJ, 1992, pp. 134–143.

[6] A. Mehrnia and A. Willson Jr., "On Optimal IFIR filter design," *Proc. of the 2004 International Symp. on Circuits and Systems (ISCAS)*, vol. 3, 23–26 May 2004, pp. 133–136.

[7] L. Lo Presti, "Efficient Modified-Sinc Filters for Sigma-Delta A/D Converters," *IEEE Trans. on Circuits and Systems-II: Analog and Digital Signal Proc.*, vol. 47, no. 11, November 2000, pp. 1204–1213.

[8] H. Oh, S. Kim, G. Choi, and Y. Lee, "On the Use of Interpolated Second-Order Polynomials for Efficient Filter Design in Programmable Downconversion," *IEEE Journal on Selected Areas in Communications*, vol. 17, no. 4, April 1999, pp. 551–560.

[9] T. Saramäki, Y. Neuvo, and S. Mitra, "Design of Computationally Efficient Interpolated FIR Filters," *IEEE Trans. on Circuits and Systems*, vol. 35, no. 1, January 1988, pp. 70–88.

Chapter 9

A Most Efficient Digital Filter: The Two-Path Recursive All-Pass Filter

Fred Harris
San Diego State University

$\mathbf{M}$any of us are introduced to digital recursive filters as mappings from analog prototype structures mapped to the sample data domain by the bilinear Z-transform. These digital filters are normally implemented as a cascade of canonic second-order filters that independently form its two poles and two zeros with two feedback and two feedforward coefficients, respectively. In this chapter we discuss an efficient alternative recursive filter structure based on simple recursive all-pass filters that use a single coefficient to form both a pole and a zero or to form two poles and two zeros.

An all-pass filter has unity gain at all frequencies and otherwise exhibits a frequency-dependent phase shift. We might then wonder that if the filter has unity gain at all frequencies, how it can form a stopband. We accomplish this by adjusting the phase in each path of a two-path filter to obtain destructive cancellation of signals occupying specific spectral bands. Thus the stopband zeros are formed by the destructive cancellation of components in the multiple paths rather than as explicit polynomial zeros. This approach leads to a wide class of very efficient digital filters that require only 25% to 50% of the computational workload of the standard cascade of canonic second order filters. These filters also permit the interchange of the resampling and filtering to obtain further workload reductions.

9.1 ALL-PASS NETWORK BACKGROUND

All-pass networks are the building blocks of every digital filter [1]. All-pass networks exhibit unity gain at all frequencies and a phase shift that varies as a function

Streamlining Digital Signal Processing: A Tricks of the Trade Guidebook, Second Edition. Edited by Richard G. Lyons.
© 2012 the Institute of Electrical and Electronics Engineers. Published 2012 by John Wiley & Sons, Inc.

of frequency. All-pass networks have poles and zeros that occur in (conjugate) reciprocal pairs. Since all-pass networks have reciprocal pole-zero pairs, the numerator and denominator are seen to be reciprocal polynomials. If the denominator is an N-th order polynomial $P_N(Z)$, the reciprocal polynomial in the numerator is $Z^N P_N(Z^{-1})$. It is easily seen that the vector of coefficients that represent the denominator polynomial is reversed to form the vector representing the numerator polynomial. A cascade of all-pass filters is also seen to be an all-pass filter. A sum of all-pass networks is not all-pass and we use these two properties to build our class of filters. Every all-pass network can be decomposed into a product of first- and second-order all-pass networks, thus it is sufficient to limit our discussion to first- and second-order filters, which we refer to as Type I and Type II, respectively. Here we limit our discussion to first- and second-order polynomials in Z and Z^2. The transfer functions of Type-I and Type-II all-pass networks are shown in (9–1) with the corresponding pole-zero diagrams shown in Figure 9–1.

$$H_1(Z) = \frac{(1+\alpha Z)}{(Z+\alpha)}, \qquad H_1(Z^2) = \frac{(1+\alpha Z^2)}{(Z^2+\alpha)}: \qquad \text{Type I}$$

$$H_2(Z) = \frac{(1+\alpha_1 Z + \alpha_2 Z^2)}{(Z^2+\alpha_1 Z + \alpha_2)}, \quad H_2(Z^2) = \frac{(1+\alpha_1 Z^2 + \alpha_2 Z^4)}{(Z^4+\alpha_1 Z^2 + \alpha_2)}: \quad \text{Type II} \tag{9–1}$$

Note that the single sample delay with Z-transform Z^{-1} (or $1/Z$) is a special case of the Type-I all-pass structure obtained by setting the coefficient α to zero. Linear phase delay is all that remains as the pole of this structure approaches the origin

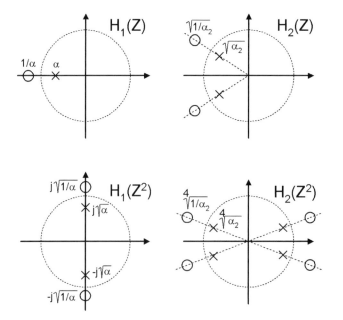

Figure 9–1   Pole-zero structure of Type-I and Type-II all-pass filters of degrees 1 and 2.

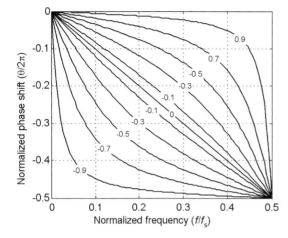

Figure 9–2a Phase response of Type-I all-pass filter, first-order polynomial in Z^{-1}, as function of coefficient α, ($\alpha = 0.9, 0.7, \ldots,$ $-0.7, -0.9$).

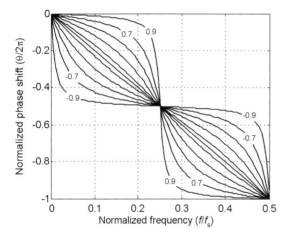

Figure 9–2b Phase response of Type-II all-pass filter, first-order polynomial in Z^{-2}, as function of coefficient α, ($\alpha = 0.9, 0.7, \ldots,$ $-0.7, -0.9$).

while its zero simultaneously approaches infinity. We use the fundamental all-pass filter (Z^{-1}) as the starting point of our design process and then develop the more general case of the Type-I and Type-II all-pass networks to form our class of filters.

A closed-form expression for the phase function of the Type-1 transfer function is obtained by evaluating the transfer function on the unit circle. The result of this exercise is shown in (9–2).

$$\phi = -2\operatorname{atan}\left[\frac{(1+\alpha)}{(1-\alpha)}\tan\left(M\frac{\theta}{2}\right)\right], \quad M = 1, 2. \tag{9–2}$$

The phase response of the first- and second-order all-pass Type-I structures is shown in Figures 9–2(a) and (b).

Note that for the first-order polynomial in Z^{-1}, the transfer function for $\alpha = 0$ defaults to the pure delay, and as expected, its phase function is linear with frequency. The phase function will vary with α and this network can be thought of, and will be used as, the generalized delay element. We observe that the phase function is anchored at its end points ($0°$ at zero frequency and $180°$ at the half sample rate) and that it warps with variation in α. It bows upward (less phase) for positive α and bows downward (more phase) for negative α. The bowing phase function permits us to use the generalized delay to obtain a specified phase shift angle at any frequency. For instance, we note that when $\alpha = 0$, the frequency for which we realize $90°$ phase shift is 0.25 (the quarter sample rate). We can determine a value of α for which the $90°$ phase shift is obtained at any normalized frequency such as at normalized frequency 0.45 ($\alpha = 0.8$) or at normalized frequency 0.05 ($\alpha = -0.73$).

9.2   IMPLEMENTING AND COMBINING ALL-PASS NETWORKS

While the Type-I all-pass network can be implemented in a number of architectures we limit the discussion to the one shown in Figure 9–3. This structure has a number of desirable implementation attributes that are useful when multiple stages are placed in cascade. We observe that the single multiplier resides in the feedback loop of the lower delay register to form the denominator of the transfer function. The single multiplier also resides in the feedforward path of the upper delay register to form the numerator of the transfer function. The single multiplier thus forms all the poles and zeros of this all-pass network and we call attention to this in the equivalent processing block to the right of the filter block diagram.

9.3   TWO-PATH FILTERS

While the M-th order all-pass filter finds general use in an M-path polyphase structure, we restrict this discussion to two-path filters. We first develop an understanding of the simplest two-path structure and then expand the class by invoking a simple set of frequency transformations. The structure of the two-path filter is presented in Figure 9–4. Each path is formed as a cascade of all-pass filters in powers of Z^2. The delay in the lower path can be placed on either side of the all-pass network. When

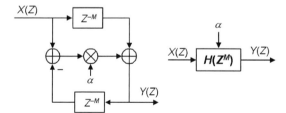

Figure 9–3   Single coefficient Type-I all-pass filter structure.

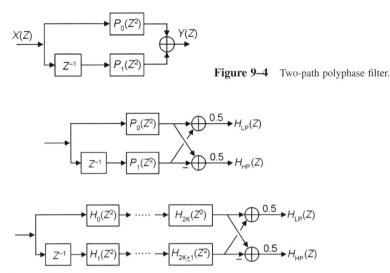

Figure 9–4   Two-path polyphase filter.

Figure 9–5   Two-path all-pass filter.

the filter is implemented as a multirate device the delay is positioned on the side of the filter operating at the higher of the two rates, where it is absorbed by the input (or output) commutator.

This deceivingly simple filter structure offers a surprisingly rich class of filter responses. The two-path all-pass structure can implement halfband lowpass and highpass filters, as well as Hilbert transform filters that exhibit minimum or non-minimum phase response. The two-path filter implements standard recursive filters such as the *Butterworth* and the *elliptical* filter. A MATLAB routine, *tony_des2*, that computes the coefficients of the two-path filter and a number of its frequency-transformed variants is available as described later in this chapter. Also, as suggested earlier, the halfband filters can be configured to embed a 1-to-2 upsampling or a 2-to-1 downsampling operation within the filtering process.

The prototype halfband filters have their 3-dB band edge at the quarter sample rate. All-pass frequency transformations applied to the two-path prototype form arbitrary bandwidth lowpass and highpass complementary filters, and arbitrary center frequency passband and stopband complementary filters. Zero packing the time response of the two-path filter, another trivial all-pass transformation, causes spectral scaling and replication. The zero-packed structure is used in cascade with other filters in iterative filter designs to achieve composite spectral responses exhibiting narrow transition bandwidths with low-order filters.

The specific form of the prototype halfband two-path filter is shown in Figure 9–5. The number of poles (or order of the polynomials) in the two paths differ by precisely one, a condition assured when the number of filter segments in the lower leg is equal to or is one less than the number of filter segments in the upper leg. The structure forms complementary lowpass and highpass filters as the scaled sum and difference of the two paths.

The transfer function of the two-path filter shown in Figure 9–5 is shown in (9–3).

$$H(Z) = P_0(Z^2) \pm Z^{-1} P_1(Z^2)$$

$$P_i(Z^2) = \prod_{k=0}^{K_i} H_{i,k}(Z^2), \quad i = 0, 1$$

$$H_{i,k}(Z^2) = \frac{1 + \alpha(i, k) Z^2}{Z^2 + \alpha(i, k)}$$

(9–3)

In particular, we can examine the simple case of two all-pass filters in each path. The transfer function for this case is shown in (9–4).

$$
\begin{aligned}
H(Z) &= \frac{1 + \alpha_0 Z^2}{Z^2 + \alpha_0} \frac{1 + \alpha_2 Z^2}{Z^2 + \alpha_2} \pm \frac{1}{Z} \frac{1 + \alpha_1 Z^2}{Z^2 + \alpha_1} \frac{1 + \alpha_3 Z^2}{Z^2 + \alpha_3} \\
&= \frac{b_0 Z^9 \pm b_1 Z^8 + b_2 Z^7 \pm b_3 Z^6 + b_4 Z^5 \pm b_4 Z^4 + b_3 Z^3 \pm b_2 Z^2 + b_1 Z^1 \pm b_0}{Z(Z^2 + \alpha_0)(Z^2 + \alpha_1)(Z^2 + \alpha_2)(Z^2 + \alpha_3)}
\end{aligned}
$$

(9–4)

We note a number of interesting properties of this transfer function, applicable to all the two-path prototype filters. The denominator roots are on the imaginary axis restricted to the interval ± 1 to assure stability. The numerator is a linear-phase FIR filter with a symmetric weight vector. As such the numerator roots must appear either on the unit circle, or if off and real, in reciprocal pairs, and if off and complex, in reciprocal conjugate quads. Thus for appropriate choice of the filter weights, the zeros of the transfer function can be placed on the unit circle, and can be distributed to obtain an equal ripple stopband response. In addition, due to the one pole difference between the two paths, the numerator must have a zero at ± 1. When the two paths are added, the numerator roots are located in the left half-plane, and when subtracted, the numerator roots are mirror imaged to the right half-plane forming lowpass and highpass filters, respectively.

The attraction of this class of filters is the unusual manner in which the transfer function zeros are formed. The zeros of the all-pass subfilters reside outside the unit circle (at the reciprocal of the stable pole positions) but migrate to the unit circle as a result of the sum or difference of the two paths. The zeros appear on the unit circle because of destructive cancellation of spectral components delivered to the summing junction via the two distinct paths, as opposed to being formed by numerator weights in the feedforward path of standard biquadratic filters. The stopband zeros are a windfall. They start as the maximum phase all-pass zeros formed concurrently with the all-pass denominator roots by a single shared coefficient and migrate to the unit circle in response to addition of the path signals. This is the reason that the two-path filter requires less than half the multiplies of the standard biquadratic filter.

Figure 9–6 presents the pole-zero diagram for this filter. The composite filter contains nine poles and nine zeros and requires two coefficients for path-0 and two coefficients for path-1. The tony_des2 design routine was used to compute weights for the ninth-order filter with −60 dB equal ripple stopband. The passband edge is

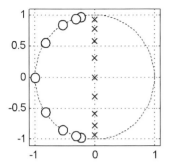

Figure 9–6 Pole-zero diagram of two-path, nine-pole, four-multiplier filter.

located at a normalized frequency of 0.25 and the stopband edge that achieved the desired 60 dB stopband attenuation is located at a normalized frequency of 0.284. This is an elliptical filter with constraints on the pole positions. The denominator coefficient vectors of the filter are listed here in decreasing powers of Z:

Path-0 Polynomial Coefficients:

Filter-0 [1 0 0.101467517]

Filter-2 [1 0 0.612422841]

Path-1 Polynomial Coefficients:

Filter-1 [1 0 0.342095596]

Filter-3 [1 0 0.867647439]

The roots presented here represent the lowpass filter formed from the two-path filter. Figure 9–7 presents the phase slopes of the two paths of this filter as well as the filter frequency response. We note that the zeros of the spectrum correspond to the zero locations on the unit circle in the pole-zero diagram.

The first subplot in Figure 9–7 presents two sets of phase responses for each path of the two-path filter. The dashed lines represent the phase response of the two paths when the filter coefficients are set to zero. In this case the two paths default to two delays in the top path and three delays in the bottom path. Since the two paths differ by one delay, the phase shift difference is precisely 180° at the half-sample rate. When the filter coefficients in each path are adjusted to their design values, the phase response of both paths assume the bowed "lazy S" curve described earlier in Figure 9–2b. Note that at low frequencies, the two phase curves exhibit the same phase profile and that at high frequencies, the two-phase curves maintain the same 180° phase difference. Thus the addition of the signals from the two paths will lead to a gain of 2 in the band of frequencies with the same phase and will result in destructive cancellation in the band of frequencies with 180° phase difference. These two bands of course are the passband and stopband, respectively. We note that the two phase curves differ by exactly 180° at four distinct frequencies as well as the

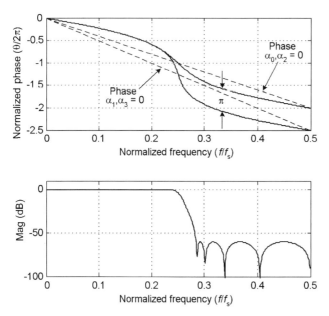

Figure 9–7   Phase slopes and frequency response of two-path, nine-pole, four-multiplier filter.

half-sample rate: those frequencies corresponding to the spectral zeros of the filter. Between these zeros, the filter exhibits stopband side lobes that, by design, are equal ripple.

9.4 LINEAR PHASE TWO-PATH HALFBAND FILTERS

We can modify the structure of the two-path filter to form filters with approximately linear phase response by restricting one of the paths to be pure delay. We accomplish this by setting all the filter coefficients in the upper leg to zero. This sets the all-pass filters in this leg to their default responses of pure delay with poles at the origin. As we pursue the solution to the phase-matching problem in the equal-ripple approximation we find that the all-pass poles must move off the imaginary axis. In order to keep real coefficients for the all-pass filters, we call on the Type-II all-pass filter structure. The lower path then contains first- and second-order filters in Z^2. We lose a design degree of freedom when we set the phase slope in one path to be a constant. Consequently, when we design an equal-ripple group delay approximation to a specified performance we require additional all-pass sections. To meet the same out-of-band attenuation and the same stopband band-edge as the nonlinear phase design of the previous section, our design routine *lineardesign*, determined that we require two first-order filters in Z^2 and three second-order filters in Z^2. This means that eight-coefficients are required to meet the specifications that in the nonlinear phase design only required four coefficients. Path-0 (the delay only path) requires 16 units of delay while the all-pass denominator coefficient vector list is presented below in decreasing powers of Z, which, along with its single delay element, form a seventeenth-order denominator.

Path-0 Polynomial Coefficients:

Delay   [zeros(1, 16)   1]

Path-1 Polynomial Coefficients:

Filter-0 [1   0   0.832280776]

Filter-1 [1   0   −0.421241137]

Filter-2 [1   0   0.67623706   0   0.23192313]

Filter-3 [1   0   0.00359228   0   0.19159423]

Filter-4 [1   0   −0.59689082   0   0.18016931]

Figure 9–8 presents the pole-zero diagram of the linear-phase all-pass filter structure that meets the same spectral characteristics as those outlined in the previous section. We first note that the filter is non-minimum phase due to the zeros outside the unit circle. We also note the near cancellation of the right half-plane pole cluster with the reciprocal zeros of the non-minimum phase zeros. Figure 9–9 presents the phase slopes of the two filter paths and the filter frequency response. We first note that the phase of the two paths is linear; consequently, the group delay is constant over the filter passband. The constant group delay matches the time delay to the peak of the impulse response, which corresponds to the 16-sample time delay of the top path. Of course the spectral zeros of the frequency response coincide with the transfer function zeros on the unit circle.

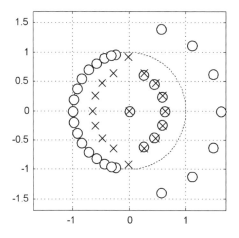

Figure 9–8   Pole-zero diagram of 2-path, 33-pole, 8-multiplier filter.

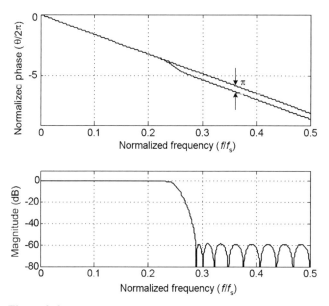

Figure 9–9   Two-path, 33-pole, 8-multiplier filter: phase slopes and frequency response.

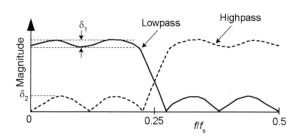

Figure 9–10   Magnitude response of lowpass and highpass halfband filter.

9.5 PASSBAND RESPONSE IN TWO-PATH HALFBAND FILTERS

The all-pass networks that formed the halfband filter exhibit unity gain at all frequencies. These are lossless filters affecting only the phase response of the signal they process. This leads to an interesting relationship between the passband and stopband ripple response of the halfband filter and, in fact, for any of the two-path filters discussed in this chapter. We noted earlier that the two-path filter presents complementary lowpass and the highpass versions of the halfband filter, the frequency responses of which are shown in Figure 9–10, where passband and stopband ripples have been denoted by δ_1 and δ_2, respectively.

The transfer functions of the lowpass and highpass filters are shown in (9–5), where $P_0(Z)$ and $P_1(Z)$ are the transfer functions of the all-pass filters in each of the two paths. The power gain of the lowpass and highpass filters is shown in (9–6).

When we form the sum of the power gains, the cross terms in the pair of products cancel and we obtain the results shown in (9–7).

$$H_{LOW}(Z) = 0.5 \cdot [P_0(Z) + Z^{-1}P_1(Z)]$$
$$H_{HIGH}(Z) = 0.5 \cdot [P_0(Z) - Z^{-1}P_1(Z)]$$

(9–5)

$$|H_{LOW}(Z)|^2 = H_{LOW}(Z)H_{LOW}(Z^{-1})$$
$$= 0.25 \cdot [P_0(Z) + Z^{-1}P_1(Z)] \cdot [P_0(Z^{-1}) + Z\, P_1(Z^{-1})]$$
$$|H_{HIGH}(Z)|^2 = H_{HIGH}(Z)H_{HIGH}(Z^{-1})$$
$$= 0.25 \cdot [P_0(Z) - Z^{-1}P_1(Z)] \cdot [P_0(Z^{-1}) - Z\, P_1(Z^{-1})]$$

(9–6)

$$|H_{LOW}(Z)|^2 + |H_{HIGH}(Z)|^2 = 0.25 \cdot [2 \cdot |P_0(Z)|^2 + 2 \cdot |P_1(Z)|^2] = 1 \quad (9–7)$$

Equation (9–7) tells us that at any frequency, the squared magnitude of the lowpass gain and the squared magnitude of the highpass gain sum to unity. This is a consequence of the filters being lossless. Energy that enters the filter is never dissipated; a fraction of it is available at the lowpass output and the rest of it is available at the highpass output. This property is the reason the complementary lowpass and highpass filters cross at their 3-dB points. If we substitute the gains at peak ripple of the lowpass and highpass filters into (9–7), we obtain (9–8), which we can rearrange and solve for the relationship between δ_1 and δ_2. The result is interesting. We learn here that the peak-to-peak in-band ripple is approximately half the square of the out-of-band peak ripple. Thus, if the out-of-band ripple is –60 dB or one part in one thousand, then the in-band peak-to-peak ripple is half of one part in one million, which is on the order of 5 μ-dB (4.34 μ-dB). The halfband recursive all-pass filter exhibits an extremely small in-band ripple. The in-band ripple response of the two-path nine-pole filter is seen in Figure 9–11.

$$[1 - \delta_1]^2 + [\delta_2]^2 = 1$$
$$[1 - \delta_1] = \sqrt{1 - \delta_2^2} \cong 1 - 0.5 \cdot \delta_2^2$$
$$\delta_1 \cong 0.5 \cdot \delta_2^2$$

(9–8)

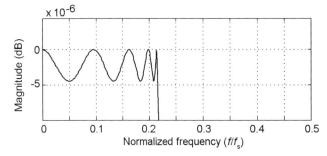

Figure 9–11   In-band ripple level of two-path, nine-pole recursive filter.

9.6 TRANSFORMING THE PROTOTYPE HALFBAND TO ARBITRARY BANDWIDTH

In the previous section we examined the design of two-path halfband filters formed from recursive all-pass first-order filters in the variable Z^2. We did this because we have easy access to the weights of this simple constrained filter, the constraint being stated in (9–5). If we include a requirement that the stopband be equal ripple, the halfband filters we examine are elliptical filters that can be designed from standard design routines. Our program *tony_des2* essentially does this in addition to the frequency transformations we are about to examine. The prototype halfband filter can be transformed to form other filters with specified (arbitrary) bandwidth and center frequency. In this section, elementary frequency transformations are introduced and their impact on the prototype architecture as well as on the system response is reviewed. In particular, the frequency transformation that permits bandwidth tuning of the prototype is introduced first. Additional transformations that permit tuning of the center frequency of the prototype filter are also discussed.

9.7 LOWPASS TO LOWPASS TRANSFORMATION

We now address the transformation that converts a lowpass halfband filter to a lowpass arbitrary-bandwidth filter. Frequency transformations occur when an existing all-pass subnetwork in a filter is replaced by another all-pass subnetwork. In particular, we present the transformation shown in (9–9).

$$\frac{1}{Z} \Rightarrow \frac{1+bZ}{Z+b}; \quad b = \frac{1-\tan(\theta_b/2)}{1+\tan(\theta_b/2)}; \quad \theta_b = 2\pi\frac{f_b}{f_S} \tag{9–9}$$

This is the generalized delay element we introduced in the initial discussion of first-order all-pass networks. We can physically replace each delay in the prototype filter with the all-pass network and then tune the prototype by adjusting the parameter "*b*". We have fielded many designs in which we perform this substitution. Some of these designs are cited in the bibliography. For the purpose of this discussion we perform the substitution algebraically in the all-pass filters comprising the two-path halfband filter, and in doing so generate a second structure for which we will develop and present an appropriate architecture.

We substitute (9–9) into the first order, in Z^2, all-pass filter introduced in (9–2) and rewritten in (9–10),

$$G(Z) = H(Z^2)\Big|_{Z \Rightarrow \frac{Z+b}{1+bZ}}$$

$$G(Z) = \frac{1+\alpha Z^2}{Z^2 + \alpha}\Big|_{Z \Rightarrow \frac{Z+b}{1+bZ}} \tag{9–10}$$

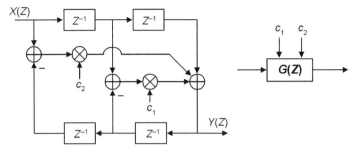

Figure 9–12   Block diagram of general second-order all-pass filter.

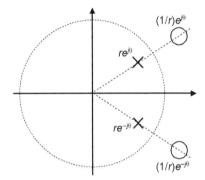

Figure 9–13   Pole zero diagram of generalized second-order all-pass filter.

After performing the indicated substitution and gathering terms, we find the form of the transformed transfer function is as shown in (9–11).

$$G(Z) = \frac{1 + c_1 Z + c_2 Z^2}{Z^2 + c_1 Z + c_2}; \quad c_1 = \frac{2b(1 + \alpha)}{1 + \alpha b^2}; \quad c_2 = \frac{b^2 + \alpha}{1 + \alpha b^2} \qquad (9–11)$$

As expected, when $b \to 0$, $c_1 \to 0$, and $c_2 \to a$, the transformed all-pass filter reverts back to the original first-order filter in Z^2. The architecture of the transformed filter, which permits one multiplier to form the matching numerator and denominator coefficient simultaneously, is shown in Figure 9–12. Also shown is a processing block $G(Z)$ that uses two coefficients, c_1 and c_2. This is seen to be an extension of the one-multiply structure presented in (9–3). The primary difference in the two architectures is the presence of the coefficient and multiplier c_1 associated with the power of Z^{-1}. This term, formerly zero, is the sum of the polynomial roots, and hence is minus twice the real part of the roots. With this coefficient being non-zero, the roots of the polynomial are no longer restricted to the imaginary axis.

   The root locations of the transformed, or generalized, second-order all-pass filter are arbitrary except that they appear as conjugates inside the unit circle, and the poles and zeros appear in reciprocal sets as indicated in Figure 9–13.

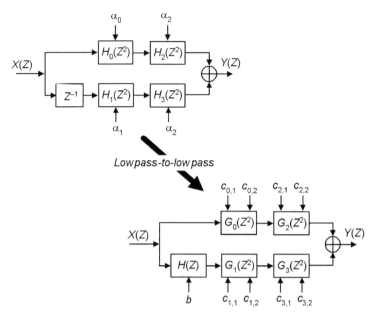

Figure 9–14   Effect on architecture of frequency transformation applied to two-path halfband all-pass filter.

The two-path prototype filter contained one or more one-multiply first-order recursive filters in Z^2 and a single delay. We effect a frequency transformation on the prototype filter by applying the lowpass-to-lowpass transformation shown in (9–10). Doing so converts the one-multiply first-order in Z^2 all-pass filter to the generalized two-multiply second-order all-pass filter and converts the delay, a zero-multiply all-pass filter to the generalized one-multiply first-order in Z all-pass filter. Figure 9–14 shows how applying the frequency transformation affects the structure of the prototype. Note that the nine-pole, nine-zero halfband filter, which is implemented with only four multipliers, now requires nine multipliers to form the same nine poles and nine zeros for the arbitrary-bandwidth version of the two-path network. This is still significantly less than the standard cascade of first- and second-order canonic filters for which the same nine-pole, nine-zero filter would require 23 multipliers.

Figure 9–15 presents the pole-zero diagram of the frequency-transformed prototype filter. The nine poles have been pulled off the imaginary axis, and the nine zeros have migrated around the unit circle to form the reduced-bandwidth version of the prototype.

Figure 9–16 presents the phase response of the two paths and the frequency response obtained by applying the lowpass-to-lowpass frequency transformation to the prototype two-path, four-multiply, halfband filter presented in Figure 9–7. The lowpass-to-lowpass transformation moved passband-edge from normalized frequency 0.25 to normalized frequency 0.1.

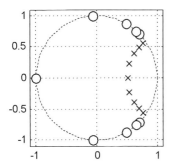

Figure 9–15   Pole-zero diagram obtained by frequency transforming halfband filter to normalized frequency 0.1.

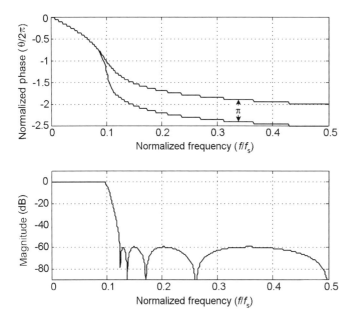

Figure 9–16   Two-path halfband filter: phase response of two paths; frequency response of filter frequency transformed to 0.1 normalized bandwidth.

9.8 LOWPASS-TO-BANDPASS TRANSFORMATION

In the previous section we examined the design of two-path, arbitrary-bandwidth lowpass filters formed from recursive all-pass first- and second-order filters as shown in Figure 9–14. We formed this filter by a transformation of a prototype halfband filter. We now address the second transformation, one that performs the lowpass-to-bandpass transformation. As in the previous section we invoke a frequency transformation wherein an existing all-pass subnetwork in a filter is replaced by another all-pass subnetwork. In particular, we now examine the transformation shown in (9–12).

$$\frac{1}{Z} \Rightarrow -\frac{1}{Z}\frac{1-cZ}{Z-c}; \quad c = \cos(\theta_C); \quad \theta_C = 2\pi\frac{f_C}{f_S} \tag{9–12}$$

This, except for the sign, is a cascade of a delay element with the generalized delay element we introduced in the initial discussion of first-order all-pass networks.

We can physically replace each delay in the prototype filter with this all-pass network and then tune the center frequency of the lowpass prototype by adjusting the parameter c. For our purposes we perform the substitution algebraically in the all-pass filters comprising the two-path predistorted arbitrary-bandwidth filter, and in doing so generate yet a third structure for which we will develop and present an appropriate architecture. We substitute (9–12) into the second-order all-pass filter derived in (9–11) and rewritten in (9–13).

$$F(Z) = G(Z)\big|_{Z \Rightarrow \frac{Z(Z-c)}{(cZ-1)}}$$

$$= \frac{(b^2+\alpha)+2b(1+\alpha)Z+(1+\alpha b^2)Z^2}{(1+\alpha b^2)+2b(1+\alpha)Z+(b^2+\alpha)Z^2}\bigg|_{Z \Rightarrow \frac{Z(Z-c)}{(cZ-1)}} \tag{9–13}$$

After performing the indicated substitution and gathering up terms, we find the form of the transformed transfer function is as shown in (9–14).

$$F(Z) = \frac{1+d_1Z+d_2Z^2+d_3Z^3+d_4Z^4}{Z^4+d_1Z^3+d_2Z^2+d_3Z+d_4}$$

$$d_1 = \frac{-2c(1+b)(1+\alpha b)}{1+\alpha b^2} \quad d_2 = \frac{(1+\alpha)(c^2(1+b)^2+2b)}{1+\alpha b^2} \tag{9–14}$$

$$d_3 = \frac{-2c(1+b)(1+\alpha b)}{1+\alpha b^2} \quad d_4 = \frac{\alpha+b^2}{1+\alpha b^2}$$

As expected, when we let $c \to 0$, d_1 and $d_3 \to 0$, while $d_2 \to c_1$ and $d_4 \to c_2$, the weights default to those of the prototype (arbitrary-bandwidth) filter. The transformation from lowpass to bandpass generates two spectral copies of the original spectrum, one each at the positive- and negative-tuned center frequency. The architecture of the transformed filter, which permits one multiplier to simultaneously form the matching numerator and denominator coefficients, is shown in Figure 9–17. Also shown is a processing block F(Z) that uses four coefficients d_1, d_2, d_3, and d_4. This is seen to be an extension of the two-multiply structure presented in Figure 9–14.

We have just described the lowpass-to-bandpass transformation that is applied to the second-order all-pass networks of the two-path filter. One additional transformation that requires attention is the lowpass-to-bandpass transformation that must be applied to the generalized delay or bandwidth-transformed delay from the prototype halfband filter. We substitute (9–12) into the first-order all-pass filter derived in (9–9) and rewritten in (9–15).

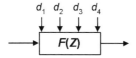

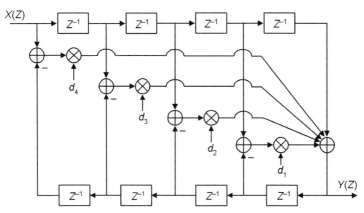

Figure 9–17   Block diagram of general fourth-order all-pass filter.

$$E(Z) = \frac{1+bZ}{Z+b}\bigg|_{Z \Rightarrow \frac{Z(Z-c)}{(cZ-1)}}$$

$$= \frac{(cZ-1)+bZ(Z-c)}{Z(Z-c)+b(cZ-1)} = \frac{-1+c(1-b)Z+bZ^2}{Z^2-c(1-b)Z-b}$$

(9–15)

As expected, when $c \to 1$, the denominator goes to $(Z+b)(Z-1)$ while the numerator goes to $(1+bZ)(Z-1)$ so that the transformed all-pass filter reverts back to the original first-order filter. The distributed minus sign in the numerator modifies the architecture of the transformed second-order filter by shuffling signs in Figure 9–13 to form the filter shown in Figure 9–18. Also shown is a processing block $E(Z)$ that uses two coefficients, e_1 and e_2.

In the process of transforming the lowpass filter to a bandpass filter we convert the two-multiply second-order all-pass filter to a four-multiply fourth-order all-pass filter, and convert the one-multiply lowpass-to-lowpass filter to a two-multiply all-pass filter. The doubling of the number of multiplies is the consequence of replicating the spectral response at two spectral centers of the real bandpass system. Note that the nine-pole, nine-zero arbitrary lowpass filter now requires 18 multipliers to form the 18 poles and 18 zeros for the bandpass version of the two-path network. This is still significantly less than the standard cascade of first- and second-order canonic filters for which the same 18-pole, 18-zero filter would require 45 multipliers. Figure 9–19 shows how the structure of the prototype is affected by applying the lowpass-to-bandpass frequency transformation.

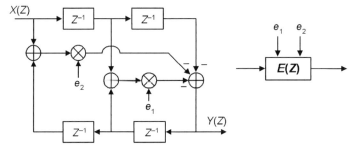

Figure 9–18 Block diagram of lowpass-to-bandpass transformation applied to lowpass-to-lowpass transformed delay element.

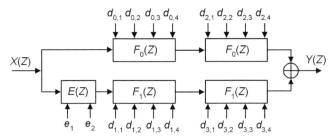

Figure 9–19 Effect on architecture of low-pass-to-band-pass frequency transformation applied to two-path arbitrary-bandwidth all-pass filter.

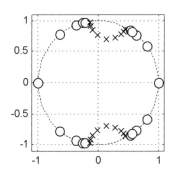

Figure 9–20 Pole-zero plot of two-path all-pass halfband filter subjected to lowpass-to-lowpass and then lowpass-to-bandpass transformations.

Figure 9–20 presents the pole-zero diagram of the frequency-transformed prototype filter. The nine poles defining the lowpass filter have been pulled to the neighborhood of the bandpass center frequency. The nine zeros have also replicated, appearing both below and above the passband frequency.

Figure 9–21 presents the phase response of the two paths and the frequency response obtained by applying the lowpass-to-bandpass frequency transformation to the prototype two-path, nine-multiply, lowpass filter presented in Figure 9–14. The one-sided bandwidth was originally adjusted to a normalized frequency of 0.1 and is now translated to a center frequency of 0.22.

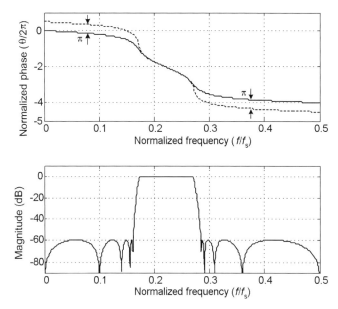

Figure 9–21   Frequency response of two-path all-pass filter subjected to lowpass-to-lowpass and then lowpass-to-bandpass transformations.

9.9 CONCLUSIONS

We have presented a class of particularly efficient recursive filters based on two-path recursive all-pass filters. The numerator and denominator of an all-pass filter have reciprocal polynomials with the coefficient sequence of the denominator reversed in the numerator. The all-pass filters described in this chapter fold and align the numerator registers with the denominator registers so that the common coefficients can be shared and thus form a pole and reciprocal zero with a single multiply. Coefficients are selected via a design algorithm to obtain matching phase profiles in specific spectral intervals with 180° phase offsets in other spectral intervals. When the time series from the two paths are added, the signals residing in the spectral intervals with 180° phase offsets are destructively cancelled.

From the transfer function perspective, the non-minimum phase zeros in the separate numerators migrate to the unit circle as a result of the interaction of the numerator-denominator cross products resulting when forming the common denominator from the separate transfer functions of the two paths. The migration to the unit circle of the essentially free reciprocal zeros, formed while building the system poles, is the reason this class of filters requires less than half the multiplies of a standard recursive filter. The destructive cancellation of the spectral regions defined for the two-path halfband filter is preserved when the filter is subjected to transformations that enable arbitrary bandwidth and arbitrary center frequencies. The one characteristic not preserved under the transformation is the ability to embed 1-to-2 or 2-to-1 multirate processing in the two-path filter. The extension of the two-path filter structure to an M-path structure with similar computational efficiency is the topic of a second presentation. The above-mentioned MATLAB routines may be obtained by request to: fharris@kahuna.sdsu.edu.

9.10   REFERENCES

[1] F. HARRIS, *Multirate Signal Processing for Communication Systems*, Prentice-Hall, Englewood Cliff, NJ, 2004.

EDITOR COMMENTS

To facilitate the reader's software modeling, below we provide the polynomial coefficients for (9–4). Expanding its denominator, we may write (9–4) as:

$$H(Z) = \frac{b_0 Z^9 \pm b_1 Z^8 + b_2 Z^7 \pm b_3 Z^6 + b_4 Z^5 \pm b_4 Z^4 + b_3 Z^3 \pm b_2 Z^2 + b_1 Z^1 \pm b_0}{Z^9 + a_2 Z^7 + a_4 Z^5 + a_6 Z^3 + a_8 Z^1}. \quad (9\text{–}16)$$

The individual lowpass and highpass paths are then described by

$$H_{\text{Low-pass}}(Z) = \frac{b_0 Z^9 + b_1 Z^8 + b_2 Z^7 + b_3 Z^6 + b_4 Z^5 + b_4 Z^4 + b_3 Z^3 + b_2 Z^2 + b_1 Z^1 + b_0}{Z^9 + a_2 Z^7 + a_4 Z^5 + a_6 Z^3 + a_8 Z^1}$$

$$(9\text{–}16')$$

and

$$H_{\text{High-pass}}(Z) = \frac{b_0 Z^9 - b_1 Z^8 + b_2 Z^7 - b_3 Z^6 + b_4 Z^5 - b_4 Z^4 + b_3 Z^3 - b_2 Z^2 + b_1 Z^1 - b_0}{Z^9 + a_2 Z^7 + a_4 Z^5 + a_6 Z^3 + a_8 Z^1}.$$

$$(9\text{–}16'')$$

The b_k and a_i coefficients in (9–16) are:

$$b_0 = \alpha_0 \alpha_2,$$

$$b_1 = \alpha_1 \alpha_3,$$

$$b_2 = \alpha_0 + \alpha_2 + \alpha_0 \alpha_1 \alpha_2 + \alpha_0 \alpha_2 \alpha_3,$$

$$b_3 = \alpha_1 + \alpha_3 + \alpha_0 \alpha_1 \alpha_3 + \alpha_1 \alpha_2 \alpha_3,$$

$$b_4 = \alpha_0 \alpha_1 + \alpha_0 \alpha_3 + \alpha_1 \alpha_2 + \alpha_2 \alpha_3 + \alpha_0 \alpha_1 \alpha_2 \alpha_3 + 1,$$

$$a_2 = \alpha_0 + \alpha_1 + \alpha_2 + \alpha_3,$$

$$a_4 = \alpha_0 \alpha_1 + \alpha_0 \alpha_2 + \alpha_0 \alpha_3 + \alpha_1 \alpha_2 + \alpha_1 \alpha_3 + \alpha_2 \alpha_3,$$

$$a_6 = \alpha_0 \alpha_1 \alpha_2 + \alpha_0 \alpha_1 \alpha_3 + \alpha_0 \alpha_2 \alpha_3 + \alpha_1 \alpha_2 \alpha_3,$$

$$a_8 = \alpha_0 \alpha_1 \alpha_2 \alpha_3.$$

Chapter 10

DC Blocker Algorithms

Randy Yates   **Richard Lyons**

Digital Signal Labs Inc. **Besser Associates**

The removal of a DC bias (a constant-amplitude component) from a signal is a common requirement in signal processing systems, thus a good DC blocking algorithm is a desirable tool to have in one's bag of signal processing tricks. In this chapter we present both a fixed-point DC blocker (using a noise-shaping trick that eliminates a signal's DC bias using fixed-point arithmetic), and a general linear-phase DC blocker network that may prove useful in various DSP applications.

10.1 FIXED-POINT DC BLOCKER

Simple fixed-point implementations of a DC blocker algorithm have a vexing quantization problem that can create more DC bias than they block. In this section we present a *leaky integrator* and a noise-shaping trick called *fraction saving* to eliminate the quantization problem when using fixed-point arithmetic.

The simplest filter that blocks DC is the digital differentiator, whose transfer function is given by

$$R(z) = 1 - z^{-1}. \qquad (10\text{--}1)$$

Having a z-plane zero at $z = 1$, the differentiator has infinite attenuation at 0 Hz and perfectly blocks DC. However, the differentiator also attenuates spectral components close to DC, as shown by the dashed curve in Figure 10–1.

The standard method of shoring up that *drooping* frequency response of a differentiator is to place a pole just inside the z-plane zero at $z = 1$ using a single-pole filter whose transfer function is:

$$S(z) = \frac{1}{1 - pz^{-1}}, \qquad (10\text{--}2)$$

Streamlining Digital Signal Processing: A Tricks of the Trade Guidebook, Second Edition. Edited by Richard G. Lyons.
© 2012 the Institute of Electrical and Electronics Engineers. Published 2012 by John Wiley & Sons, Inc.

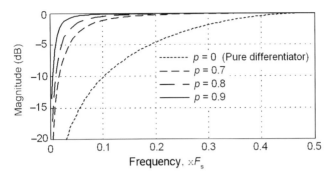

Figure 10–1   Cascaded differentiator/integrator frequency response.

where p is a real pole, and $0 < p < 1$. Equation (10–2) describes a leaky integrator, a nonideal integrator that leaks some energy away rather than perfectly integrating DC. The integrator's response to a DC input will not be an ever-increasing output, but rather an output that increases for a time and then levels off. The term *leaky* is carried over from analog design in which the capacitor used to implement an integrator was imperfect due to the flow of leakage current. The cascaded differentiator/integrator transfer function is given by

$$H(z) = R(z)S(z) = \frac{1 - z^{-1}}{1 - pz^{-1}}. \tag{10–3}$$

The location of the pole $z = p$ of this nonlinear-phase filter presents a tradeoff between the filter's bandwidth and time-domain transient response. The magnitude responses of $H(z)$, for various values of p, are shown in Figure 10–1.

Cascaded filters implementing (10–3) work fine in a floating-point number system and have been described in the literature [1]–[2]. Next we discuss a potential problem in fixed-point implementations of (10–3).

10.2 FIXED-POINT IMPLEMENTATION

In order to understand the effects of fixed-point arithmetic on the DC blocker algorithm in (10–3), let's consider the differentiator and leaky integrator sections independently using the network in Figure 10–2(a). Assume we are using 16-bit input and output binary words as indicated in the figure.

The differentiator's difference equation is

$$w[n] = x[n] - x[n-1]. \tag{10–4}$$

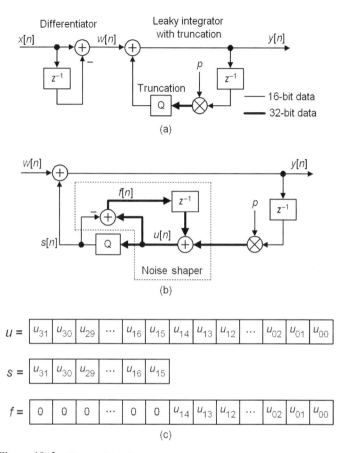

Figure 10–2 Fixed-point DC blocking filter: (a) simple structure; (b) optimized integrator structure; (c) bit-wise representation of u, s, and f.

Note that, at most, the computation in (10–4) will require one more bit than the input (17 bits total) in order to avoid possible overflow, which happens when the input changes sign and the magnitude of the difference is greater than 32768. However, barring that case, we can implement the differentiator without requiring extra precision in the intermediate computation and thus avoid the associated requantization to 16 bits at the output. Also note that the differentiator is nonrecursive, so quantization effects, if they do occur, are not circulated back through the differentiator.

The leaky integrator has a difference equation of

$$y[n] = p \cdot y[n-1] + w[n]. \tag{10–5}$$

Because the goal is to implement (10–5) on a fixed-point (integer) processor, its terms must each be represented as an integer and the operations must be performed using integer arithmetic. If we want the $x[n]$ input and $y[n]$ output to be scaled identically, then we must quantize (truncate) the 32-bit product result in (10–5) to 16 bits. It is precisely this quantization that introduces a potentially significant DC offset error into the $y[n]$ output, and the closer the pole is to the unit circle the larger the potential DC error.

Like any quantizer, the quantization's DC error can be shaped by placing feedback around the quantizer. In order to implement our noise-shaping trick, all that is required is to *save the fraction* of the quantizer input and feed it back in the next update. Hence the term *fraction saving*. This process is shown in Figure 10–2(b). Note that sequence $f[n] = u[n] - s[n]$ is the fractional part of $u[n]$ in two's complement arithmetic after quantization by truncation. This can be illustrated as follows: let u be represented bit-wise as shown in Figure 10–2(c). Notice that s is u with the 15 least-significant bits truncated. So $f = u - s$ is the fractional part of u.

Using the fraction saving scheme in Figure 10–2(b) guarantees that the 16-bit $y[n]$ output will have a DC bias of zero [4]. (Detailed descriptions of noise-shaping can be found in references [5], [6].) In the next section we describe a computationally efficient linear-phase DC blocking network.

10.3 LINEAR-PHASE DC BLOCKER

Another approach to eliminate the DC bias of a signal is to compute the moving average of a signal and subtract that average value from the signal as shown in Figure 10–3(a). The delay element in the figure is a simple delay line, having a length equal to the averager's group delay, enabling us to time-synchronize the averager's $v[n]$ output with the $x[n]$ input in preparation for the subtraction operation.

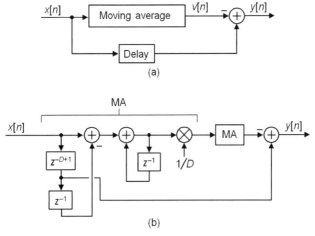

Figure 10–3   Linear phase DC blocker: (a) moving average subtraction method; (b) dual-MA implementation.

The most computationally efficient form of a D-point moving averager (MA) is the network whose transfer function is defined by

$$H_{MA}(z) = \frac{1}{D} \cdot \frac{1-z^{-D}}{1-z^{-1}}. \qquad (10\text{-}6)$$

The second ratio in (10–6) is merely a recursive running sum comprising a D-length comb filter followed by a digital integrator. If D is an integer power of two the $1/D$ scaling in (10–6) can be performed using a binary right shift by $\log_2(D)$ bits.

However, if D is an integer power of two then the MA's group delay is not an integer number of samples, making the synchronization of the delayed $x[n]$ and $v[n]$ difficult. To solve that problem we use two cascaded D-point MAs as shown in Figure 10–3(b). Because the dual-MA has an integer group delay of $D - 1$ samples, our trick is to tap off the first averager's delay line eliminating the bottom-path delay element in Figure 10–3(a).

The magnitude response of our dual-MA DC blocker, for $D = 32$, is shown in Figure 10–4(a). In that figure we show the details of this DC blocker's passband with its peak–peak ripple of 0.42 dB. The frequency axis value of 0.5 corresponds to a cyclic frequency of half the input signal's F_s sample rate. This DC blocker has the desired infinite attenuation at 0 Hz.

What we've created then is a linear-phase, multiplierless, DC blocking network having a narrow transition region near 0 Hz. It's worth noting that standard tapped delay-line, linear phase, FIR filter designs using least-squares error minimization, or the Parks-McClellan method, require more than one hundred taps to approximate our $D = 32$ DC blocking filter's performance

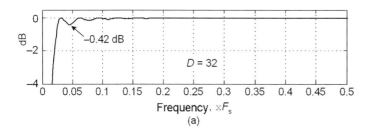

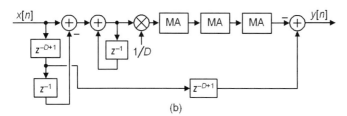

Figure 10–4   DC blocker: (a) $D = 32$ dual-MA passband performance; (b) quad-MA implementation.

On a practical note, the MAs in Figure 10–3(b) contain integrators that can experience data overflow. (An integrator's gain is infinite at DC!) Using two's complement fixed-point arithmetic avoids integrator overflow errors if we ensure that the number of integrator (accumulator) bits are at least:

$$\text{accumulator bits} = \text{number of bits in} q[n] + \lceil \log_2(D) \rceil \qquad (10\text{–}7)$$

where $q[n]$ is the input sequence to an accumulator, and $\lceil k \rceil$ means: if k is not an integer, round it up to the next larger integer.

For a narrower transition region width, in the vicinity of 0 Hz, than that shown in Figure 10–4(a), we can set D to a larger integer power of two, however, this will not reduce the DC blocker's passband ripple.

At the expense of three additional delay lines, and four new addition operations per output sample, we can implement the DC blocker shown in Figure 10–4(b). That quad-MA implementation, having a group delay of $2D - 2$ samples, yields an improved passband peak–peak ripple of only 0.02 dB as well as a reduced-width transition region relative to the dual-MA implementation. The DC blocker in Figure 10–4(b) contains four $1/D$ scaling operations that, of course, can be combined and implemented as a single binary right shift by $4 \cdot \log_2(D)$ bits.

10.4   CONCLUSIONS

We presented a non-linear-phase, but computationally efficient, DC blocking filter that achieves ideal operation when output data quantization is used. In addition we described an alternate DC blocking filter that, at the expense of larger data memory, exhibits a linear phase frequency response.

10.5   REFERENCES

[1] C. DICK and F. HARRIS, "FPGA Signal Processing Using Sigma-Delta Modulation," *IEEE Signal Proc. Magazine*, vol. 17, no. 1, January 2000, pp. 200–235.

[2] K. SHENOI, *Digital Signal Processing in Communications Systems*. Chapman & Hall, New York, 1994, p. 275.

[3] A. BATEMAN, "Implementing a Digital AC Coupling Filter," *GlobalDSP Magazine*, February 2003. [Online: http://www.globaldsp.com/index.asp.]

[4] R. BRISTOW-JOHNSON, *DSP Trick: Fixed-Point DC Blocking Filter With Noise-Shaping*. [Online: http://www.dspguru.com/comp.dsp/tricks/alg/dcblock.htm.]

[5] P. AZIZ, H. SORENSEN, and J. Van Der SPIEGEL, "An Overview of Sigma-Delta Converters," *IEEE Signal Proc. Magazine*, vol. 13, no. 1, January 1996, pp. 61–84.

[6] G. BOURDOPOULOS, A. PNEVMATIKAKIS, V. ANASTASSOPOULOS, and Theodore L. DELIYANNIS, *Delta-Sigma Modulators: Modeling, Design and Applications*. London: Imperial College Press, 2003, pp. 36–40.

Chapter 11

Precise Variable-Q Filter Design

Shlomo Engelberg
Jerusalem College of Technology

$\mathbf{T}$he quality factor, Q, of a bandpass filter is the ratio of the filter's center frequency over its bandwidth. As such, the value Q is a measure of the sharpness of a bandpass filter. Variable-Q filters are filters whose Q can easily be controlled, and these filters have applications in instrumentation, adaptive filtering, audio bandpass and notch filtering, and sonar and ultrasonic systems.

In this chapter we describe two ways to implement a variable-Q digital filter. First we present a simple variable-Q filter and then illustrate why its frequency response is less than ideal. Next we show a trick to modify that filter making its performance nearly ideal. As always, the choice of which filter to use is made based on the user's application.

11.1 A SIMPLE VARIABLE-Q FILTER

The block diagram of a simple variable-Q narrow bandpass filter is given in Figure 11–1(a) where the feedback path contains a notch filter. The frequency responses of several bandpass filters having various Q values are illustrated in Figure 11–1(b).

The simplest (real) notch filter with a notch frequency of F_o Hz is the filter whose transfer function is given by

$$H_N(z) = 1 - 2\cos(2pF_o/F_s)z^{-1} + z^{-2} \tag{11–1}$$

where F_s is the input signal sample rate in Hz. This notch filter attenuates signals whose frequencies are in the vicinity of F_o Hz and allows other frequencies to pass.

Streamlining Digital Signal Processing: A Tricks of the Trade Guidebook, Second Edition. Edited by Richard G. Lyons.
© 2012 the Institute of Electrical and Electronics Engineers. Published 2012 by John Wiley & Sons, Inc.

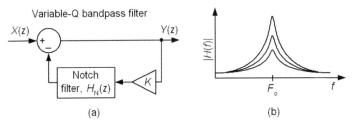

Figure 11–1   Variable-Q filtering: (a) block diagram of a simple variable-Q filter; (b) filter responses for several values of K.

Using the notch filter in the feedback path, the frequency response of the Figure 11–1(b) variable-Q bandpass filter can be represented by

$$H_{BP}(f) = \frac{1}{1 + K \cdot H_N(f)}. \qquad (11\text{--}2)$$

For large values of K, it is clear that when the frequency magnitude response of the notch filter is near zero, the frequency response of the bandpass filter will be very near one. At frequencies where the notch filter response is not near zero, the frequency response of the bandpass filter should be less than one. The bandpass filter in Figure 11–1(b) will tend to pass a narrow band of frequencies in the neighborhood of F_o Hz. The larger K is, the higher the Q and the narrower the filter's passband will be.

The transfer function of the simple variable-Q bandpass filter in Figure 11–1(b) is given by

$$H_{BP}(z) = \frac{\alpha}{1 - \alpha K [2 \cos(2\pi F_o / F_s)] z^{-1} + \alpha K z^{-2}} \qquad (11\text{--}3)$$

where $\alpha = 1/(K + 1)$. In implementing the $H_{BP}(z)$ filter one need only store the previous two values of the output. This is a very simple formula and it can be implemented using a simple microcontroller. The Q of this bandpass filter is controlled by K, and the filter is stable for all positive values of K [1].

In Figure 11–2 we plot the frequency response, the solid curve, of the bandpass filter when $F_s = 5000$ samples/sec, $K = 10$, and the desired center frequency is $F_o = F_s/4 = 1.25$ kHz. This filter has an ideal passband gain of unity (0 dB), just as expected.

11.2   THE PROBLEM

A problem occurs when we set the center frequency of our simple variable-Q filter to a frequency other than $F_s/4$. For example, consider a bandpass filter with the same

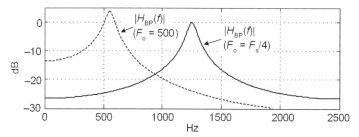

Figure 11–2   Magnitude responses of the simple variable-Q filter when $F_o = F_s/4 = 1.25$ kHz (solid line) and when $F_o = 500$ Hz (dashed line).

parameters as earlier save that the desired center frequency is now set to 500 Hz. The magnitude response plot of the filter is shown as the dashed curve in Figure 11–2. There we find a problem: yes, the simple variable-Q bandpass filter's gain is the desired value of unity (0 dB) at 500 Hz, but the filter has a gain of greater than unity (>0 dB) at a frequency just above 500 Hz. And that gain peak is located, by definition, at the center frequency of the filter. So our simple bandpass filter has a peak gain greater than the desired value of unity, and its center frequency is not the desired 500 Hz.

Interestingly enough, it can be shown that our simple bandpass filter's gain may be greater than unity when the real part of the notch filter's frequency response is negative, and it is this unpleasant behavior that we wish to correct.

11.3 THE SOLUTION

There is a way to cure the problem with the dashed curve in Figure 11–2 but, as is so often the case, it requires complicating the filter somewhat. Our trick is to use the standard lowpass-to-lowpass digital filter transformation technique [1]–[3], but rather than moving the edge of the passband of a prototype lowpass filter using the transformation, we move the center frequency of our ideal variable-Q bandpass filter from $F_s/4$ to our desired center frequency F_D. The standard lowpass-to-lowpass transformation requires that the variable z in the transfer function of the prototype filter be replaced by the function

$$\frac{z-a}{1-az} \qquad (11\text{--}4)$$

where the value of a is given by

$$a = \frac{\tan(\pi/4 - \pi F_D/F_s)}{\sin(2\pi F_D/F_s) + \cos(2\pi F_D/F_s) \cdot \tan(\pi/4 - \pi F_D/F_s)}. \qquad (11\text{--}5)$$

This transformation stretches the response of the prototype filter but does not change its magnitude, thus eliminating the undesirable peaking seen in the dashed curve of Figure 11–2.

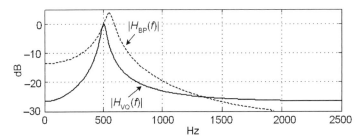

Figure 11–3   Magnitude responses of the original ($|H_{BP}(f)|$) and improved ($|H_{VQ}(f)|$) variable-Q filters when $F_o = 500$ Hz.

Automating the calculation of (11–5) is simple and takes only a few lines of code. Our final variable-Q filter's transfer function has the form

$$H_{VQ}(z) = \frac{1 - 2az^{-1} + a^2 z^{-2}}{(\beta + Ka^2) - 2a(\beta + K)z^{-1} + (a^2\beta + K)z^{-2}} \tag{11–6}$$

where $\beta = 1 + K$, and a satisfies (11–5).

Implementing this filter requires storing the preceding two values of the filter's output *and* the preceding two values of the filter's input. We are trading the simplicity of our first design for improved accuracy. Repeating our last example with $F_s = 5000$ samples/sec, $K = 10$, $F_{old} = F_s/4 = 1.25$ kHz, and the desired center frequency is $F_{new} = 500$ Hz, we find that $a = 0.2065$. Using this design, we find that the frequency response of the improved filter, shown as the solid $|H_{VQ}(f)|$ curve in Figure 11–3, is nearly ideal. (For ease of comparison, we have included the simple bandpass filter's magnitude response in Figure 11–3 as the dashed curve.)

Using the improved bandpass filter design method, defined by (11–6), we have

- exact control over the filter's center frequency,
- forced the filter's passband gain to always be unity (0 dB),
- improved the filter's frequency response symmetry.

As with all recursive filters implemented using a fixed-point number format, having finite-width coefficient values, the precise locations of the variable-Q filter's poles should be examined to ensure that they reside within the z-domain's unit circle.

11.4 CONCLUSIONS

We showed how to analyze and design simple variable-Q bandpass filters. The simplest variable-Q filter suffers from two shortcomings. Its center frequency is not exactly the planned-for center frequency, and the amplitude at the center frequency is not exactly the planned-for amplitude of unity. We presented a slightly more complicated filter in (11–6) that meets all the variable-Q filter's desired specifications quite precisely.

11.5   REFERENCES

[1] S. Engelberg, *Digital Signal Processing: An Experimental Approach.* Springer, London, 2008, pp. 151–153, 168–170.
[2] J. Proakis and D. Manolakis, *Digital Signal Processing: Principles, Algorithms, and Applications,* 3rd ed. Prentice Hall, Upper Saddle River, NJ, 1996, pp. 698–700.
[3] A. Oppenheim and R. Schafer, *Discrete-Time Signal Processing,* 2nd ed. Prentice Hall, Englewood Cliffs, NJ, 1989, pp. 430–435.
[4] Alen Docef, Assoc. Prof., Virginia Commonwealth University School of Engineering, Private Communication, September 2008.

EDITOR COMMENTS

There is a simplification to this chapter's expression (11–5) [4]. Here is the story. From references [1]–[3], the standard lowpass-to-lowpass digital filter transformation process defines (11–5)'s parameter a to be:

$$a = \frac{\tan(\pi F_o/F_s - \pi F_D/F_s)}{\sin(2\pi F_D/F_s) + \cos(2\pi F_D/F_s)\tan(\pi F_o/F_s - \pi F_D/F_s)}$$

where F_o is the critical frequency of the original filter, F_D is the desired critical frequency of the transformed filter, and F_s is the filters' sample rate. This chapter's "trick" is to set $F_o = F_s/4$, yielding the (11–5) expression for parameter a repeated here as:

$$a = \frac{\tan(\pi/4 - \pi F_D/F_s)}{\sin(2\pi F_D/F_s) + \cos(2\pi F_D/F_s)\cdot \tan(\pi/4 - \pi F_D/F_s)}. \tag{11–5}$$

The denominator of (11–5) can be expressed in the form:

$$\sin(2x) + \cos(2x)\cdot \tan(\pi/4 - x) = \frac{\sin(2x)\cos(\pi/4 - x) + \cos(2x)\sin(\pi/4 - x)}{\cos(\pi/4 - x)}. \tag{11–I}$$

Realizing that $\sin(\alpha)\cdot\cos(\beta) + \cos(\alpha)\cdot\sin(\beta) = \sin(\alpha + \beta)$, and $\sin(\theta) = \cos(\pi/2-\theta)$, (11–I) becomes:

$$\frac{\sin(2x + \pi/4 - x)}{\cos(\pi/4 - x)} = \frac{\sin(\pi/4 + x)}{\cos(\pi/4 - x)} = \frac{\cos(\pi/4 - x)}{\cos(\pi/4 - x)} = 1 \tag{11–II}$$

Because (11–II) is the denominator of (11–5), we obtain a simplified version of (11–5-simple) as:

$$a = \tan(\pi/4 - \pi F_D/F_s). \tag{11–5-simple}$$

Chapter 12

Improved Narrowband Lowpass IIR Filters in Fixed-Point Systems

Richard Lyons
Besser Associates

$\mathbf{D}$ue to their resistance to quantized-coefficient errors, traditional second-order infinite impulse response (IIR) filters are the fundamental building blocks in computationally efficient high-order IIR digital filter implementations. However, when used in fixed-point number systems, the inherent properties of quantized-coefficient second-order IIR filters do not readily permit their use in narrowband lowpass filtering applications. Narrowband lowpass IIR filters have traditionally had a bad reputation; for example, MATLAB's Signal Processing Toolbox documentation warns: "All classical IIR lowpass filters are ill-conditioned for extremely low cutoff frequencies."

This chapter presents a neat trick to overcome the shortcomings of narrowband second-order lowpass IIR filters, with no increase in filter coefficient bit widths and no increase in the number of filter multiplies per output sample.

12.1 MOTIVATION

Narrowband lowpass IIR filters are difficult to implement because of intrinsic limitations on their z-plane pole locations. Let's quickly review the restrictions on the z-plane pole locations of a standard second-order IIR filter whose structure is shown in Figure 12–1(a).

Streamlining Digital Signal Processing: A Tricks of the Trade Guidebook, Second Edition. Edited by Richard G. Lyons.
© 2012 the Institute of Electrical and Electronics Engineers. Published 2012 by John Wiley & Sons, Inc.

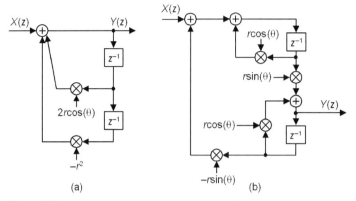

Figure 12–1   Second-order IIR filters: (a) standard form; (b) coupled form.

Such an IIR filter, having a transfer function given by

$$H(z) = \frac{1}{1 - 2r\cos(\theta)z^{-1} + r^2 z^{-2}} = \frac{1}{(1 - re^{j\theta}z^{-1})(1 - re^{-j\theta}z^{-1})}, \qquad (12\text{–}1)$$

has a pair of conjugate poles located at radii of r, at angles of $\pm\theta$ radians. In fixed-point implementations, quantizing the $2r\cos(\theta)$ and r^2 coefficients restricts the possible pole locations [1], [2]. On the z-plane, a pole can only reside at the intersection of a vertical line defined by the quantized value of $2r\cos(\theta)$ and a concentric circle whose radius is defined by the square root of the quantized value of r^2. For example, Figure 12–2 shows the first quadrant of possible z-plane pole locations when five magnitude bits are used to represent the filter's two coefficients. Notice the irregular spacing of those permissible pole locations. (Due to trigonometric symmetry, the pole locations in the other three quadrants of the z-plane are mirror images of those shown in Figure 12–2.)

12.2 THE PROBLEM

So here's the problem: if we use floating-point software to design a very narrowband lowpass IIR filter (implemented as cascaded second-order filters) having poles residing in the shaded area near $z = 1$, subsequent quantizing of the designed filter coefficients to five magnitude bits will make the poles shift to one of the locations shown by the dots on the border of the shaded region in Figure 12–2. Unfortunately that pole shifting, inherent in the Figure 12–1(a) IIR filter implementation due to coefficient quantization, prevents us from realizing the desired narrowband lowpass filter. We can always reduce the size of the shaded forbidden zone near $z = 1$ in Figure 12–2 by increasing the number of bits used to represent the second-order filters' coefficients. However, in some filter implementation scenarios increasing coefficient word bit widths may not be a viable option.

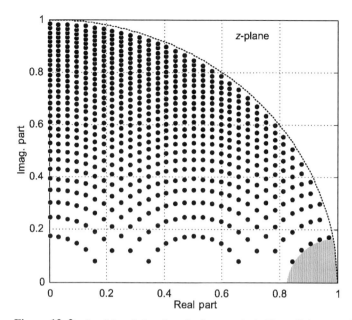

Figure 12–2   Possible pole locations for five magnitude-bit coefficient quantization.

One solution to the above problem is to use a so-called *coupled-form* IIR filter (also called the Gold-Rader filter [3]) structure, shown in Figure 12–1(b), having a transfer function given by

$$H_{cf}(z) = \frac{r\sin(\theta)z^{-1}}{1 - 2r\cos(\theta)z^{-1} + r^2 z^{-2}}. \qquad (12–2)$$

Because the coupled-form filter's quantized coefficients in Figure 12–1(b) are linear in $r\cos(\theta)$ and $r\sin(\theta)$, its possible pole locations are on a regularly-spaced grid on the z-plane defined by $z = r\cos(\theta) + jr\sin(\theta)$. This enables us to build second-order narrowband lowpass IIR filters with poles in the desired shaded region of Figure 12–2. Unfortunately, the coupled-form filter requires twice the number of multiplies needed by the standard second-order IIR filter in Figure 12–1(a).

The remainder of this chapter describes a novel narrowband lowpass IIR filter structure, proposed by Harris and Loudermilk, having poles residing in the shaded region of Figure 12–2 with no additional multiplication operations beyond those needed for a standard second-order IIR filter [4].

12.3   AN IMPROVED NARROWBAND LOWPASS IIR FILTER

The improved lowpass IIR filter is created by replacing each unit-delay element in a standard second-order IIR filter with multiple unit-delay elements as shown in

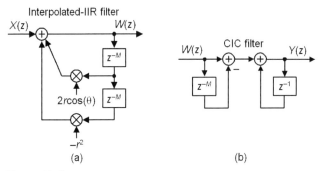

Figure 12–3   Filters: (a) single section interpolated-IIR filter; (b) first-order CIC filter.

Figure 12–3(a). The astute reader will notice that this scheme is reminiscent of interpolated finite impulse response (IFIR) filters where each delay element in a FIR filter's tapped-delay line is replaced by multiple delay elements [5]. For this reason we will call the filter in Figure 12–3(a) an *interpolated infinite impulse response* (interpolated-IIR) filter. The M-length delay lines, where M is a positive integer, in the interpolated-IIR filter shifts a standard IIR filter's conjugate poles, originally located at $z = re^{\pm j\theta}$ (in the vicinity of $z = 1$), to the new locations of

$$z_{\text{new}} = \sqrt[M]{r} \cdot e^{\pm j\theta/M}. \tag{12–3}$$

That is, the new conjugate pole locations are at radii of the Mth root of r, at angles of $\pm\theta/M$ radians. Those interpolated-IIR filter pole locations reside in the desired shaded region of Figure 12–2 without using more bits to represent the original IIR filter's coefficients.

   If the original Figure 12–1(a) second-order IIR filter contains feedforward coefficients, those coefficients are also delayed by M-length delay lines.

12.4 INTERPOLATED-IIR FILTER EXAMPLE

Here we show an example of an interpolated-IIR filter in action. With f_s representing a filter's input signal sample rate in hertz, assume we wish to implement a recursive lowpass filter whose one-sided passband width is $0.005f_s$ with a stopband attenuation greater than 60 dB. If we choose to set $M = 4$, then we start our interpolated-IIR filter design by first designing an standard IIR filter having a one-sided passband width of $M \cdot 0.005f_s = 0.02f_s$. Using our favorite IIR filter design software (for an elliptic IIR filter in this example), we obtain a fifth-order prototype IIR filter. Partitioning that fifth-order prototype IIR filter into two second-order and one single-order IIR filter sections, all in cascade and having coefficients represented by 12-bit words, yields the frequency magnitude response shown in Figure 12–4(a).

   Next, replacing the unit-delay elements in the filter sections with $M = 4$ unit-delay elements results in the frequency magnitude response shown in Figure 12–

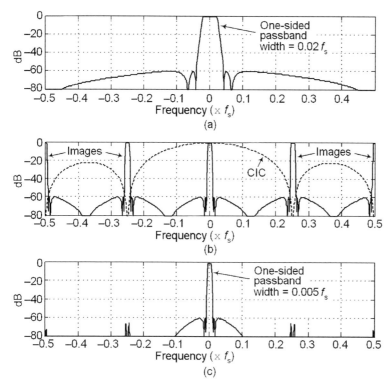

Figure 12–4   Frequency magnitude responses: (a) original IIR prototype filter; (b) zero-packed interpolated-IIR filter and CIC filters (dashed); (c) final narrowband 12-bit coefficient filter.

4(b). There we see the multiple narrowband passband images induced by the $M = 4$ length delay lines of the interpolated-IIR filter. Our final job is to attenuate those unwanted passband images. We can do so by following the cascaded increased-delay IIR filter sections with a cascaded integrator-comb (CIC) filter, whose structure is shown in Figure 12–3(b) [6], [7]. (The CIC filter is computationally advantageous because it requires no multiplications.) To satisfy our desired 60 dB stopband attenuation requirement, we use a second-order CIC filter—two first-order CIC filters in cascade—to attenuate the passband images in Figure 12–4(b). The result of our design is the interpolated-IIR and CIC filter combination whose composite frequency magnitude response meets our filter requirements as shown Figure 12–4(c).

12.5   CONCLUSION

We discussed the limitations encountered when using traditional second-order quantized-coefficient IIR filters to perform narrowband lowpass filtering. Next we showed how Gold and Rader resolved the problem, albeit with an increased computational cost of doubling the number of multiplies per filter output sample. Finally we described, and then demonstrated, the interpolated-IIR filter proposed by Harris and Loudermilk to overcome the shortcomings of traditional lowpass IIR filters. The interpolated-IIR filter provides improved lowpass filter performance while requiring no increase in filter coefficient bit widths and no additional multiply operations beyond a traditional IIR filter. When it comes to narrowband IIR filters, there's a new sheriff in town.

12.6 REFERENCES

[1] J. PROAKIS and D. MANOLAKIS, *Digital Signal Processing: Principles, Algorithms, and Applications*, 3rd ed. Prentice Hall, Upper Saddle River, NJ, 1996, pp. 572–576.

[2] A. OPPENHEIM and R. SCHAFER, *Discrete-Time Signal Processing*, 2nd ed. Prentice Hall, Englewood Cliffs, NJ, 1989, pp. 382–386.

[3] B. GOLD and C. RADER, "Effects of Parameter Quantization on the Poles of a Digital Filter." *Proceedings IEEE*, vol. 55, May 1967, pp. 688–689.

[4] F. HARRIS and W. LOWDERMILK, "Implementing Recursive Filters with Large Ratio of Sample Rate to Bandwidth," in *Conference Record of the Forty-First Asilomar Conference on Signals, Systems and Computers*, Pacific Grove, CA, November 4–7, 2007, pp. 1149–1153.

[5] R. LYONS, "Interpolated Narrowband Lowpass FIR Filters," *IEEE Signal Processing Magazine*, DSP Tips and Tricks Column, vol. 20, no. 1, January 2003, pp. 50–57.

[6] E. HOGENAUER, "An Economical Class of Digital Filters for Decimation and Interpolation," *IEEE Trans. Acoust. Speech and Signal Proc.*, vol. ASSP–29, April 1981, pp. 155–162.

[7] R. LYONS, "Understanding Cascaded Integrator-Comb Filters," *Embedded Systems Programming Magazine*, vol. 18, no. 4, April 2005, pp. 14–27. [Online: http://www.embedded.com/showArticle. jhtml?articleID=160400592.]

Chapter 13

Improving FIR Filter Coefficient Precision

Zhi Shen
Huazhong University of Science and Technology

There is a method for increasing the precision of fixed-point coefficients used in linear-phase finite impulse response (FIR) filters to achieve improved filter performance. The improved performance is accomplished without increasing either the number of coefficients or coefficient bitwidths. At first, such a process does not seem possible, but this chapter shows exactly how this novel filtering process works.

13.1 TRADITIONAL FIR FILTERING

To describe our method of increasing FIR filter coefficient precision, let's first recall a few characteristics of traditional linear-phase tapped-delay line FIR filter operation.

Consider an FIR filter whose impulse response is shown in Figure 13–1(a). For computational efficiency reasons (reduced number of multipliers), we implement such filters using the *folded* tapped-delay line structure shown in Figure 13–1(c) [1].

The filter's b_k floating-point coefficients are listed in the second column of Figure 13–1(b). When quantized to an 8-bit two's complement format, those coefficients are the decimal integers and binary values shown in the third and fourth columns, respectively, in Figure 13–1(b).

Compared with b_4, the other coefficients are smaller, especially the outer coefficients such as b_0 and b_8. Because of the fixed bitwidth quantization, many high-order bits of the low-amplitude coefficients, the bold underscored bits in the fourth column of Figure 13–1(b), are the same as the sign bit. These bits are wasted because they have no effective (no weight) in the calculations. If we can remove those wasted bits (consecutive bits adjacent to, and equal to, the sign bit), and replace them with

Streamlining Digital Signal Processing: A Tricks of the Trade Guidebook, Second Edition. Edited by Richard G. Lyons.
© 2012 the Institute of Electrical and Electronics Engineers. Published 2012 by John Wiley & Sons, Inc.

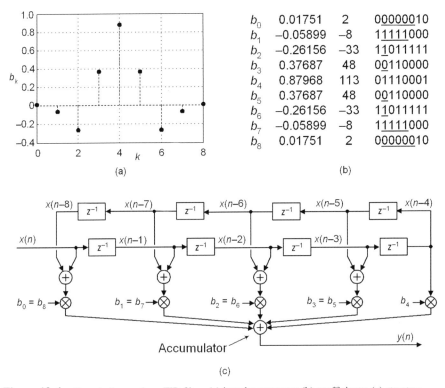

b_0	0.01751	2	00000010
b_1	−0.05899	−8	11111000
b_2	−0.26156	−33	11011111
b_3	0.37687	48	00110000
b_4	0.87968	113	01110001
b_5	0.37687	48	00110000
b_6	−0.26156	−33	11011111
b_7	−0.05899	−8	11111000
b_8	0.01751	2	00000010

(a)   (b)

(c)

Figure 13–1 Generic linear-phase FIR filter: (a) impulse response; (b) coefficients; (c) structure.

more significant coefficient bits, we will obtain improved numerical precision for the low-amplitude beginning and ending coefficients.

Replacing a low-amplitude coefficient's wasted bits with more significant bits is the central point of our FIR filtering trick (of course some filter architecture modification is needed, as we shall see). So let's have a look at a generic example of what we call a "serial" implementation of our trick.

13.2 SERIAL IMPLEMENTATION

As a simple example of replacing wasted bits, we list the Figure 13–1(b) b_k coefficients as the floating-point numbers in the upper left side of Figure 13–2. Assume we quantize the maximum-amplitude coefficient, b_4, to eight bits. In this FIR filter trick we quantize the lower-amplitude coefficients to larger bitwidths than the maximum coefficient (b_4) as shown on the upper right side of Figure 13–2. (The algorithm used to determine those variable bitwidths is discussed later in this chapter.) Next we eliminate the appropriate wasted bits, the bold underscored bits in the lower left side of Figure 13–2, to arrive at our final 8-bit coefficients shown on the lower right side of Figure 13–2.

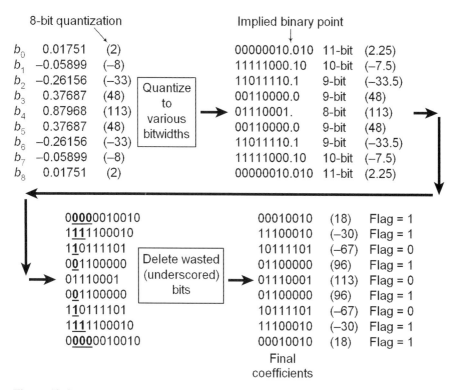

Figure 13–2   Filter coefficients for serial implementation.

Appended to each coefficient is a flag bit that indicates whether that coefficient used one more quantization bit than the previous, next larger, coefficient.

Now, you may say: "Stop! You can't do this. The outer coefficients are left-shifted, so they are enlarged, and the product accumulations are changed. Using these modified coefficients, the filter results will be wrong!" Don't worry, we correct the filter results by modifying the way we accumulate products. Let's see how.

The coefficients, and flag bits, from Figure 13–2 are used in the serial implementation shown in Figure 13–3. The data registers in Figure 13–3 represent the folded delay-line elements in Figure 13–1(c). This implementation is called *serial* because there is only one multiplier and, when a new $x(n)$ input sample is to be processed, we perform a series of multiplications and accumulations (using multiple clock cycles) to produce a single $y(n)$ filter output sample.

For an N-tap FIR filter, where N is odd, due to our folded delay-line structure only $(N + 1)/2$ coefficients are stored in the coefficient ROM (read-only memory). (When N is even, $N/2$ coefficients are stored.) Crucial to this FIR filter trick is that when processing a new $x(n)$ input sample, the largest coefficient, b_4, is applied to the multiplier prior to the first accumulation. Following that is the next smaller

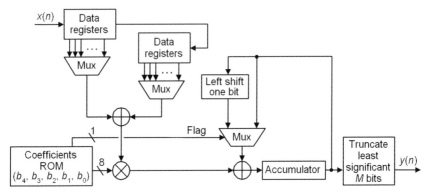

Figure 13–3   Serial implementation with 8-bit coefficients.

coefficient, b_3, and so on. In other words, in this serial implementation the coefficient sequence applied to the multiplier, for each $x(n)$ input sample, is in the order of the largest to the smallest coefficient.

Given these properties, when a new $x(n)$ sample is to be processed we clear the current accumulator value and multiply the sum of the appropriate data registers by the b_4 coefficient. That product is then added to the accumulator. On the next clock cycle we multiply the sum of the appropriate data registers by the b_3 coefficient. If the flag bit of the b_3 coefficient is one, we left-shift the current accumulator value and then the current multiplier's output is added to the shifted accumulator value. (If the current coefficient's flag bit is zero the accumulator word is not shifted prior to an addition.) We continue these multiplications, possible left shifts, and accumulations for the remaining b_2, b_1, and b_0 coefficients.

The left shifting of an accumulator value is the key to this entire FIR filter trick. To minimize truncation errors due to right shifting a multiplier output word, we preserve precision by left-shifting the previous accumulator word.

To maintain our original FIR filter's gain, after the final accumulation we truncate the final accumulator value by discarding its least significant M bits, where M is the summation of the flag bits in the ROM memory, to produce a $y(n)$ output sample. Now let's have a look at an actual FIR filter example.

13.3 SERIAL METHOD EXAMPLE

Suppose we want to implement a lowpass filter whose cutoff frequency is $0.167f_s$ and whose stopband begins at $0.292f_s$, where f_s is the input data sample rate. If the filter has 29 taps (coefficients), and is implemented with floating point coefficients, its frequency magnitude response will be that shown by the solid curve in Figure 13–4(a). Anticipating a hardware implementation using an Altera field-programmable gate array (FPGA) having 9-bit multipliers, when using coefficients that have been quantized to 9-bit lengths in a traditional manner (with no wasted coefficient bits

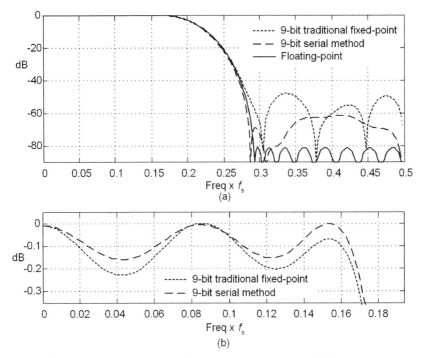

Figure 13–4 Lowpass serial method filter frequency responses: (a) full frequency range; (b) passband detail.

removed), the filter's frequency magnitude response is the dotted curve in Figure 13–4(a).

When we use our FIR filter trick's serial implementation, with its enhanced-precision 9-bit coefficients (not counting the flag bit) obtained in the manner shown in Figure 13–2, the filter's frequency magnitude response is the dashed curve in Figure 13–4(a). We see in the figure that, relative to the traditional fixed-point implementation, the serial method provides:

• Improved stopband attenuation
• Reduced transition region width
• Improved passband ripple performance

All of these improvements occur without increasing the bitwidths of our filter's multiplier or coefficients, nor the number of coefficients. Because it preserves the impulse response symmetry of the original floating-point filter, the serial implementation filter exhibits phase linearity [2].

It is possible to improve upon the stopband attenuation of our compressed-coefficient serial method FIR filter. We do so by implementing what we call the *parallel method*.

13.4 PARALLEL IMPLEMENTATION

In the above serial method of filtering, adjacent filter coefficients were quantized to a precision differing by no more than one bit. That's because we use a single flag bit to control the one-bit shifting of the accumulator word prior to a single accumulation. In the parallel method, described now, adjacent coefficients can be quantized to a precision differing by more than one bit. Figure 13–5 shows an example of our parallel method's coefficient quantization process.

Again we list the Figure 13–1(b) b_k coefficients as the floating-point numbers in the upper left side of Figure 13–5. In this parallel method, however, notice that the expanded quantized b_1 and b_2 words differ by more than one bit in the upper right side of Figure 13–5. Coefficients b_2 and b_6 are quantized to 9 bits while the b_1 and b_7 coefficients are quantized to 12 bits. While we only deleted some of the wasted coefficient bits in Figure 13–2, in our parallel method all the wasted coefficient bits are deleted. As such, our final 8-bit coefficients are those listed in the lower right side of Figure 13–5.

All of us are familiar with the operation known as *bit extension*—the process of extending the bit length of a binary word without changing its value or sign. With

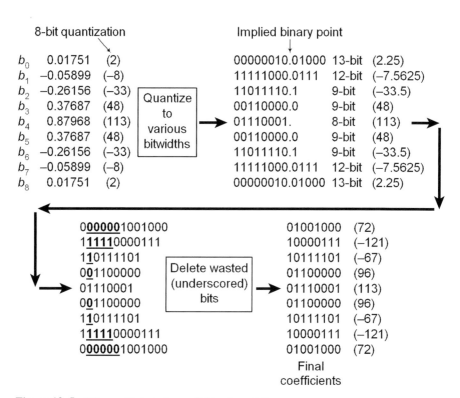

Figure 13–5   Filter coefficients for parallel implementation.

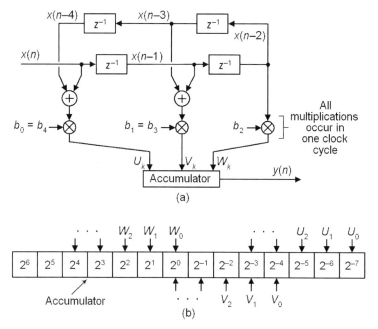

Figure 13–6   Parallel method implementation: (a) filter structure; (b) accumulator organization.

that process in mind, we can refer to our trick's operation of removing wasted bits as *bit compression*.

Because no flag bits are used in the parallel method, this filtering method is easiest to implement using FPGAs with their flexible multi-data bus routing capabilities.

For example, consider the filter structure shown in Figure 13–6(a) where we perform the three multiplications in parallel (in a single clock cycle), and that is why we use the phrase *parallel method*. Instead of shifting the accumulator word to the left as we did in the serial method, here we merely reroute the multiplier outputs to the appropriate bit positions as they are added to the accumulator word as shown in Figure 13–6(b). In our hypothetical Figure 13–6 example, if there were four wasted bits deleted from the high-precision b_1 coefficient then the V_k product is shifted to the right by four bits, relative to the W_k product bits, before being added to the accumulator word. If there were seven wasted bits deleted from the high-precision b_0 coefficient then the U_k product is shifted to the right by seven bits, relative to the W_k product bits, before being added to the accumulator word.

13.5   PARALLEL METHOD EXAMPLE

With the solid curve, Figure 13–7 shows our parallel method's performance in implementing the desired lowpass filter used in the above serial method implementation

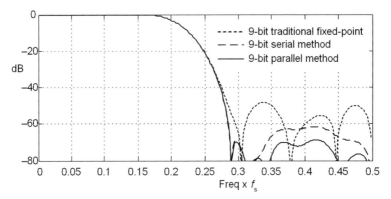

Figure 13–7   Traditional fixed-point, serial method, and parallel method filter frequency responses.

example. For comparison, we have also included the 9-bit traditional fixed point (no bit compression) and the serial method magnitude responses in Figure 13–7.

The enhanced precisions of the parallel method's quantized coefficients, beyond their serial method precisions, yield improved filter performance. The parallel method of our FIR filter trick achieves a stopband attenuation improvement of 21 dB beyond the traditional fixed point implementation—again, without increasing the bitwidths of our filter's multipliers or coefficients, nor the number of coefficients.

13.6   COMPUTING COEFFICIENT BITWIDTHS

Determining the bitwidths of the quantized filter coefficients in our DSP trick depends on whether you are implementing the serial or the parallel filtering method.

13.7   SERIAL METHOD COEFFICIENT QUANTIZATION

In the serial method, let's assume we want our ROM to store coefficients whose bitwidths are integer B (not counting the flag bit).

The steps in computing the integer ROM coefficients for the serial method are as follows:

> **Step 1:** Set a temporary scale factor variable to SCALE = 1, and temporary bitwidth integer variable to $K = B - 1$. Apply the following quantization steps to the largest-magnitude original b_k floating-point coefficient (for example, b_4 in the upper left side of Figure 13–2).
>
> **Step 2:** If the b_k floating-point coefficient being quantized and all the remaining unquantized coefficients are less than the value SCALE/2, set SCALE = SCALE/2, set $K = K + 1$, and set the current coefficient's flag bit to Flag = 1. If the b_k floating-point coefficient being quantized or any of the

remaining unquantized coefficients are equal to or greater than SCALE/2, variables SCALE and K remain unchanged, and set the current coefficient's flag bit to Flag = 0.

Step 3: Multiply the b_k floating-point coefficient being quantized by 2^K and round the result to the nearest integer. That integer is our final value saved in ROM.

Step 4: Repeat Steps 2 and 3 for all the remaining original unquantized b_k floating-point coefficients, in sequence from the remaining largest-magnitude to the remaining smallest-magnitude coefficient.

Table 13–1 illustrates the serial method quantization steps for the floating-point coefficients in Figure 13–2.

13.8 PARALLEL METHOD COEFFICIENT QUANTIZATION

In the parallel method, let's assume we want our ROM to store coefficients whose bitwidths are integer B. (For example, in the lower right side of Figure 13–5, $B = 8$.) Next, let's define an optimum *magnitude* range, R, as

$$0.5 \le R < 1. \tag{13–1}$$

The steps in computing the integer ROM coefficients for the parallel method are as follows:

Step 1: Repeatedly multiply an original b_k floating-point coefficient (the upper left side of Figure 13–5) by two until the magnitude of the result resides in the optimum magnitude range R. Denote the number of necessary multiply-by-two operations as Q.

Step 2: Multiply the original b_k floating-point coefficient by 2^{B+Q-1} (the minus one in the exponent accounts for the final coefficient's sign bit) and round the result to the nearest integer. That integer is our final value saved in ROM.

Step 3: Repeat Steps 1 and 2 for all the remaining original b_k floating-point coefficients.

13.9 CONCLUSIONS

We introduced two novel methods for improving the precision of the fixed-point coefficients of FIR filters. Using the modified (compressed) coefficients, we achieved enhanced filter performance, while maintaining phase linearity, without increasing the bitwidths of our filter multiplier or coefficients, nor the number of coefficients. The so-called serial method of filtering is compatible with traditional programmable DSP chip and FPGA processing, while the parallel method is most appropriate with an FPGA implementation.

Table 13–1 Serial Method Quantization Example

Coefficient being quantized	Current SCALE	ALL un-quantized coefficients less than SCALE /2?	New SCALE	K	ROM equivalent coefficient bitwidth	Flag bit	Left shift and round
$b_4 = 0.87968$	1	No $|b_4| > 1/2$	1	7	8	0	$B_4 = \text{round}[b_4 \cdot 2^{\wedge}7] = 113$
$b_3 = 0.37687$	1	Yes $|b_3|, |b_2|, |b_1|, |b_0| < 1/2$	0.5	8	9	1	$B_3 = \text{round}[b_3 \cdot 2^{\wedge}8] = 96$
$b_2 = -0.26156$	0.5	No $|b_2| > 0.5/2$	0.5	8	9	0	$B_2 = \text{round}[b_2 \cdot 2^{\wedge}8] = -67$
$b_1 = -0.05899$	0.5	Yes $|b_1|, |b_0| < 0.5/2$	0.25	9	10	1	$B_1 = \text{round}[b_1 \cdot 2^{\wedge}9] = -30$
$b_0 = 0.01751$	0.25	Yes $|b_0| < 0.25/2$	0.125	10	11	1	$B_0 = \text{round}[b_0 \cdot 2^{\wedge}10] = 18$

13.10   REFERENCES

[1] R. LYONS, *Understanding Digital Signal Processing*, 3rd ed. Prentice Hall, Upper Saddle River, NJ, 2010, pp. 702–703.

[2] J. PROAKIS, and D. MANOLAKIS, *Digital Signal Processing-Principles, Algorithms, and Applications*, 3rd ed. Prentice Hall, Upper Saddle River, NJ, 1996, pp. 620–621.

Part Two

Signal and Spectrum Analysis Tricks

Chapter 14

Fast, Accurate Frequency Estimators

Eric Jacobsen   **Peter Kootsookos**
Anchor Hill Communications   **Emuse Technologies Ltd.**

$\mathbf{T}$he problem of estimating the frequency of a tone, contaminated with noise, appears in communications, audio, medical, instrumentation, and a host of other applications. Naturally, the fundamental tool for such analysis is the discrete Fourier transform (DFT) or its efficient cousin the fast Fourier transform (FFT). A well-known tradeoff exists between the amount of time needed to collect data, the number of data points collected, the type of time-domain window used, and the resolution that can be achieved in the frequency domain [1]–[7]. This chapter presents computationally simple estimators that provide substantial refinement of the frequency estimation of tones based on DFT samples without the need for increasing the DFT size.

14.1 SPECTRAL PEAK LOCATION ALGORITHMS

An important distinction between the "resolution" in the frequency domain and the accuracy of frequency estimation should be clarified. Typically when the term *resolution* is used in the context of frequency estimation the intent is to describe the 3-dB width of the $\sin(x)/x$ response in the frequency domain. The *resolution* is affected by N, the number of data points collected; the type of window used; and the sample rate. One of the primary uses of the resolution metric is the ability to resolve closely spaced tones within an estimate. The frequency estimation problem, while it can be affected by the resolution, only seeks to find, as accurately as possible, the location of the peak value of the $\sin(x)/x$ spectral envelope, regardless of its width or other characteristics. This distinction is important since improving the resolution is often a common avenue taken by someone wishing to improve the frequency estimation capability of a system. Improving the resolution is typically

Streamlining Digital Signal Processing: A Tricks of the Trade Guidebook, Second Edition. Edited by Richard G. Lyons.
© 2012 the Institute of Electrical and Electronics Engineers. Published 2012 by John Wiley & Sons, Inc.

137

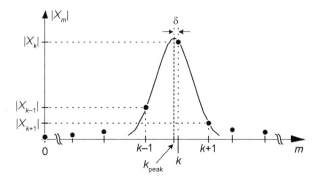

Figure 14–1 DFT magnitude samples of a spectral tone.

costly in computational burden or latency and this burden is often unnecessary if the only goal is to improve the frequency estimate of an isolated tone.

The general concept in using spectral peak location estimators is to estimate the frequency of the spectral peak, k_{peak} in Figure 14–1, based on the three X_{k-1}, X_k, and X_{k+1} DFT samples. If we estimated k_{peak} to be equal to the k index of the largest DFT magnitude sample, the maximum error in k_{peak} would be half the width of the DFT bin. Using the frequency-domain peak sample, X_k, and one or two adjacent samples allows some simple best- or approximate-fit estimators to be used to improve the estimate of the peak location. In this material, each estimator provides a fractional correction term, δ, which is added to the integer peak index, k, to determine a *fine* k_{peak} estimate of the $\sin(x)/x$ main lobe peak location using

$$k_{peak} = k + \delta \qquad (14\text{–}1)$$

$$f_{tone} = k_{peak} f_s / N \qquad (14\text{–}1')$$

where f_s is the time data sample rate in Hz and N is the DFT size. Note that δ can be positive or negative, and that k_{peak} need not be an integer. This refinement of the original bin-location estimate can be surprisingly accurate even in low signal-to-noise ratio (SNR) conditions.

Figure 14–1 shows the basic concept where the main lobe of the $\sin(x)/x$ response in the frequency domain traverses three samples. These three samples can be used with simple curve-fit techniques to provide an estimate of the peak location between bins, δ, which can then be added to the peak index, k, to provide a fine estimate of the tone frequency. An example of such an estimator is

$$\delta = (|X_{k+1}| - |X_{k-1}|)/(4|X_k| - 2|X_{k-1}| - 2|X_{k+1}|) \qquad (14\text{–}2)$$

using three DFT magnitude samples [8], [9]. This estimator is simple, but it is statistically biased, and performs poorly in the presence of noise.

Jacobsen suggested some simple changes that improve the performance of estimator (14–2) [4]. For example, (14–3) uses the complex DFT values of X_k, X_{k+1}, and X_{k-1}. Taking the real part of the computed result provides significantly improved spectral peak location estimation accuracy and eliminates the statistical bias of (14–2).

$$\delta = -\text{Re}[(X_{k+1} - X_{k-1})/(2X_k - X_{k-1} - X_{k+1})] \qquad (14\text{–}3)$$

Because the magnitude calculations in (14–2) are nontrivial, (14–3) provides a potential for computation reduction as well.

Quinn contributed two efficient estimators with good performance [1], [3]. Quinn's first estimator is:

$$\alpha_1 = \text{Re}(X_{k-1}/X_k) \qquad (14\text{–}4)$$

$$\alpha_2 = \text{Re}(X_{k+1}/X_k) \qquad (14\text{–}5)$$

$$\delta_1 = \alpha_1/(1-\alpha_1) \qquad (14\text{–}6)$$

$$\delta_2 = -\alpha_2/(1-\alpha_2) \qquad (14\text{–}7)$$

If $\delta_1 > 0$ and $\delta_2 > 0$, then $\delta = \delta_2$, otherwise $\delta = \delta_1$.

Quinn's second estimator [3] performs better but includes a number of transcendental function computations that take it out of the realm of computationally efficient estimators.

An estimator provided by MacLeod [2] requires the computation of a square root but is otherwise fairly simple. That estimator begins with

$$r = X_k \qquad (14\text{–}8)$$

$$R_n = \text{Re}(X_n r^*). \qquad (14\text{–}9)$$

That is, create a real-valued vector R_n that is the real part of the result of the Fourier coefficient vector X_n made up of coefficients X_{k-1} through X_{k+1}, times the conjugate of the spectral peak coefficient, X_k. This phase-aligns the terms relative to the peak so that the result gets a strong contribution in the real part from each of the coefficients. Then

$$d = (R_{k-1} - R_{k+1})/(2 * R_k + R_{k-1} + R_{k+1}) \qquad (14\text{–}10)$$

$$\delta = (\text{sqrt}(1 + 8d^2) - 1)/4d \qquad (14\text{–}11)$$

While the above estimators work well for analysis when a rectangular time-domain window is applied to the DFT input samples, it is often beneficial or necessary to use nonrectangular windowing. Estimators that allow nonrectangular time-domain windows include those of Grandke [5], Hawkes [10], and Lyons [11].

Grandke proposed a simple estimator assuming the use of a Hanning window, using only the peak sample, X_k, and its largest-magnitude adjacent sample. Letting X_k be the peak sample, the estimator is then simply:
When $|X_{k-1}| > |X_{k+1}|$,

$$\alpha = |X_k|/|X_{k-1}| \qquad (14\text{–}12)$$

Table 14–1   Correction Scaling Values for Four Common Window Types for the (14–14) and (14–15) Estimators

Window	P	Q
Hamming	1.22	0.60
Hanning	1.36	0.55
Blackman	1.75	0.55
Blackman-Harris (3-term)	1.72	0.56

$$\delta = (\alpha - 2)/(\alpha + 1). \tag{14–12'}$$

When $|X_{k-1}| < |X_{k+1}|$,

$$\alpha = |X_{k+1}|/|X_k| \tag{14–13}$$

$$\delta = (2\alpha - 1)/(\alpha + 1). \tag{14–13'}$$

Hawkes proposed the estimator (14–14), similar to (14–2) with a scaling term, P, which can be adjusted for different window applications.

$$\delta = P(|X_{k+1}| - |X_{k-1}|)/(|X_k| + |X_{k-1}| + |X_{k+1}|) \tag{14–14}$$

Inspired by (14–3) and (14–14), Lyons has suggested estimator (14–15) with a window-specific scaling term Q [11].

$$\delta = \text{Re}[Q(X_{k-1} - X_{k+1})/(2X_k + X_{k-1} + X_{k+1})] \tag{14–15}$$

Table 14–1 shows the scaling values for the (14–14) and (14–15) estimators for certain common windowing functions.

14.2 ALGORITHM PERFORMANCE

Simulation of the above estimators yields some interesting results. While the estimators differ in computational efficiency, their performance also varies and some performance metrics vary with the tone offset within the bin. Thus the menu of estimators available that would perform best in a certain application depends on the nature of the system. In general the estimators used with nonrectangular windows provide less accuracy and have more highly biased outputs than those used with rectangular windows [5], [10], [11], but their performance is better than that with the rectangular-windowed estimators when applied to DFT samples from nonrectangular-windowed data.

The performance of the best estimators is captured in Figures 14–2 and 14–3. A tone was simulated with peak location varying from bin 9 to bin 10 at small increments. The root mean squared error (RMSE) of the estimators for tones at bin 9.0 and bin 9.5 are shown in Figure 14–2. Some thresholding of the FFT peak location detection is seen in Figure 14–2(b) at low SNRs, which is independent of the fine

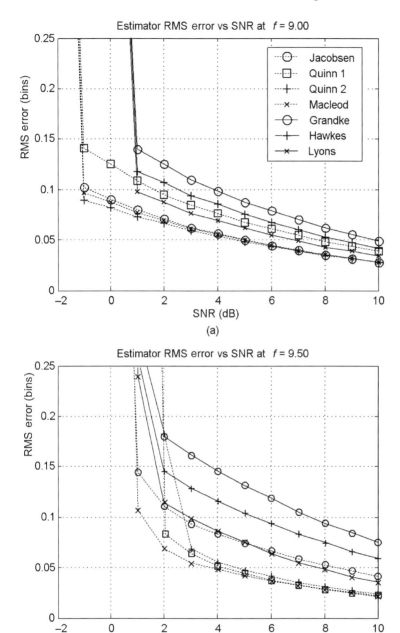

Figure 14–2   RMS error performance in AWGN for the indicated estimators: (a) tone at bin 9; (b) tone centered between bin 9 and bin 10.

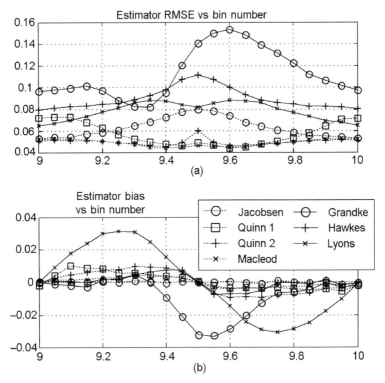

Figure 14–3   Interbin performance in AWGN with SNR = 1.4 dB: (a) RMSE; (b) bias.

estimator used. The estimators of Grandke, Hawkes, and Lyons were tested with a Hanning window applied while the remainder were tested with rectangular windowing.

It can be seen that the relative locations of some of the estimator's performance curves change depending on the tone offset within the bin. This is further seen in Figure 14–3, where the RMSE and bias at ≈1.4 dB SNR demonstrates that for some estimators the error is worse when the tone is aligned on a bin and for others it is worse when the tone is between bins. In Figure 14–4 the estimator performances are shown in no noise.

All of the estimators shown are unbiased except for Grandke (14–12) and (14–13), Hawkes (14–14), and Lyons (14–15).

The relative merits of the various estimators are captured in Table 14–2. All of the estimators use a number of adds or subtracts, but these are typically trivial to implement in logic or software and so are not considered significant for computational complexity comparison. The relative complexity of multiplies, divides, magnitude computations, or transcendental functions via computation or lookup-table depends highly on the specific implementation and are enumerated here for consideration.

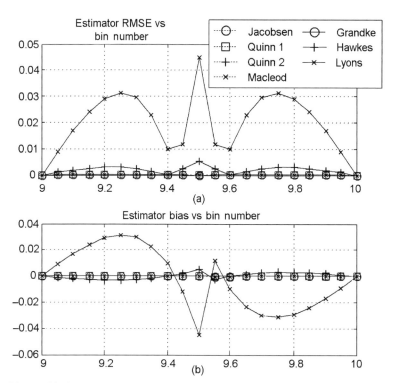

Figure 14–4   Interbin performance in no noise: (a) RMSE; (b) bias.

Table 14–2   Algorithm Computational and Performance Comparison

Equ	Mult.	Div.	Mag.	RMSE	Bias?	Remarks
(14–2)	0	1	3	High at low SNR	Yes	Not considered due to poor performance
(14–3)	4	1	0	Medium, increasing with bin offset	No	Very good bias performance
(14–4)–(14–5)	4	0	0	Medium, decreasing with bin offset	No	
n/a	8	7	0	Low	No	Quinn's second estimator [3]*.
n/a	1(3)	2	0	Very low	No	MacLeod's estimator [2]**.
(14–12)–(14–13)	0	2	2	High, increasing with bin offset	No	Worst RMSE, Hanning only
(14–14)	1	1	3	High	Yes	Good for windowed applications
(14–15)	5	1	0	High	Yes	Good for windowed applications

* Requires four logarithmic computations.
** Requires one square root computation.

14.3 CONCLUSIONS

Interbin tone peak location estimation for isolated peaks can be performed with a number of effective estimators in the presence or absence of noise or nonrectangular windowing. Which estimator is most appropriate for a particular application depends on the type of windowing used, system sensitivity to bias, location-dependent RMS error, and computational complexity. Macleod's and Quinn's second estimator perform especially well but may be computationally burdensome in complexity-sensitive applications. The Jacobsen estimator provides a good complexity-performance tradeoff, requiring only a single divide but at the expense of a small performance reduction, especially for tones between bins. Lyons' estimator is similar for windowed applications, requiring only an additional multiplication compared with Jacobsen's but at the expense of some bias. Since estimator bias is predictable it can often be removed simply with an additional arithmetic operation.

14.4 REFERENCES

[1] B.G. QUINN, "Estimating Frequency by Interpolation Using Fourier Coefficients," *IEEE Trans. Signal Processing*, vol. 42, no. 5, May 1994, pp. 1264–1268.

[2] M. MACLEOD, "Fast Nearly ML Estimation of the Parameters of Real or Complex Single Tones or Resolved Multiple Tones," *IEEE Trans. Signal Processing*, vol. 46, No. 1, January 1998, pp. 141–148.

[3] B. QUINN, "Estimation of Frequency, Amplitude and Phase from the DFT of a Time Series," *IEEE Trans. Signal Processing*, vol. 45, No. 3, March 1997, pp. 814–817.

[4] E. JACOBSEN, "On Local Interpolation of DFT Outputs," [Online: http://www.ericjacobsen.org/FTinterp.pdf.]

[5] T. GRANDKE, "Interpolation Algorithms for Discrete Fourier Transforms of Weighted Signals," *IEEE Trans. Instrumentation and Measurement*, vol. IM-32, June 1983, pp 350–355.

[6] V. JAIN et al., "High-Accuracy Analog Measurements via Interpolated FFT," *IEEE Trans. Instrumentation and Measurement*, vol. IM-28, June 1979, pp. 113–122.

[7] D. RIFE and R. BOORSTYN, "Single-Tone Parameter Estimation from Discrete-Time Observations," *IEEE Trans. Information Theory*, vol. IT-20, September 1974, pp. 591–598.

[8] W. PRESS, et al., *Numerical Recipes in C*, Cambridge University Press, Cambridge, 1992, Chapter 10.

[9] P. VOGLEWEDE, "Parabola Approximation for Peak Determination," Global DSP Magazine, May 2004.

[10] K. HAWKES, "Bin Interpolation," *Technically Speaking*, ESL Inc., January, 1990, pp. 17–30.

[11] R. LYONS, *Private communication*, August 30, 2006.

EDITOR COMMENTS

Estimators (14–3) and (14–15) have the advantage that DFT magnitude calculations, with their computationally costly square root operations, are not required as is necessary with some other spectral peak location estimators described in this chapter. However, the question arises, "How do we determine the index k of the largest-magnitude DFT sample, $|X_k|$, in Figure 14–1 without computing square roots to obtain DFT magnitudes?" The answer is we can use the complex sample magnitude estimation algorithms, requiring no square root computations, described in Chapter 25.

The following shows the (14–3) and (14–15) estimators in real-only terms. If we express the complex spectral samples, whose magnitudes are shown in Figure 14–1, in rectangular form as:

$$X_{k-1} = X_{k-1,\text{real}} + jX_{k-1,\text{imag}},$$

$$X_k = X_{k,\text{real}} + jX_{k,\text{imag}},$$

$$X_{k+1} = X_{k+1,\text{real}} + jX_{k+1,\text{imag}},$$

we can express estimator (14–3) using real-only values, and eliminate its minus sign, as

$$\delta_{(3)\,\text{real-only}} = \frac{R_{\text{num}}R_{\text{den}} + I_{\text{num}}I_{\text{den}}}{R_{\text{den}}^2 + I_{\text{den}}^2} \qquad (14\text{–}16)$$

where

$$R_{\text{num}} = X_{k-1,\text{real}} - X_{k+1,\text{real}}$$

$$I_{\text{num}} = X_{k-1,\text{imag}} - X_{k+1,\text{imag}}$$

$$R_{\text{den}} = 2X_{k,\text{real}} - X_{k-1,\text{real}} - X_{k+1,\text{real}}$$

$$I_{\text{den}} = 2X_{k,\text{imag}} - X_{k-1,\text{imag}} - X_{k+1,\text{imag}}.$$

Thus (14–16) is the actual real-valued arithmetic needed to compute (14–3). In a similar manner we can express estimator (14–15) using real-only values as

$$\delta_{(15)\,\text{real-only}} = \frac{Q(R_{\text{num}}R_{\text{den}} + I_{\text{num}}I_{\text{den}})}{R_{\text{den}}^2 + I_{\text{den}}^2}$$

where

$$R_{\text{num}} = X_{k-1,\text{real}} - X_{k+1,\text{real}}$$

$$I_{\text{num}} = X_{k-1,\text{imag}} - X_{k+1,\text{imag}}$$

$$R_{\text{den}} = 2X_{k,\text{real}} + X_{k-1,\text{real}} + X_{k+1,\text{real}}$$

$$I_{\text{den}} = 2X_{k,\text{imag}} + X_{k-1,\text{imag}} + X_{k+1,\text{imag}}.$$

Chapter 15

Fast Algorithms for Computing Similarity Measures in Signals

James McNames
Portland State University

This chapter describes fast algorithms that compute *similarity measures* between contiguous segments of a discrete-time one-dimensional signal $x(n)$ and a single template or pattern $p(n)$, where N_x is the duration of the observed signal $x(n)$ in units of samples and N_p is the duration of the template $p(n)$ in units of samples. Similarity measures are used to detect occurrences of significant intermittent events that occur in the observed signal. The occurrences are detected by finding the sample times in which a segment of the signal is similar to the template. For example, Figure 15–1(a) shows the output sequence of a noisy microelectrode recording and Figure 15–1(b) shows the cross-correlation of that sequence with template matching. It is much easier to detect spikes as the peaks above the noise floor (the dots) in the cross-correlation than it is in the original recorded sequence.

This type of event detection is called *template matching*, and it is used in many signal and image processing applications. Some examples of similarity measure applications are QRS and arrhythmia detection in electrocardiograms [1], labeling of speech [2], object detection in radar [3], matching of patterns in text strings [4], and detection of action potentials in extracellular recordings of neuronal activity [5].

Direct calculation of similarity measures normally requires computation that scales (the number of computations increases) linearly with the duration of the template, N_p. Although fast algorithms for computing similarity measures in signals have been discovered independently by many signal processing engineers, there is no publication that describes these algorithms in a common framework. Most of the

Streamlining Digital Signal Processing: A Tricks of the Trade Guidebook, Second Edition. Edited by Richard G. Lyons.
© 2012 the Institute of Electrical and Electronics Engineers. Published 2012 by John Wiley & Sons, Inc.

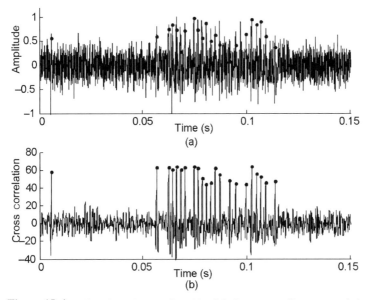

Figure 15–1   Microelectrode recording: (a) original sequence; (b) cross-correlation result.

similarity measures used in practice are based on second-order statistics, which include averages, products, and squares of the terms. These statistics include sample means, cross-correlations, and squared differences. The aims of this chapter are to demonstrate how nonlinear similarity measures based on these second-order statistics can be computed efficiently using simple components and to estimate the template size N_p at which the fast algorithms outperform direct implementations.

Although no one has published fast algorithms for one-dimensional signals, many publications describe fast algorithms for template matching in images. Image template matching differs fundamentally from signal template matching in several important respects. The size of the template, in units of pixels, usually representing an object that one wishes to locate within the image, is often on the same scale as the size of the image, which limits the efficiency of fast-filtering algorithms based on block processing that can be used in one-dimensional signals. In image processing applications the template is often not just shifted (in the x-y directions), but rotated, scaled in size, and scaled in amplitude. Rotation has no analogous operation in one-dimensional signal processing. In both image and signal processing applications, the detection of templates with various scales or amplitudes is usually treated by selecting a similarity measure that is invariant to changes in scale.

Most of the similarity measures of interest in template-matching applications are nonlinear and based on second-order statistics of the template and the signal segment. Direct implementations of these measures require computation that scales linearly with the duration of the template. If the signal length N_x is much greater than the template length ($N_x \gg N_p$), the similarity measure must be calculated over the entire possible range ($0 \le n \le N_x - N_p$), and the similarity measures can be expressed in terms of second-order components, then it is possible to calculate these measures much more quickly with running sums and fast finite impulse response (FIR) filtering algorithms than is possible with direct calculation.

15.1   SIMILARITY MEASURE COMPONENTS

In order to describe our similarity measure algorithms in a common framework, Table 15–1 lists the *components* that can be computed to evaluate various similarity measures. Later we will define these components and present algorithms for computing these components in the most efficient way possible. The range of each sum in this table is $i = 0$ to N_p. The weighting function $w(i)^2$ permits the user to control the relative influence of each term in the template.

15.2   SIMILARITY MEASURES

Four similarity measures are discussed in the following sections. Each section defines the measure, describes its relevant properties, cites some example applications, and gives a mathematically equivalent expression in terms of the components listed in Table 15–1. Generalizations to other measures based on second-order statistics, such as normalized cross-correlation, cross-covariance, and normalized mean squared error, are straightforward.

Cross-Correlation

The sample cross-correlation of the template and the signal is defined as

$$r(n) = \frac{1}{N_p} \sum_{i=0}^{N_p-1} x(n+i)p(i) \tag{15–1}$$

The primary advantage of the cross-correlation is that it can be computed efficiently [6]. It is widely used for matched filter and eigenfilter applications [7].

Cross-correlation is the most popular similarity measure in signal processing applications because it can be calculated with traditional linear filter architectures

Table 15–1   Similarity Measure Components

$s_x(n) = \sum_i x(n+i)$	$s_{x^2}(n) = \sum_i x(n+i)^2$
$s_{x^2w^2}(n) = \sum_i x(n+i)^2 w(i)^2$	$s_{xp}(n) = \sum_i x(n+i)p(i)$
$s_{xpw^2}(n) = \sum_i x(n+i)p(i)w(i)^2$	$s_p(n) = \sum_i p(i)$
$s_{p^2} = \sum_i p(i)^2$	$s_{p^2w^2}(n) = \sum_i p(i)^2 w(i)^2$

by treating the template as the impulse response of an anticausal tapped-delay line FIR filter. The cross-correlation can be expressed in terms of the Table 15–1 second-order components as

$$r(n) = \frac{1}{N_p} s_{xp}(n) \qquad (15\text{–}2)$$

Mean Squared Error

The mean squared error is defined as

$$\zeta(n) = \frac{1}{N_p} \sum_{i=0}^{N_p-1} [x(n+i) - p(i)]^2 \qquad (15\text{–}3)$$

This similarity measure is popular for many applications because it has several useful properties. It is invariant to identical shifts in the amplitude of the signal and template means, is related to the maximum likelihood estimate of the template location, and maximizes the peak signal-to-noise ratio [8]–[12].

The mean squared error can be expressed in terms of the Table 15–1 second-order components as

$$\zeta(n) = \frac{1}{N_p} \sum_{i=0}^{N_p-1} x(n+i)^2 - 2x(n+i)p(i) + p(i)^2 \qquad (15\text{–}4)$$

$$\zeta(n) = \frac{1}{N_p} \left[s_{x^2}(n) - 2s_{xp}(n) + s_{p^2} \right] \qquad (15\text{–}5)$$

The same decomposition in terms of the components in Table 15–1 has been used in image processing applications for both analysis and fast computation [6], [9], [12].

Some signal analysts prefer to use the *root mean squared error* (RMSE), the square root of the above mean squared error (MSE), as a similarity measure. In most applications the MSE conveys the same similarity information as the RMSE, but the MSE does not require the computationally costly square root operation demanded by RMSE calculations.

Weighted Mean Squared Error

The mean squared error in the previous section can be generalized to facilitate relative weighting of each term in the sum,

$$\zeta_w(n) = \frac{1}{N_p} \sum_{i=0}^{N_p-1} (w(i)[x(n+i) - p(i)])^2 \qquad (15\text{–}6)$$

where w is a vector of weights. This similarity measure may be appropriate if the occurrence of the template in the signal also affects the signal variance and for detecting patterns that vary. Under some general assumptions, this similarity measure can also be used for maximum likelihood estimation of the template locations, as has been done for images [11]. The weighted mean squared error can be expressed as

$$\zeta_w(n) = \frac{1}{N_p} \sum_{i=0}^{N_p-1} \left[w(i)x(n+i) - w(i)p(i) \right]^2 \tag{15-7}$$

and then expanded in the same manner as (15–3),

$$\zeta_w(n) = s_{x^2 w^2}(n) - 2s_{xpw^2}(n) + s_{p^2 w^2} \tag{15-8}$$

Centered Correlation Coefficient

The centered correlation coefficient is defined as

$$\rho_c(n) = \frac{\sum_{i=0}^{N_p-1} \left[x(n+i) - \bar{x}(n) \right]\left[p(i) - \bar{p} \right]}{\left(\sum_{i=0}^{N_p-1} \left[x(n+i) - \bar{x}(n) \right]^2 \right)\left(\sum_{i=0}^{N_p-1} \left[p(i) - \bar{p} \right]^2 \right)} \tag{15-9}$$

where $\bar{x}(n)$ and $\bar{p}$ are the sample means of the signal segment and template,

$$\bar{x}(n) = \frac{1}{N_p} \sum_{i=0}^{N_p-1} x(n+i) \quad \bar{p} = \frac{1}{N_p} \sum_{i=0}^{N_p-1} p(i) \tag{15-10}$$

The first term in the denominator of (15–9) is the scaled sample variance of the signal segment, $(N_p - 1)\sigma_x^2(n)$. It can be expressed as

$$(N_p - 1)\sigma_x^2(n) = \sum_{i=0}^{N_p-1} \left[x(n+i) - \bar{x}(n) \right]^2 \tag{15-11}$$

$$= s_{x^2}(n) - \frac{1}{N_p} s_x(n)^2 \tag{15-12}$$

In the image processing literature, this similarity measure is sometimes called the "normalized cross-correlation" [10], [13]. It is invariant to changes in the mean and scale of both the signal segment and template. It is also normalized and bounded, $-1 \le \rho_c(n) \le 1$. It can be expressed in terms of the Table 15–1 second-order components as

$$\rho_c(n) = \frac{s_{x\tilde{p}}(n)}{\sqrt{s_{x^2}(n) - \dfrac{1}{N_p} s_x(n)^2 s_{\tilde{p}^2}}} \qquad (15\text{–}13)$$

where the centered template is defined as

$$\tilde{p}(i) = p(i) - \bar{p} \qquad (15\text{–}14)$$

and

$$s_{\tilde{p}^2} = \sum_{i=0}^{N_p-1} \tilde{p}(i)^2 \qquad (15\text{–}15)$$

15.3 FAST CALCULATION OF SIMILARITY MEASURE COMPONENTS

Since the last three similarity measure components in Table 15–1, s_p, s_{p^2}, and $s_{p^2w^2}$, do not scale with the signal length N_x, the computational cost of computing these components is insignificant compared with the first five components. The following sections describe fast techniques for computing the first five Table 15–1 components.

Running Sums

The components $s_x(n)$ and $s_{x^2}(n)$ are the same operation applied to $x(n)$ and $x(n)^2$, respectively. Because these are simple sums, they can be computed with a running sum,

$$s_x(n) = s_x(n-1) + x(n+N_p-1) - x(n-1) \qquad (15\text{–}16)$$

$$s_{x^2}(n) = s_{x^2}(n-1) + x(n+N_p-1)^2 - x(n-1)^2 \qquad (15\text{–}17)$$

where we define $x(n) = 0$ for $n < 0$. The computational cost of this operation is two additions per output sample.

Similar schemes have been used in image processing based on precomputed summed area and summed squares tables [12]–[14]. It is also possible to calculate these components with fast FIR filters, which compute the convolution as multiplication in the frequency domain using the FFT, by convolving $x(n)$ and $x(n)^2$ with an impulse response of ones [15], but this is less efficient than running sums.

A disadvantage of running sums is that the quantization errors accumulate in the same manner as a random walk, and the quantization error variance scales linearly with N_x. This effect can be reduced by computing the sum directly every N_r samples. This increases the computational cost to $\lfloor N_p + 2(N_r - 1) \rfloor / N_r$ per sample. Compared with a direct computation of $s_x(n)$ and $s_{x^2}(n)$, the computational cost of a running sum is reduced by nearly a factor of N_r as compared with a direct computation of the sum at every sample time.

Fast FIR Filters

The components $s_{x^2w^2}(n)$, $s_{xp}(n)$, and $s_{xpw^2}(n)$ defined in Table 15–1 can each be expressed as a convolution of a signal with a tapped-delay line FIR filter impulse response,

$$s_{x^2w^2}(n) = x(n)^2 * w(-n)^2 \qquad (15\text{–}18)$$

$$s_{xp}(n) = x(n) * p(-n) \qquad (15\text{–}19)$$

$$s_{xpw^2}(n) = x(n) * p(-n)w(-n)^2 \qquad (15\text{–}20)$$

where * denotes convolution. Each impulse response has a finite duration of N_p and is anticausal with a range of $-(N_p - 1) \leq n \leq 0$.

Expressing these components as a convolution with an FIR filter makes it possible to apply any of the fast filtering algorithms that have been developed over the past 40 years for a variety of hardware architectures. For large pattern lengths, say $N_p > 100$, the fastest algorithms take advantage of the well-known convolution property of the discrete Fourier transform (DFT) and the efficiency of the FFT to minimize computation.

15.4   RESULTS

Floating Point Operations

Table 15–2 lists the number of multiplication/division operations (M), addition/subtraction operations (A), and square root operations (S) per sample for direct implementations of each of the four similarity measures.

Table 15–3 lists the number of FIR filtering operations (F) and other arithmetic operations per sample for fast implementations of each of the four similarity measures.

Table 15–2   Computational Cost of a Direct Implementation of Each Similarity Measure

Similarity measure	Eqn.	M	A	S
$r(n)$	(15–1)	$N_p + 1$	$N_p - 1$	–
$\xi(n)$	(15–3)	$N_p + 1$	$2N_p - 1$	–
$\xi_w(n)$	(15–6)	$2N_p$	$2N_p - 1$	–
$\rho_c(n)$	(15–9), (15–10)	$2N_p + 3$	$5N_p - 3$	1

Table 15–3   Computational Cost of a Fast Implementation of Each Similarity Measure

Similarity measure	Eqn.	F	M	A	S
$r(n)$	(15–2)	1	–	–	–
$\xi(n)$	(15–5)	1	3	5	–
$\xi_w(n)$	(15–8)	2	2	2	–
$\rho_c(n)$	(15–13)	1	5	7	–

These tables do not include the cost of computing s_p, s_{p^2}, and $s_{p^2 w^2}$, since $N_x \gg N_p$ and the cost of computing these components does not scale with N_x. These tables assume three additions per sample for each running sum, which corresponds to a reset interval of approximately $N_r = N_p$. Note that these tables only include basic arithmetic operations. Other operations that are not accounted for here, such as memory access and conditional branching, can have a large impact on performance [16], [17].

Simulation Example

To demonstrate an example of the possible reduction in computational cost achievable by using running sums and fast FIR filters, we compared the direct and fast implementations of each similarity measure on a personal computer (Pentium 4, 2 GHz processor, 512 MB of RAM). The direct implementations and running sums were written in C and compiled with the Visual C/C++ optimizing compiler (Microsoft, Version 12.00.8168) for 80 × 86. The fast FIR filtering was performed by the overlap-add method implemented in the fftfilt() function in MATLAB (MathWorks, Version 7.0.0.19920). The block length was chosen automatically by MATLAB to minimize the total number of floating point operations (flops). The source code used to generate these results is available at http://bsp.pdx.edu.

Figure 15–2 shows the average execution time required per sample for each of the four measures. The signal length was $N_x = 2^{18} = 262{,}144$ samples. The execution time for each measure and template length was averaged over 20 runs.

The range of execution times (max–min) in a set of 20 runs never exceeded 0.24 µs/sample. In each case, the mean absolute error between the direct and fast implementations was never greater than 5.4ε, where ε is the distance from a floating point representation of 1.0 to the next larger floating point number. Thus, quantization did not significantly reduce the accuracy of the similarity measures computed with the fast implementations.

15.5   CONCLUSIONS

Popular similarity measures based on second-order statistics can be expressed in terms of basic components and computed with fast algorithms. These techniques are

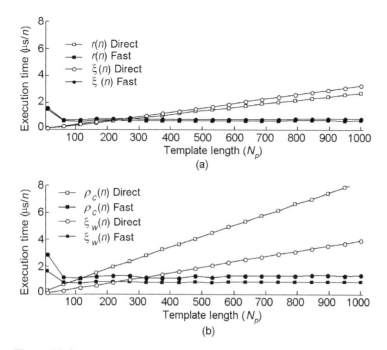

Figure 15–2   Similarity measures' average execution times required per sample (μs/n).

most suitable for detection applications in which the similarity measure must be calculated for every sample in a given signal segment [1]–[5]. For template lengths of greater than 200, these fast algorithms are much more computationally efficient than direct implementations.

15.6 REFERENCES

[1] A. BOLLMANN, K. SONNE, H. ESPERER, I. TOEPFFER, J. LANGBERG, and H. KLEIN, "Non-invasive Assessment of Fibrillatory Activity in Patients with Paroxysmal and Persistent Atrial Fibrillation Using the Holter ECG," *Cardiovascular Research*, vol. 44, 1999, pp. 60–66.

[2] S. HANNA and A. CONSTANTINIDES, "An Automatic Segmentation and Labelling Algorithm for Continuous Speech Signals," *Digital Signal Processing*, vol. 87, 1987, pp. 543–546.

[3] A. OLVER and L. CUTHBERT, "FMCW Radar for Hidden Object Detection," *IEE Proceedings-F Radar and Signal Processing*, vol. 135, no. 4, August 1988, pp. 354–361.

[4] M. ATALLAH, F. CHYZAK, and P. DUMAS, "A Randomized Algorithm for Approximate String Matching," *Algorithmica*, vol. 29, 2001, pp. 468–486.

[5] I. BANKMAN, K. JOHNSON, and W. SCHNEIDER, "Optimal Detection, Classification, and Superposition Resolution in Neural Waveform Recordings," *IEEE Transactions on Biomedical Engineering*, vol. 40, no. 8, August 1993, pp. 836–841.

[6] L. BROWN, "A Survey of Image Registration Techniques," *ACM Computing Surveys*, vol. 24, no. 4, December 1992, pp. 325–376.

[7] E. HALL, R. KRUGER, S. DWYER, III, D. HALL, R. McLAREN, and G. LODWICK, "A Survey of Prepocessing and Feature Extraction Techniques for Radiographic Images," *IEEE Transactions on Computers*, vol. 20, no. 9, September 1971, pp. 1032–1044.

[8] K. McGILL and L. DORFMAN, "High-Resolution Alignment of Sampled Waveforms," *IEEE Transactions on Biomedical Engineering*, vol. 31, no. 6, June 1984, pp. 462–468.

[9] W. PRATT, *Digital Image Processing*, 2nd ed. John Wiley & Sons, Inc., 1991.

[10] R. GONZALEZ and R. WOODS, *Digital Image Processing*. Addison-Wesley, 1992.

[11] C. OLSON, "Maximum-Likelihood Image Matching," *IEEE Transactions on Pattern Analysis and Machine Intelligence*, vol. 24, no. 6, June 2002, pp. 853–857.

[12] S. KILTHAU, M. DREW, and T. MÖLLER, "Full Search Content Independent Block Matching Based on the Fast Fourier Transform," in *2002 International Conference on Image Processing*, vol. 1, 2002, pp. 669–672.

[13] K. BRIECHLE and U. HANEBECK, "Template Matching Using Fast Normalized Cross Correlation," *Proceedings of SPIE—The International Society for Optical Engineering*, vol. 4387, 2001, pp. 95–102.

[14] H. SCHWEITZER, J. BELL, and F. WU, "Very Fast Template Matching," *Computer Vision—ECCV 2002*, vol. 2353, 2002, pp. 358–372.

[15] M. UENOHARA and T. KANADE, "Use of Fourier and Karhunen-Loeve Decomposition for Fast Pattern Matching with A Large Set of Templates," *IEEE Transactions on Pattern Analysis and Machine Intelligence*, vol. 19, no. 8, August 1997, pp. 891–898.

[16] Z. MOU and P. DUHAMEL, "Short-Length FIR Filters and Their Use in Fast Nonrecursive Filtering," *IEEE Transactions on Signal Processing*, vol. 39, no. 6, June 1991, pp. 1322–1332.

[17] A. GACIC, M. PÜSCHEL, and J. MOURA, "Fast Automatic Software Implementations of FIR filters," in *International Conference on Acoustics, Speech and Signal Processing (ICASSP'03)*, vol. 2, April 2003, pp. 541–544.

Chapter 16

Efficient Multi-tone Detection

Vladimir Vassilevsky
Abvolt Ltd.

$\mathbf{T}$his chapter presents the DSP tricks employed to build a computationally efficient multi-tone detection system implemented without multiplications, and with minimal data memory requirements. More specifically, we describe the detection of incoming dial tones, the validity checking to differentiate valid tones from noise signals, and the efficient implementation of the detection system. While our discussion focuses on dual-tone multifrequency (DTMF) telephone dial tone detection, the processing tricks presented may be employed in other multi-tone detection systems.

16.1 MULTI-TONE DETECTION

Multi-tone detection is the process of detecting the presence of spectral tones, each of which has the frequencies $\omega_1, \omega_2, \ldots \omega_k$, where $k = 8$ in our application. A given combination of tones is used to represent a symbol of information, so the detection system's function is to determine what tones are present in the $x(n)$ input signal. A traditional method for multi-tone detection is illustrated in Figure 16–1.

As shown in this figure, the incoming $x(n)$ multi-tone signal is multiplied by the frequency references $\exp(j\omega_k t)$ for all possible multi-tone frequencies ω_k, downconverting any incoming tones so they become centered at zero hertz. Next, the complex $u_k(n)$ products are lowpass filtered. Finally, the magnitudes of the complex lowpass-filtered sequences are logically compared to determine which of the ω_k tones are present. For the multi-tone detection system in Figure 16–1, the DSP tricks employed are as follows:

- Use a 1-bit signal representation so multiplications are replaced with exclusive-or operations;

- Group signal samples into 8-bit words so eight signal samples are manipulated simultaneously in one clock cycle;

Streamlining Digital Signal Processing: A Tricks of the Trade Guidebook, Second Edition. Edited by Richard G. Lyons.
© 2012 the Institute of Electrical and Electronics Engineers. Published 2012 by John Wiley & Sons, Inc.

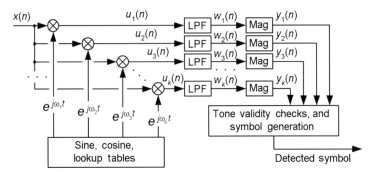

Figure 16–1   Traditional multi-tone detection process.

- Perform multiplier-free complex downconversion using a lookup table (LUT);
- Perform multiplier-free lowpass filtering;
- Perform computation-free elimination of quantization DC bias error;
- Perform multiplier-free complex sample magnitude estimation.

16.2 COMPLEX DOWNCONVERSION

As a first trick to simplify our implementation, the $x(t)$ input signal is converted from an analog signal to a 1-bit binary signal, and the $\exp(j\omega_k t)$ reference frequencies are quantized to a 1-bit representation. Because of this step, the downconversion multiplication of 1-bit data by a 1-bit reference frequency sample can be performed using a simple exclusive-or (XOR) operation. To improve computational efficiency the incoming 1-bit data samples from the comparator, arriving at an 8 kHz sample rate, are collected into one 8-bit byte. The 8-bit XOR operations, one XOR for the sine part and one XOR for the cosine part of the complex reference frequencies, are performed to process eight $x(n)$ samples simultaneously. The logical zeros in the result of XOR operations correspond to the situation when the input is in-phase with an $\exp(j\omega_k t)$ reference, and the logical ones of the XOR result indicate that the input is in quadrature phase with an $\exp(j\omega_k t)$ reference.

The number of zeros and ones in the XOR results, which comprise the real and imaginary parts of the $u_k(n)$ sequences, can be found by counting the nonzero bits directly. However, to enhance execution speed, we use a LUT indexed by the XOR result instead of the direct count of ones and zeros.

The XOR lookup table entries, optimized to provide the best numeric accuracy in conjunction with the follow-on lowpass filters, take advantage of the full numeric range $(-128 \ldots +127)$ of a signed 8-bit word. If N is the number of ones in an 8-bit XOR result, the LUT entries are computed as follows:

$$\text{Table}[N] = \text{Round}[29.75(N-4)+8] \qquad (16\text{–}1)$$

Table 16–1   XOR Lookup Table

N	LUT entry, $u_k(n)$
0	−111
1	−81
2	−52
3	−22
4	+8
5	+38
6	+68
7	+97
8	+127

where Round [·] denotes rounding to the nearest integer. Using (16–1) for our application, the XOR LUT contains the entries listed in Table 16–1. Because XOR results are 8-bit words, the XOR LUT has 256 entries where each entry is one of the nine values in Table 16–1 depending on the number of logic ones in an XOR result. The multiplication factor 29.75 in (16–1) was chosen to partition the LUT entry range, −128 . . . +127, into nine intervals while accommodating the +8 term. (The constant term in (16–1) equal to +8 will be explained shortly.)

The reference frequency samples (square waves actually) are also stored in LUTs. Ideally, the size of the sine and cosine LUTs should be equal to the least common period for all DTMF frequencies. However, the least common period of the DTMF frequencies is one second. For the sample rate of 8 kHz, the reference frequency LUT sizes are equal to 16k (2^{14}). Because tables of this size cannot be realized in low-end microcontrollers, the values of the frequencies were modified to fit the least common period into a smaller table. We found that the common period of 32 milliseconds is a good compromise between the frequency accuracy and the size of the LUT. In this case the size of the LUT is equal to 512 bytes, and the difference between the LUT frequencies and the DTMF standard frequencies is 10 Hz or less. This mismatch does not affect the operation of the multi-tone detector.

The LUT of 512 bytes for the reference frequencies may be too large for some applications. Numerically controlled oscillator (NCO) sine and cosine generation can be used as an alternative [2], [3]. In that scenario, typical 16-bit NCOs require an extra 16 bytes of random-access memory (RAM) and create an additional computing workload on the order of the three million operations per second (MIPS); however, the NCO method frees the 512 bytes of ROM.

16.3 LOWPASS FILTERING

The $w_k(n)$ product sequences must be lowpass filtered. The next trick is that we apply the $w_k(n)$ sequences to a bank of computationally efficient lowpass filters defined by

$$w_k(n) = w_k(n-1) + [u_k(n) - w_k(n-1)]/16. \qquad (16\text{–}2)$$

This economical filter requires only a single data memory storage location and no filter coefficient storage. To enhance execution speed, we implement the divide by 16 with an arithmetic right-shift by 4 bits. The filter outputs are computed with the precision of 8 bits.

The astute reader may recognize (16–2) as the computationally efficient *exponential averaging* lowpass filter. That filter, whose frequency magnitude response is shown in Figure 16–2(a), exhibits nonlinear phase, but that is of no consequence in our application. The filter's weighting factor of 1/16 is determined by the filter's necessary 3-dB bandwidth of 10 Hz, as mandated by the accuracy of the entries stored in the reference frequencies lookup table.

When the arithmetic right-shift is executed in the lowpass filters, a round to the nearest integer operation is performed to improve precision by adding 8 to the number before making the right-shift by 4 bits. Instead of repeatedly adding 8 to each filter output sample, the trick we use next to eliminate all these addition operations is to merely add a constant value of 8 to the XOR LUT. This scheme accounts for the constant 8 term in (16–1).

16.4 MAGNITUDE APPROXIMATION

Every 16 ms, the magnitudes of all the complex $w_k(n)$ sequences are computed by obtaining the real and imaginary parts of $w_k(n)$ and using the approximation

$$
\begin{aligned}
|w_k(n)| \approx y_k(n) = \max\{|\text{real}[w_k(n)]|, |\text{imag}[w_k(n)]|\} \\
+ \min\{|\text{real}[w_k(n)]|, |\text{imag}[w_k(n)]|\}/4
\end{aligned}
\tag{16–3}
$$

to estimate the magnitude of the complex $w_k(n)$ sequences. This magnitude approximation trick requires no multiplications because the multiplication by 1/4 is implemented using an arithmetic right-shift by 2 bits. The maximum magnitude estimation error of (16–3), shown in Figure 16–2(b), is equal to 11.6%, which is acceptable for the operation of the DTMF decoder.

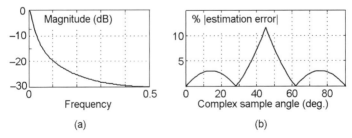

Figure 16–2 Detector performance: (a) lowpass filter response; (b) approximation error of the $w_k(n)$ magnitude.

16.5 VALIDITY CHECKING

The computed magnitudes are processed to decide if valid DTMF tones are present and, if so, the DTMF symbol is decoded. First, the three maximum magnitudes are found and arranged in descending order: $M_1 \geq M_2 \geq M_3$. Then the validity conditions are checked using the following rules:

1. The two frequencies corresponding to M_1 and M_2 must fall into the two different touch-tone telephone frequency groups (the <1 kHz group and the >1 kHz group).
2. Magnitude M_2 must be greater than the absolute threshold value T_1.
3. Ratio M_2/M_3 must be greater than the relative threshold value T_2.

The threshold values T_1 and T_2 depend on the real-world system's characteristics such as the signal sample rate, the analog signal's spectrum, the analog filtering prior to digital processing, the probability density of the signal and the noise, the desired probability of false alarm, and the probability of detection. As such, the threshold values are found empirically while working with the hardware implementation of the multi-tone detector. Note that if we normalize the ideal single-tone maximum amplitude at the output to unity, the value of the threshold T_1 will be on the order of the absolute value range of the XOR LUT divided by four. As such, in this application T_1 is approximately 120/4 = 30. The relative threshold T_2 does not depend on tone amplitudes. A typical value for T_2 is approximately 2.

Also note that the third conditional check above does not require a division operation. That condition is verified if $M_2 \geq T_2 M_3$. To enhance speed, the multiplication by the constant threshold T_2 is implemented by arithmetic shift and addition operations. Finally, if all three validity conditions are satisfied, the DTMF tones are considered to be valid. This validity checking allows us to distinguish true DTMF tones from speech, the silence in speech pauses, noise, or other signals.

16.6 IMPLEMENTATION ISSUES

The multi-tone detection algorithm requires roughly 1000 bytes of read-only memory (ROM), and 64 bytes of read/write memory (RAM). The algorithm requires no analog-to-digital (A/D) converter, no multiply operations, and its processing workload is quite low, equal to only 0.5 MIPS. (For comparison, the DTMF detection system described in [1] requires 24 MIPS of processing power on a fixed-point DSP chip.) As such, fortunately, it is possible to implement the DTMF multi-tone detector/decoder using a low-cost 8-bit microcontroller such as Atmel's AVR, Microchip Technology's PIC, Freescale's HC08, or the x51 family of microcontrollers.

16.7 OTHER CONSIDERATIONS

The performance of the above algorithm is satisfactory for most practical DTMF detection applications. However, in some cases more robust processing is needed.

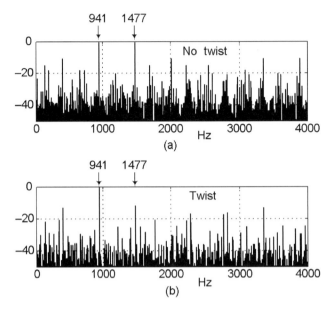

Figure 16–3   One-bit comparator output spectra: (a) no frequency twist; (b) with frequency twist.

For example, if the DTMF signal has become distorted by the processing of a low-bitrate speech coder/decoder, the above algorithm may not operate properly. The weakness of the algorithm is its sensitivity to *frequency twist* and other noise interference. Frequency twist occurs when the power levels of the two detected tones differ by 4–8 dB, and the sensitivity in this case is due to the input signal being digitized to 1 bit by the comparator (a highly nonlinear procedure).

To illustrate this behavior, Figure 16–3(a) shows the output spectra of the 1-bit comparator for the input tones of 944 and 1477 Hz when no frequency twist occurs. There we can see the mix of the input tones and the miscellaneous nonlinear spectral products. If the amplitudes of both input tones are equal, the nonlinear spectral artifacts are at least 10 dB lower then the desired signals. Figure 16–3(b) also shows the output spectra of the comparator with frequency twist. If the amplitudes of the two input tones differ by 6 dB or more (twist), then the strongest nonlinear spectral products have approximately equal amplitude as that of the weakest of the two input tones. Some of the products are falling into the bandwidth of the lowpass filters. Note also that a difference in the levels of input tones equal to 6 dB causes that difference to be 12 dB after the comparator, which limits the performance of the multi-tone detector.

As such, we caution the reader that 1-bit quantization is not always appropriate for signal detection systems. Sadly, there are no hard-and-fast rules to identify the cases when 1-bit quantization can be used.

For the detection of spectrally simple, high-signal-to-noise-ratio, stable-amplitude over-sampled signals, the 1-bit quantization may be applicable. However, because the statistical analysis of 1-bit quantization errors is complicated, careful modeling must be conducted to evaluate the nonlinear performance and spectral aliasing effects of 1-bit quantization.

If improved performance beyond that of 1-bit quantization is required, a similar algorithm can be applied for processing $x(n)$ input signals from an A/D converter. The multi-bit $x(n)$ signal's multiplication by the 1-bit reference frequencies can, again, be performed using simple 16-bit addition and subtraction operations. The A/D converter-based algorithm is able to decode the DTMF under any conditions; however, the trade-off is that the processing has to be done for every sample, not at eight samples simultaneously. This increases the computing burden to several MIPS.

Finally, note that the DTMF detection algorithm presented here is not strictly compliant with the ITU/Bellcore standards [4]. However, this is not detrimental in most practical cases.

16.8 REFERENCES

[1] M. FELDER, J. MASON, and B. EVANS, "Efficient Dual-Tone Multifrequency Detection Using the Nonuniform Discrete Fourier Transform," *IEEE Signal Processing Letters*, vol. 5, no. 7, July 1998, pp. 160–163.
[2] Analog Devices Inc., "A Technical Tutorial on Digital Signal Synthesis," [Online: http://www.analog.com/UploadedFiles/Tutorials/450968421DDS_Tutorial_rev12-2-99.pdf.]
[3] L. CORDESSES, "Direct Digital Synthesis: A Tool for Periodic Wave Generation," *IEEE Signal Processing Magazine*, DSP Tips & Tricks column, vol. 21, no. 2, July 2005, pp. 50–54.
[4] R. FREEMAN, *Reference Manual for Telecommunications Engineering*. Wiley-Interscience, New York, 2002.

EDITOR COMMENTS

One-bit representation (quantization) of signals is an interesting and tricky process. For additional examples of systems using one-bit signal quantization see Chapters 5 and 35.

The weighting factor of the lowpass filters (exponential averagers) in this chapter was 1/16 in order to achieve a 3-dB bandwidth of 10 Hz when the filter input sample rate is 1 kHz. The following shows how to obtain that weighting factor.

We can compute the appropriate value of an exponential averager's weighting factor, W, to achieve any desired filter 3-dB bandwidth. If f_c is the desired positive *cutoff* frequency in Hz where the frequency magnitude response is 3 dB below the averager's zero-Hz response, then the value of W needed to achieve such an f_c cutoff frequency is

$$W = \cos(\Omega) - 1 + \sqrt{\cos^2(\Omega) - 4\cos(\Omega) + 3} \qquad (16\text{–}4)$$

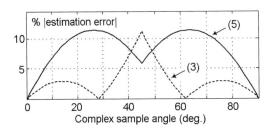

Figure 16–4   Magnitude approximation errors in percent.

where $\Omega = 2\pi f_c/f_s$ and f_s is the filter's input sample rate in Hz. So when $f_c = 10$ Hz and $f_s = 1$ kHz, expression (16–4) yields a desired weighting factor of $W = 0.061$. Because $1/16 = 0.0625$ is very close to the desired value for W, a weighting factor of $1/16$ was acceptable for use in (16–2).

To elaborate on the lowpass filters' complex output sample magnitude estimation algorithm, the two magnitude estimation schemes appropriate for use in this application are the above expression (16–3) and a similar algorithm defined by

$$|w_k(n)| \approx y_k(n) = \max\{|\mathrm{real}[w_k(n)]|, |\mathrm{imag}[w_k(n)]|\}$$
$$+ \min\{|\mathrm{real}[w_k(n)]|, |\mathrm{imag}[w_k(n)]|\}/2. \qquad (16\text{–}5)$$

The difference between (16–3) and (16–5) is their second terms' scaling factors. For both scaling factors, the multiplication can be implemented with binary right-shifts. Both expressions have almost identical maximum error as shown in Figure 16–4. However, (16–3) has the lowest average error, making it the optimum choice with respect to minimizing the average magnitude estimation error. Complex sample magnitude approximation algorithms similar to (16–3) and (16–5) having improved accuracy at the expense of additional shift and add/subtract operations are described in Chapter 25.

Chapter 17

Turning Overlap-Save into a Multiband, Mixing, Downsampling Filter Bank

Mark Borgerding
3dB Labs, Inc.

In this chapter, we show how to extend the popular overlap-save fast convolution filtering technique to create a flexible and computationally efficient bank of filters, with frequency translation and decimation implemented in the frequency domain. In addition, we supply some tips for choosing appropriate fast Fourier transform (FFT) size.

Fast convolution is a well-known and powerful filtering technique. All but the shortest finite impulse response (FIR) filters can be implemented more efficiently in the frequency domain than when performed directly in the time domain. The longer the filter impulse response is, the greater the speed advantage of fast convolution.

When more than one output is filtered from a single input, some parts of the fast convolution algorithm are redundant. Removing this redundancy increases fast convolution's speed even more. Sample rate change by decimation (downsampling) and frequency translation (mixing) techniques can also be incorporated efficiently in the frequency domain. These concepts can be combined to create a flexible and efficient bank of filters. Such a filter bank can implement mixing, filtering, and decimation of multiple arbitrary channels much faster than direct time-domain implementation.

17.1 SOMETHING OLD AND SOMETHING NEW

The necessary conditions for vanilla-flavored fast convolution are covered pretty well in the literature. However, the choice of FFT size is not. Filtering multiple

Streamlining Digital Signal Processing: A Tricks of the Trade Guidebook, Second Edition. Edited by Richard G. Lyons.
© 2012 the Institute of Electrical and Electronics Engineers. Published 2012 by John Wiley & Sons, Inc.

165

channels from the same forward FFT requires special conditions not detailed in textbooks. To downsample and shift those channels in the frequency domain requires still more conditions.

The first section is meant to be a quick reminder of the basics before we extend the *overlap-save* (OS) fast convolution technique. If you feel comfortable with these concepts, skip ahead. On the other hand, if this review does not jog your memory, check your favorite DSP book for "fast convolution," "overlap-add" (OA), "overlap-save," or "overlap-scrap" [1]–[5].

17.2 REVIEW OF FAST CONVOLUTION

The convolution theorem tells us that multiplication in the frequency domain is equivalent to convolution in the time domain [1]. Circular convolution is achieved by multiplying two discrete Fourier transforms (DFTs) to effect convolution of the time sequences the transforms represent. By using the FFT to implement the DFT, the computational complexity of circular convolution is approximately $O(N\log_2 N)$ instead of $O(N^2)$, as in direct linear convolution. Although very small FIR filters are most efficiently implemented with direct convolution, fast convolution is the clear winner as the FIR filters get longer. Conventional wisdom places the efficiency crossover point at 25–30 filter coefficients. The actual value depends on the relative strengths of the platform in question (CPU pipelining, zero-overhead looping, memory addressing modes, etc.). On a desktop processor with a highly optimized FFT library, the value may be as low as 16. On a fixed-point DSP with a single-cycle multiply-accumulate instruction, the efficiency crossover point can be greater than 50 coefficients.

Fast convolution refers to the blockwise use of circular convolution to accomplish linear convolution. Fast convolution can be accomplished by overlap-add or overlap-save methods. Overlap-save is also known as "overlap-scrap" [5]. In OA filtering, each signal data block contains only as many samples as allows circular convolution to be equivalent to linear convolution. The signal data block is zero padded prior to the FFT to prevent the filter impulse response from "wrapping around" the end of the sequence. OA filtering adds the input-on transient from one block with the input-off transient from the previous block.

In OS filtering, shown in Figure 17–1, no zero-padding is performed on the input data, thus the circular convolution is not equivalent to linear convolution. The portions that wrap around are useless and discarded. To compensate for this, the last part of the previous input block is used as the beginning of the next block. OS requires no addition of transients, making it faster than OA. The OS filtering method is recommended as the basis for the techniques outlined in the remainder of this discussion. The nomenclature "FFT$_N$" in Figure 17–1 indicates that an FFT's input sequence is zero-padded to a length of N samples, if necessary, before performing the N-point FFT.

For clarity, the following table defines the symbols used in this material.

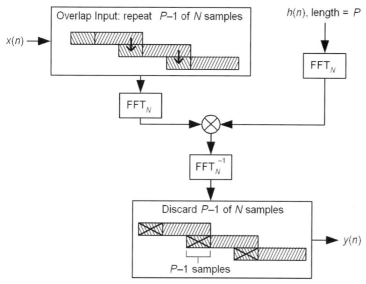

Figure 17–1   Overlap-save (OS) filtering, $y(n) = x(n) * h(n)$.

Symbol Conventions

$x(n)$	Input data sequence
$h(n)$	FIR filter impulse response
$y(n) = x(n) * h(n)$	Convolution of $x(n)$ and $h(n)$
L	Number of new input samples consumed per data block
P	Length of $h(n)$
$N = L + P - 1$	FFT size
$V = N/(P - 1)$	Overlap factor, (FFT length)/(filter transient length)
D	Decimation factor

17.3 CHOOSING FFT SIZE: COMPLEXITY IS RELATED TO FILTER LENGTH AND OVERLAP FACTOR

Processing a single block of input data that produces L outputs incurs a computational cost related to $N\log(N) = (L + P - 1) \log(L + P - 1)$. The computational cost per sample is related to $(L + P - 1) \log(L + P - 1)/L$. The filter length P is generally a fixed parameter, so choosing L such that this equation is minimized gives the theoretically optimal FFT length. It may be worth mentioning that the log base is the radix of the FFT. The only difference between different based logarithms is a scaling factor. This is irrelevant to "big O" scalability, which generally excludes constant scaling factors.

Larger FFT sizes are more costly to perform, but they also produce more usable (non-wraparound) output samples. In theory, one is penalized more for choosing too small an overlap factor, V, than too large. In practice, the price of large FFTs paid in computation and/or numerical accuracy may suggest a different conclusion.

"In theory there is no difference between theory and practice. In practice there is."

—Yogi Berra (American philosopher and occasional baseball manager)

Some other factors to consider while deciding the overlap factor for fast convolution filtering:

- **FFT speed.** It is common for actual FFT computational cost to differ greatly from theory (e.g., due to memory caching).
- **FFT accuracy.** The numerical accuracy of fast convolution filtering is dependent on the error introduced by the FFT-to-inverse FFT round trip. For floating point implementations, this may be negligible, but fixed point processing can lose significant dynamic range in these transforms.
- **Latency.** Fast convolution filtering process increases delay by at least L samples. The longer the FFT, the longer the latency.

While there is no substitute for benchmarking on the target platform, in the absence of benchmarks, choosing a power-of-2 FFT length about four times the length of the FIR filter is a good rule of thumb.

17.4   FILTERING MULTIPLE CHANNELS: REUSE THE FORWARD FFT

It is often desirable to apply multiple filters against the same input sample sequence. In these cases, the advantages of fast convolution filtering become even greater. The computationally expensive operations are the forward FFT, the inverse FFT, and the multiplication of the frequency responses. The forward FFT needs to be computed just once. This is roughly a "Buy one filter—get one 40% off" sale.

In order to realize this computational cost savings for two or more filters, all filters must have the same impulse response length. This condition can always be achieved by zero padding the shorter filters. Alternately, the engineer may redesign the shorter filter(s) to make use of the additional coefficients without increasing the computational workload.

17.5   FREQUENCY DOMAIN DOWNSAMPLING: ALIASING IS YOUR FRIEND

The most intuitive method for reducing sample rate in the frequency domain is to simply perform a smaller inverse FFT using only those frequency bins of interest. This is not a 100% replacement for time domain downsampling. This simple method will cause ripples (Gibbs phenomenon) at FFT buffer boundaries caused by the multiplication in the frequency domain by a rectangular window.

The sampling theorem tells us that sampling a continuous signal aliases energy at frequencies higher than the Nyquist rate (half the signal sample rate) back into the baseband spectrum (below the Nyquist rate). This is equally true for decimation of a digital sequence as it is for analog-to-digital conversion [6]. Aliasing is a natural part of downsampling. In order to accurately implement downsampling in the frequency domain, it is necessary to preserve this behavior. It should be noted that, with a suitable anti-aliasing filter, the energy outside the selected bins might be negligible. This discussion is concerned with the steps necessary for equivalence. The designer should decide how much aliasing, if any, is necessary.

Decimation (i.e., downsampling) can be performed exactly in the frequency domain by coherently adding the frequency components to be aliased [7]. The following octave/MATLAB code demonstrates how to swap the order of an inverse DFT and decimation.

```
% Make up a completely random frequency spectrum
Fx = randn(1,1024) + i*randn(1,1024);
% Time-domain decimation - inverse transform then
decimate
x_full_rate = ifft(Fx);
x_time_dom_dec = x_full_rate(1:4:1024); % Retain every
fourth sample
% Frequency-domain decimation, alias first, then
inverse transform
Fx_alias = Fx(1:256) + Fx(257:512) + Fx(513:768) +
Fx(769:1024);
x_freq_dom_dec = ifft(Fx_alias)/4;
```

The sequences x_time_dom_dec and x_freq_dom_dec are equal to each other. The above sample code assumes a complex time-domain sequence for generality. The division by 4 in the last step accounts for the difference in scaling factors between the inverse FFT sizes. As various FFT libraries handle scaling differently, the designer should keep this in mind during implementation. It's worth noting that this discussion assumes the FFT length is a multiple of the decimation rate D. That is, N/D must be an integer.

To implement time-domain decimation in the frequency domain as part of fast convolution filtering, the following conditions must be met.

1. The FIR filter order must be a multiple of the decimation rate D.

$$P - 1 = K_1 D$$

2. The FFT length must be a multiple of the decimation rate D.

$$L + P - 1 = K_2 D$$

where

- D is the decimation rate or the least common multiple of the decimation rates for multiple channels
- K_1 and K_2 are integers

Note if the overlap factor V is an integer, then the first condition implies the second. It is worth noting that others have also explored the concepts of rate conversion in overlap-add/save [8]. Also note that decimation by large primes can lead to FFT inefficiency. It may be wise to decimate by such factors in the time domain.

17.6 MIXING AND OS FILTERING: ROTATE THE FREQUENCY DOMAIN FOR COARSE MIXING

Mixing, or frequency shifting, is the multiplication of an input signal by a complex sinusoid [1]. It is equivalent to convolving the frequency spectrum of an input signal with the spectrum of a sinusoid. In other words, the frequency spectrum is shifted by the mixing frequency.

It is possible to implement time-domain mixing in the frequency domain by simply rotating the DFT sequence, but there are limitations:

1. The precision with which one can mix a signal by rotating a DFT sequence is limited by the resolution of the DFT.

2. The mixing precision is limited further by the fact that we don't use a complete buffer of output in fast convolution. We use only L samples. We must restrict the mixing to the subset of frequencies whose periods complete in those L samples. Otherwise phase discontinuities occur. That is, one can only shift in multiples of V bins.

The number of "bins" to rotate is

$$N_{rot} = round\left(\frac{Nf_r}{Vf_s}\right) \cdot V \qquad (17\text{--}1)$$

where f_r is the desired mixing frequency and f_s is the sampling frequency. The second limitation may be overcome by using a bank of filters corresponding to different phases. However, this increase in design/code complexity probably does not outweigh the meager cost of multiplying by a complex phasor.

If coarse-grained mixing is unacceptable, mixing in the time domain is a better solution. The general solution to allow multiple channels with multiple mixing frequencies is to postpone the mixing operation until the filtered, decimated data is back in the time domain.

If mixing is performed in the time domain:

- All filters must be specified in terms of the input frequency (i.e., nonshifted) spectrum.

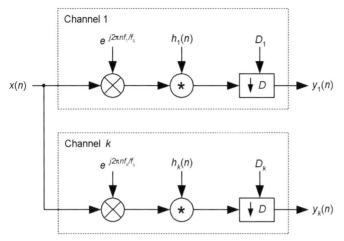

Figure 17–2   Conceptual model of a filter bank.

- The complex sinusoid used for mixing the output signal must be created at the output rate.

17.7 PUTTING IT ALL TOGETHER

By making efficient implementations of conceptually simple tools we help ourselves to create simple designs that are as efficient as they are easy to describe. Humans are affected greatly by the simplicity of the concepts and tools used in designing and describing a system. We owe it to ourselves as humans to make use of simple concepts whenever possible. ("Things should be described as simply as possible, but no simpler."—A. Einstein.) We owe it ourselves as engineers to realize those simple concepts as efficiently as possible.

The familiar and simple concepts shown in Figure 17–2 may be used for the design of mixed, filtered, and decimated channels. The design may be implemented more efficiently using the equivalent structure shown in Figure 17–3.

17.8 FOOD FOR THOUGHT

Answering a question or exploring an idea often leads to more questions. Along the path to understanding the concepts detailed in this chapter, various side paths have been glimpsed but not fully explored. Here are a few such side paths:

- Highly decimated channels are generally filtered by correspondingly narrow bandwidth filters. Those frequency bins whose combined power falls below a given threshold may be ignored without adversely affecting output. Skipping the multiplications and additions associated with filtering and aliasing those bins can speed processing at the expense of introducing arbitrarily low

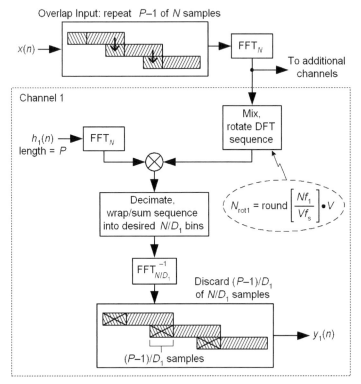

Figure 17–3   Overlap-save filter bank.

error energy. Some researchers have modeled the error created by forcing *all* frequency bins outside the decimated baseband to zero a cyclostationary process [7]. By including *some* of the bins, the error can be made arbitrarily low and compensation is made unnecessary.

- If channels of highly differing rates and filter orders are required, it may be more efficient to break up the structure using partitioned convolution or multiple stages of fast convolution. What is the crossover point? Could such designs effectively use intermediate FFT stages?

- What application, if any, does multiple fast convolution filtering have with regard to number theoretic transforms (NTT)?

17.9 CONCLUSIONS

We outlined considerations for implementing multiple overlap-save channels with decimation and mixing in the frequency domain, as well as supplying recommendations for choosing FFT size. We also provided implementation guidance to streamline this powerful multichannel filtering, downconversion, and decimation process.

17.10   REFERENCES

[1] A. OPPENHEIMER and R. SCHAFER, *Discrete-Time Signal Processing*. Prentice Hall, Upper Saddle River, NJ, 1989.

[2] L. RABINER and B. GOLD, *Theory and Application of Digital Signal Processing*. Prentice Hall, Englewood Cliffs, NJ, 1975.

[3] R. LYONS, *Understanding Digital Signal Processing*, 2nd ed. Prentice Hall, Upper Saddle River, New Jersey, 2004.

[4] S. ORFANIDIS, *Introduction to Signal Processing*. Prentice Hall, Englewood Cliffs, NJ, 1995.

[5] M. FRERKING, *Digital Signal Processing in Communication Systems*. Chapman & Hall, New York, 1994.

[6] R. CROCHIERE and L. RABINER, *Multirate Digital Signal Processing*. Prentice Hall, Englewood Cliffs, NJ, 1983.

[7] M. BOUCHERET, I. MORTENSEN, and H. FAVARO, "Fast Convolution Filter Banks for Satellite Payloads with On-board Processing," *IEEE Journal on Selected Areas in Communications*, February 1999.

[8] S. MURAMATSU and H. KIYA, "Extended Overlap-Add and -Save methods for Multirate Signal Processing," *IEEE Trans. on Signal Processing*, September 1997.

EDITOR COMMENTS

One important aspect of this overlap-save fast convolution scheme is that the FFT indexing bit-reversal problem inherent in some hardware-FFT implementations is not an issue here. If the identical FFT structures used in Figure 17–1 produce $X(m)$ and $H(m)$ having bit-reversed indices, the multiplication can still be performed directly on the scrambled $H(m)$ and $X(m)$ sequences. Next an appropriate inverse FFT structure can be used that expects bit-reversed input data. That inverse FFT then provides an output sequence whose time-domain indexing is in the correct order.

An implementation issue to keep in mind: the complex amplitudes of the standard radix-2 FFT's output samples are proportional to the FFT size, N. As such, we can think of the FFT as having a gain of N. (That's why the standard inverse FFT has a *scaling*, or *normalizing*, factor of $1/N$.) So the product of two FFT outputs, as in our fast convolution process, will have a gain proportional to N^2 and the inverse FFT has a normalizing gain reduction of only $1/N$. Thus, depending on the forward and inverse FFT software being used, the fast convolution filtering method may have an overall gain that is not unity. The importance of this possible nonunity gain depends, of course, on the numerical format of the data as well as the user's filtering application. We won't dwell on this subject here because it's so dependent on the forward and inverse FFT software routines being used. We'll merely suggest that this normalization topic be considered during the design of any fast convolution system.

Chapter 18

Sliding Spectrum Analysis

Eric Jacobsen   **Richard Lyons**

Anchor Hill Communications   **Besser Associates**

The standard method for spectrum analysis in DSP is the discrete Fourier transform (DFT), typically implemented using a fast Fourier transform (FFT) algorithm. However, there are applications that require spectrum analysis only over a subset of the N center frequencies of an N-point DFT. A popular, as well as efficient, technique for computing sparse DFT results is the Goertzel algorithm, which computes a single complex DFT spectral bin value for every N input time samples. This chapter describes *sliding spectrum analysis* techniques whose spectral bin output rates are equal to the input data rate, on a sample-by-sample basis, with the advantage that they require fewer computations than the Goertzel algorithm for real-time spectral analysis. In applications where a new DFT output spectrum is desired every sample, or every few samples, the *sliding DFT* is computationally simpler than the traditional radix-2 FFT. We'll start our discussion by providing a brief review of the Goertzel algorithm, and use its behavior as a yardstick to evaluate the performance of the sliding DFT technique. Following that, we will examine stability issues regarding the sliding DFT implementation as well as review the process of frequency-domain convolution to accomplish time-domain windowing. Finally, a modified sliding DFT structure is proposed that provides improved computational efficiency.

18.1 GOERTZEL ALGORITHM

The Goertzel algorithm, used in dual-tone multifrequency decoding and PSK/FSK modem implementations, is commonly used to compute DFT spectra [1]–[4]. The algorithm is implemented in the form of a second-order IIR filter as shown in Figure 18–1. This filter computes a single DFT output (the kth bin of an N-point DFT) defined by

Streamlining Digital Signal Processing: A Tricks of the Trade Guidebook, Second Edition. Edited by Richard G. Lyons.
© 2012 the Institute of Electrical and Electronics Engineers. Published 2012 by John Wiley & Sons, Inc.

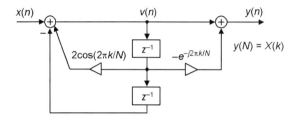

Figure 18–1   IIR filter implementation of the Goertzel algorithm.

$$X(k) = \sum_{n=0}^{N-1} x(n)e^{-j2\pi nk/N}. \tag{18–1}$$

The filter's $y(n)$ output is equal to the DFT output frequency coefficient, $X(k)$, at the time index $n = N$. For emphasis, we remind the reader that the filter's $y(n)$ output is not equal to $X(k)$ at any time index when $n \neq N$. The frequency-domain index k is an integer in the range $0 \leq k \leq N - 1$. The derivation of this filter's structure is readily available in the literature [5]–[7].

The z-domain transfer function of the Goertzel filter is

$$H_G(z) = \frac{1 - e^{-j2\pi k/N} z^{-1}}{1 - 2\cos(2\pi k/N)z^{-1} + z^{-2}} \tag{18–2}$$

with a single z-domain zero located at $z = e^{-j2\pi k/N}$ and conjugate poles at $z = e^{\pm j2\pi k/N}$ as shown in Figure 18–2(a). The pole/zero pair at $z = e^{-j2\pi k/N}$ cancel each other. The frequency magnitude response, provided in Figure 18–2(b), shows resonance centered at a normalized frequency of $2\pi k/N$, corresponding to a cyclic frequency $k \cdot f_s/N$ hertz (where f_s is the signal sample rate).

We remind the reader that while the typical Goertzel algorithm description in the literature specifies the frequency resonance variable k in (18–2) and Figure 18–1 to be an integer (making the Goertzel filter's output equivalent to an N-point DFT bin output), variable k can in fact be *any* value between 0 and $N - 1$ giving us full flexibility in specifying a Goertzel filter's resonance frequency.

While the Goertzel algorithm is derived from the standard DFT equation, it's important to realize that the filter's frequency magnitude response is not the $\sin(x)/(x)$-like response of a single-bin DFT. The Goertzel filter is a complex resonator having an infinite-length unit impulse response, $h(n) = e^{j2\pi nk/N}$, and that's why its magnitude response is so narrow. The time-domain difference equations for the Goertzel filter are

$$v(n) = 2\cos(2\pi k/N)v(n-1) - v(n-2) + x(n). \tag{18–3a}$$

$$y(n) = v(n) - e^{-j2\pi k/N} v(n-1). \tag{18–3b}$$

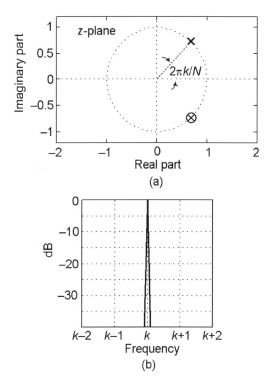

Figure 18–2   Goertzel filter: (a) z-domain pole/zero locations; (b) frequency magnitude response.

An advantage of the Goertzel filter in calculating an N-point $X(k)$ DFT bin is that (18–3a) is implemented N times while (18–3b), the feedforward path in Figure 18–1, need only be computed once after the arrival of the Nth input sample. Thus, for real $x(n)$ the filter requires $N + 2$ real multiplies and $2N + 1$ real adds to compute an N-point $X(k)$. However, when modeling the Goertzel filter if the time index begins at $n = 0$, the filter must process $N + 1$ time samples with $x(N) = 0$ to compute $X(k)$. Now let's look at the sliding DFT process.

18.2   SLIDING DISCRETE FOURIER TRANSFORM (SDFT)

The sliding DFT (SDFT) algorithm performs an N-point DFT on time samples within a sliding-window as shown in Figure 18–3. In this example the SDFT initially computes the DFT of the $N = 16$ time samples in Figure 18–3(a). The time window is then advanced one sample, as in Figure 18–3(b), and a new N-point DFT is calculated. The value of this process is that each new DFT is efficiently computed directly from the results of the previous DFT. The incremental advance of the time window for each output computation is what leads to the name *sliding DFT* or *sliding-window DFT*.

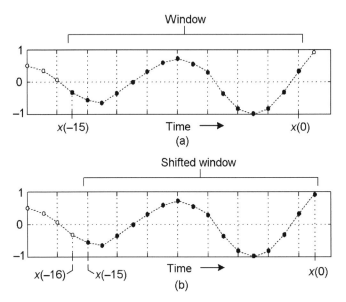

Figure 18–3   Signal windowing for two 16-point DFTs: (a) data samples in the first computation; (b) second computation samples.

The principle used for the SDFT is known as the *DFT shifting theorem*, or the *circular shift property* [8]. It states that if the DFT of a windowed (finite-length) time-domain sequence is $X(k)$, then the DFT of that sequence, circularly shifted by one sample, is $X(k)e^{j2\pi k/N}$. Thus the spectral components of a shifted time sequence are the original (unshifted) spectral components multiplied by $e^{j2\pi k/N}$, where k is the DFT bin of interest. We use this shift principal to express our sliding DFT process as

$$S_k(n) = e^{j2\pi k/N}[S_k(n-1) + x(n) - x(n-N)] \tag{18–4}$$

where $S_k(n)$ is the new spectral component and $S_k(n-1)$ is the previous spectral component. The subscript k reminds us that the spectra are those associated with the kth DFT bin.

Equation (18–4), whose derivation is provided in the appendix to this chapter, reveals the value of this process in computing real-time spectra. We calculate $S_k(n)$ by phase shifting the sum of the previous $S_k(n-1)$ with the difference between the current $x(n)$ sample and the $x(n-N)$ sample. The difference between the $x(n)$ and $x(n-N)$ samples can be computed once for each n and used for each $S_k(n)$ computation. So the SDFT requires only one complex multiply and two real adds per output sample.

The computational complexity of each successive N-point output is then $O(N)$ for the sliding DFT compared with $O(N^2)$ for the DFT and $O[N\log_2(N)]$ for the FFT. Unlike the DFT or FFT, however, due to its recursive nature the sliding DFT output

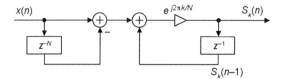

Figure 18–4   Single-bin sliding DFT filter structure.

must be computed for each new input sample. If a new N-point DFT output is required only every N inputs, the sliding DFT requires $O(N^2)$ computations and is equivalent to the DFT. When output computations are required every M input samples, and M is less than $\log_2(N)$, the sliding DFT can be computationally superior to traditional FFT implementations even when all N DFT outputs are required.

Equation (18–4) leads to the single-bin SDFT filter structure shown in Figure 18–4.

The single-bin SDFT algorithm is implemented as an IIR filter with a comb filter followed by a complex resonator [9]. (If you want to compute all N DFT spectral components, N resonators with $k = 0$ to $N - 1$ will be needed, all driven by a single comb filter.) The comb filter delay of N samples forces the filter's transient response to be $N - 1$ samples in length, so the output will not reach steady state until the $S_k(N)$ sample. In practical applications the algorithm can be initialized with zero-input and zero-output. The output will not be valid, or equivalent to (18–1)'s $X(k)$, until N input samples have been processed. The z-domain transfer function for the kth bin of the sliding DFT filter is

$$H_{\text{SDFT}}(z) = \frac{e^{j2\pi k/N}(1 - z^{-N})}{1 - e^{j2\pi k/N}z^{-1}}. \tag{18-5}$$

This complex filter has N zeros equally spaced around the z-domain's unit circle, due to the N-delay comb filter, as well as a single pole canceling the zero at $z = e^{j2\pi k/N}$ as shown in Figure 18–5(a). The SDFT filter's complex $h(n)$ unit impulse response is shown in Figure 18–5(b) for the example where $k = 2$ and $N = 20$.

Because of the comb subfilter, the SDFT filter's complex sinusoidal unit impulse response is finite in length—truncated in time to N samples—and that property makes the frequency magnitude response of the SDFT filter identical to the $\sin(Nx)/\sin(x)$ response of a single DFT bin centered at a normalized frequency of $2\pi k/N$.

We've encountered a useful property of the SDFT that's not widely known, but is important. If we change the SDFT's comb filter feedforward coefficient (in Figure 18–4) from -1 to $+1$, the comb's zeros will be rotated counterclockwise around the unit circle by an angle of π/N radians. This situation, for $N = 8$, is shown on the right side of Figure 18–6(a). The zeros are located at angles of $2\pi(k + 1/2)/N$ radians. The $k = 0$ zeros are shown as solid dots. Figure 18–6(b) shows the zeros locations for an $N = 9$ SDFT under the two conditions of the comb filter's feedforward coefficient being -1 and $+1$.

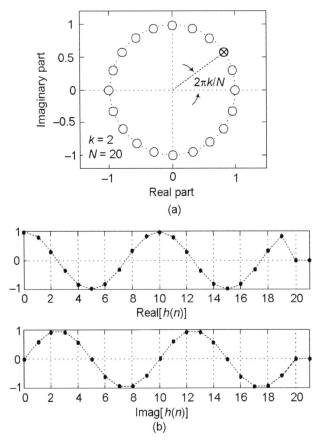

Figure 18–5   Sliding DFT characteristics for $k = 2$ and $N = 20$: (a) impulse response; (b) pole/zero locations.

   This alternate situation is useful, we can now expand our set of spectrum analysis center frequencies to more than just N angular frequency points around the unit circle. The analysis frequencies can be either $2\pi k/N$ or $2\pi(k + 1/2)/N$, where integer k is in the range $0 \leq k \leq N - 1$. Thus we can build an SDFT analyzer that resonates at any one of $2N$ frequencies between 0 and f_s Hz. Of course, if the comb filter's feedforward coefficient is set to +1, the resonator's feedforward coefficient must be $e^{j2\pi(k+1/2)/N}$ to achieve pole/zero cancellation.

   One of the attributes of the SDFT is that once an $S_k(n - 1)$ is obtained, the number of computations to calculate $S_k(n)$ is fixed and independent of N. A computational workload comparison between the Goertzel and SDFT filters is provided later in this discussion. Unlike the radix-2 FFT, the SDFT's N can be any positive integer giving us greater flexibility to *tune* the SDFT's center frequency by defining integer k such that $k = N \cdot f_i / f_s$, when f_i is a frequency of interest in Hz. In addition,

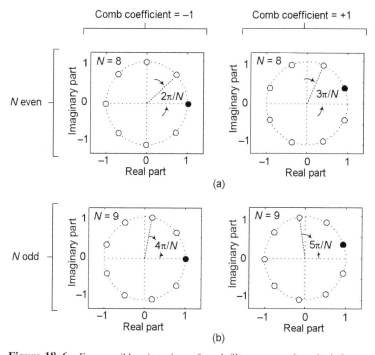

Figure 18–6   Four possible orientations of comb filter zeros on the unit circle.

the SDFT requires no bit-reversal processing as does the FFT. Like Goertzel, the SDFT is especially efficient for narrowband spectrum analysis.

For completeness, we mention that a radix-2 *sliding FFT* technique exists for computing all N bins of $X(k)$ in (18–1) [10], [11]. This method is computationally attractive because it requires only N complex multiplies to update the N-point FFT for all N bins; however, it requires $3N$ memory locations ($2N$ for data and N for twiddle coefficients). Unlike the SDFT, the radix-2 sliding FFT scheme requires address bit-reversal processing and restricts N to be an integer power of two.

18.3 SDFT STABILITY

The SDFT filter is only marginally stable because its pole resides on the z-domain's unit circle. If filter coefficient numerical rounding error is not severe, the SDFT is bounded-input-bounded-output stable. Filter instability can be a problem, however, if numerical coefficient rounding causes the filter's pole to move outside the unit circle. We can use a damping factor r to force the pole to be at a radius of r inside the unit circle and guarantee stability using a transfer function of

$$H_{\text{SDFT,gs}}(z) = \frac{re^{j2\pi k/N}(1 - r^{N}z^{-N})}{1 - re^{j2\pi k/N}z^{-1}}$$

$$(18–6)$$

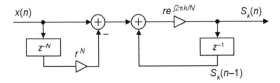

Figure 18–7   Guaranteed-stable sliding DFT filter structure.

with the subscript "gs" meaning guaranteed-stable. The stabilized feedforward and feedback coefficients become $-r^N$ and $re^{j2\pi k/N}$, respectively. The difference equation for the stable SDFT filter becomes

$$S_k(n) = re^{j2\pi k/N}[S_k(n-1) + x(n) - r^N x(n-N)] \qquad (18\text{–}7)$$

with the stabilized-filter structure shown in Figure 18–7.

Using a damping factor as in Figure 18–7 guarantees stability, but the $S_k(n)$ output, defined by

$$X_{r<1}(k) = \sum_{n=0}^{N-1} x(n)r^{(N-n)}e^{-j2\pi nk/N} \qquad (18\text{–}8)$$

is no longer exactly equal to the kth bin of an N-point DFT in (18–1). While the error is reduced by making r very close to (but less than) unity, a scheme does exist for eliminating that error completely once every N output samples at the expense of additional conditional logic operations [12]. Determining whether the damping factor r is necessary for a particular SDFT application requires careful empirical investigation.

Another stabilization method worth consideration for fixed-point applications is decrementing the largest component (either real or imaginary) of the filter's $e^{j2\pi k/N}$ feedback coefficient by one least significant bit. This technique can be applied selectively to problematic output bins and is effective in combating instability due to rounding errors that result in finite-precision $e^{j2\pi k/N}$ coefficients having magnitudes greater than unity.

Like the DFT, the SDFT's output is proportional to N, so in fixed-point binary implementations the designer must allocate sufficiently wide registers to hold the computed results.

18.4   TIME-DOMAIN WINDOWING IN THE FREQUENCY DOMAIN

The spectral leakage of the SDFT can be reduced by the standard concept of windowing the $x(n)$ input time samples. However, windowing by time-domain multiplication would compromise the computational simplicity of the SDFT. Alternatively,

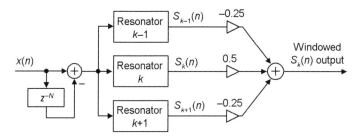

Figure 18–8   Three-resonator structure to compute three SDFT bin results, and a three-point convolution.

we can implement a time-domain window by means of frequency-domain convolution.

Spectral leakage reduction performed in the frequency domain is accomplished by convolving adjacent $S_k(n)$ values with the DFT of a window function. For example, the DFT of a Hanning window comprises only three non-zero values, −0.25, 0.5, and −0.25. As such we can compute a Hanning-windowed $S_k(n)$, the kth DFT bin, with a three-point convolution using

$$\text{Hanning-windowed } S_k(n) = -0.25 \cdot S_{k-1}(n) + 0.5 \cdot S_k(n) - 0.25 \cdot S_{k+1}(n). \quad (18\text{–}9)$$

Figure 18–8 shows this process where the comb filter stage need only be implemented once. Thus a Hanning window can be implemented by binary right shifts and two complex adds for each SDFT bin, making the Hanning window attractive in ASIC and FPGA implementations where single-cycle hardware multiplies are costly. If a gain of four is acceptable, then only two left shifts (one for the real part and one for the imaginary parts of $S_k(n)$) and two complex adds are required using

$$\text{Hanning-windowed } S_k(n) = -S_{k-1}(n) + 2 \cdot S_k(n) - S_{k+1}(n). \quad (18\text{–}10)$$

The Hanning window is a member of a category called $cos^\alpha(x)$ *window functions* [13], [14]. These functions are also known as *generalized cosine windows* because their N-point time-domain samples are defined as

$$w(n) = \sum_{m=0}^{\alpha-1} (-1)^m a_m \cos(2\pi mn/N) \quad (18\text{–}11)$$

where $n = 0, 1, 2, \ldots, N - 1$, and the integer α specifies the number of terms in the window's time function. These window functions are attractive for frequency domain convolution because their DFTs contain only a few non-zero samples. The frequency domain vectors of various $cos^\alpha(x)$ window functions follow the form $(1/2) \cdot (a_2, -a_1, 2a_0, -a_1, a_2)$, with a few examples presented in Table 18–1. Additional $cos^\alpha(x)$ window functions are described in the literature [14].

Table 18–1 $\cos^\alpha(x)$ Windows, Frequency Domain Coefficients

Window function:	a_0	a_1	a_2
Rectangular	1.0	–	–
Hanning, ($\alpha = 2$)	0.5	0.5	–
Blackman, ($\alpha = 3$)	0.42	0.5	0.08
Exact Blackman, ($\alpha = 3$)	0.4265907	0.4965606	0.0768487
Hamming	0.54	0.46	–

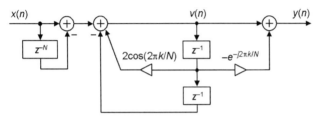

Figure 18–9   Structure of the sliding Goertzel DFT filter.

18.5   SLIDING GOERTZEL DFT

We can reduce the number of multiplications required in the SDFT by creating a new pole/zero pair in its $H_{\mathrm{SDFT}}(z)$ system function [15]. This is done by multiplying the numerator and denominator of $H_{\mathrm{SDFT}}(z)$ in (18–5) by the factor $(1 - e^{-j2\pi k/N}z^{-1})$, yielding

$$
\begin{aligned}
H_{\mathrm{SG}}(z) &= \frac{(1-e^{-j2\pi k/N}z^{-1})(1-z^{-N})}{(1-e^{-j2\pi k/N}z^{-1})(1-e^{j2\pi k/N}z^{-1})} \\
&= \frac{(1-e^{-j2\pi k/N}z^{-1})(1-z^{-N})}{1-2\cos(2\pi k/N)z^{-1}+z^{-2}}
\end{aligned}
\tag{18–12}
$$

where the subscript "SG" means sliding Goertzel. The filter block diagram for $H_{\mathrm{SG}}(z)$ is shown in Figure 18–9, where this new filter is recognized as the standard Goertzel filter preceded by a comb filter. The sliding Goertzel DFT filter, unlike the standard Goertzel filter, has a finite-duration impulse response identical to that shown in Figure 18–5(b), for $k = 2$ and $N = 20$.

Of course, unlike the traditional Goertzel filter in Figure 18–1, the sliding Goertzel DFT filter's complex feedforward computations must be performed for each input time sample. The sliding Goertzel filter's $\sin(Nx)/\sin(x)$ frequency magnitude response, for $k = 2$ and $N = 20$, is provided in Figure 18–10(a). The asymmetrical frequency response is defined by the filter's N zeros equally spaced around the z-domain's unit circle in Figure 18–10(b) due to the N-delay comb filter, as well as an

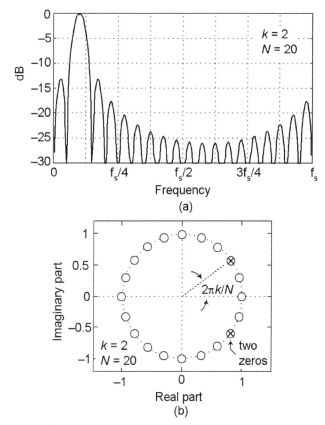

Figure 18–10   Sliding Goertzel filter for $N = 20$ and $k = 2$: (a) frequency magnitude response; (b) z-domain pole/zero locations.

additional (uncanceled) zero located at $z = e^{-j2\pi k/N}$ on account of the $(1 - e^{-j2\pi k/N}z^{-1})$ factor in the $H_{SG}(z)$ transfer function's numerator. In addition, the filter has conjugate poles canceling zeros at $z = e^{+j2\pi k/N}$.

The sliding Goertzel DFT filter is of interest because its computational workload is less than that of the SDFT. This is because the $v(n)$ samples in Figure 18–9 are real-only due to the real-only feedback coefficients. A single-bin DFT computational comparison, for real-only inputs, is provided in Table 18–2. For real-time processing requiring spectral updates on a sample by sample basis the sliding Goertzel method requires fewer multiplies than either the SDFT or the traditional Goertzel algorithm.

18.6 CONCLUSIONS

The sliding DFT process for spectrum analysis was presented, and shown to be more efficient than the popular Goertzel algorithm for sample-by-sample DFT bin

Table 18–2   Single-Bin DFT Comparison

Method	Single $S_k(n)$ computation: Real multiplies:	Real adds:	Next $S_k(n + 1)$ computation: Real multiplies:	Real adds:
DFT	$2N$	$2N$	$2N$	$2N$
Goertzel	$N + 2$	$2N + 1$	$N + 2$	$2N + 1$
Sliding DFT	$4N$	$4N$	4	4
Sliding Goertzel	$N + 2$	$3N + 1$	3	4

computations. The sliding DFT provides computational advantages over the traditional DFT or FFT for many applications requiring successive output calculations, especially when only a subset of the DFT output bins are required. Methods for output stabilization as well as time-domain data windowing by means of frequency-domain convolution were also discussed. A modified sliding DFT algorithm, called the sliding Goertzel DFT, was proposed to further reduce computational workload.

18.7 REFERENCES

[1] M. FELDER, J. MASON, and B. EVANS, "Efficient Dual-Tone Multi-frequency Detection Using the Non-uniform Discrete Fourier Transform," *IEEE Signal Processing Lett.*, vol. 5, July 1998, pp. 160–163.

[2] ANALOG DEVICES INC., *ADSP-2100 Family User's Manual*, 3rd ed. Chapter 14, vol. 1, 1995, Norwood, MA. [Online: http://www.analog.com/Analog_Root/static/library/dspManuals/Using_ADSP-2100_Vol1_books.html.]

[3] S. GAY, J. HARTUNG, and G. SMITH, "Algorithms for Multi-channel DTMF Detection for the WERDSP32 Family," *in Proceedings on the International Conference on ASSP*, pp. 1134–1137, 1989.

[4] K. BANKS, "The Goertzel Algorithm," *Embedded Systems Programming Magazine*, September 2002, pp. 34–42.

[5] G. GOERTZEL, "An Algorithm for the Evaluation of Finite Trigonometric Series," *American Math. Monthly*, vol. 65, 1958, pp. 34–35.

[6] J. PROAKIS and D. Manolakis, *Digital Signal Processing: Principles, Algorithms, and Applications*, 3rd ed., Prentice Hall, Upper Saddle River, NJ, 1996, pp. 480–481.

[7] A. OPPENHEIM, R. SCHAFER, and J. BUCK, *Discrete-Time Signal Processing*, 2nd ed. Prentice Hall, Upper Saddle River, NJ, 1996, pp. 633–634.

[8] T. SPRINGER, "Sliding FFT Computes Frequency Spectra in Real Time," *EDN Magazine*, September 29, 1988, pp. 161–170.

[9] L. RABINER and B. GOLD, *Theory and Application of Digital Signal Processing*, Prentice Hall, Upper Saddle River, NJ, 1975, pp. 382–383.

[10] B. FARHANG-BOROUJENY and Y. LIM, "A Comment on the Computational Complexity of Sliding FFT," *IEEE Trans. Circuits and Syst. II*, vol. 39, no. 12, December 1992, pp. 875–876.

[11] B. FARHANG-BOROUJENY and S. GAZOR, "Generalized Sliding FFT and Its Application to Implementation of Block LMS Adaptive Filters," *IEEE Trans. Sig. Proc.*, vol. 42, no. 3, March 1994, pp. 532–538.

[12] S. DOUGLAS and J. SOH, "A Numerically-Stable Sliding-Window Estimator and Its Application to Adaptive Filters," *Proc. 31st Annual Asilomar Conf. on Signals, Systems, and Computers*, Pacific Grove, CA, vol. 1, November 1997, pp. 111–115.

[13] F. Harris, "On the Use of Windows for Harmonic Analysis with the Discrete Fourier Transform," *Proc. IEEE*, vol. 66, January 1978, pp. 51–84.

[14] A. Nuttall, "Some Windows with Very Good Sidelobe Behavior," *IEEE Trans.*, vol. 29, no. 1, February 1981, pp. 84–91.

[15] K. Larson, Texas Instruments Inc., *Private communication*, November 2002.

18.8 APPENDIX

The derivation of the general SDFT can be understood starting with the definition of the DFT of a contiguous subset of a longer sequence. For this derivation we consider a DFT of length N computed on a subset of an input sequence with length of at least $N + q + 1$ where q is the index of the start of the DFT window in the input sequence. The additional time unit increment is merely to accommodate the slide to the next window.

We start the derivation with the definition of the kth DFT bin for a window starting at the qth element of the input sequence:

$$X(k,q) = \sum_{n=0}^{N-1} x(n+q)e^{-j2\pi nk/N}. \qquad (18\text{–}A1)$$

The transform of the $(q + 1)$th window, the next DFT in the sliding sequence, is then:

$$X(k,q+1) = \sum_{n=0}^{N-1} x(n+q+1)e^{-j2\pi nk/N}. \qquad (18\text{–}A2)$$

Substituting $p = n+1$, so the range of p is 1 to N and we have:

$$X(k,q+1) = \sum_{p=1}^{N} x(p+q)e^{-j2\pi(p-1)k/N}. \qquad (18\text{–}A3)$$

Next we change the summation by formally expressing the Nth component separately and adding the $p = 0$ case to the summation, and then subtracting it formally:

$$X(k,q+1) = \sum_{p=0}^{N-1} x(p+q)e^{-j2\pi(p-1)k/N} + x(q+N)e^{-j2\pi(N-1)k/N} - x(q)e^{j2\pi k/N}. \qquad (18\text{–}A4)$$

The exponential terms can be factored as follows:

$$X(k,q+1) = e^{j2\pi k/N}\left[\sum_{p=0}^{N-1} x(p+q)e^{-j2\pi pk/N} + x(q+N)e^{-j2\pi Nk/N} - x(q)\right]. \qquad (18\text{–}A5)$$

Because $e^{-j2\pi Nk/N} = 1$, (18–A5) can be rewritten as:

$$X(k, q+1) = e^{j2\pi k/N} \left[\sum_{p=0}^{N-1} x(p+q)e^{-j2\pi pk/N} + x(q+N) - x(q) \right]. \qquad (18\text{–}A6)$$

Notice that the summation is the DFT of the qth time window, but (18–A6) is the DFT of the $(q + 1)$th window. The sliding DFT can therefore be expressed as:

$$X(k, q+1) = e^{j2\pi k/N}[X(k, q) + x(q+N) - x(q)]. \qquad (18\text{–}A7)$$

The individual, kth, frequency bins for the $(q + 1)$th SDFT window can therefore be computed from the qth window bins and the time-domain inputs by:

$$S_k(n) = e^{j2\pi k/N}[S_k(n-1) + x(n) - x(n-N)]. \qquad (18\text{–}A8)$$

The above derivation provides exact equivalence to the traditional DFT computation with a recursive algorithm that reduces the computational for successive sliding windows.

_____

_____

EDITOR COMMENT

A variation of this sliding DFT process is discussed in Chapter 21.

Chapter 19

Recovering Periodically Spaced Missing Samples

Andor Bariska
Institute of Data Analysis and Process Design, Zurich University of Applied Sciences

There are times when we need to process time-domain sequences that have been corrupted due to missing samples of a desired signal sequence. This chapter presents a powerful technique to recover (compute) periodically spaced missing samples of a time domain sequence.

19.1 MISSING SAMPLES RECOVERY PROCESS

To clarify our missing samples problem, Figure 19–1 shows a corrupted $x_c[n]$ time sequence, represented by the black dots. What we desire is an $x_o[n]$ sequence representing correct sampling of the continuous $x_o(t)$ signal at the same sample rate as $x_c[n]$. We represent the *missing* samples of $x_o[n]$, which are shown by the white circles, using zero-valued samples in $x_c[n]$. We call the black dot samples of $x_c[n]$ riding on the $x_o(t)$ curve *available samples*, and we refer to $x_c[n]$'s zero-valued black-dot samples as *missing samples*.

As shown in Figure 19–1, the available and missing samples in $x_c[n]$ form a periodic pattern, meaning that $x_c[n]$ contains a repetitive pattern of two available samples, one missing sample, one available sample, and one missing sample. Each repeating T-length period contains $A = 3$ available samples and $M = 2$ missing samples where, of course, $T = A + M = 5$.

Let us assume that $x_o(t)$ is bandlimited to B Hz, and that a sampling rate of F_s Hz was used to generate the sequence $x_o[n]$, that is, $x_o[n] = x_o(n/F_s)$. Again, our missing sample recovery problem is to compute the desired $x_o[n]$ signal from the

Streamlining Digital Signal Processing: A Tricks of the Trade Guidebook, Second Edition. Edited by Richard G. Lyons.
© 2012 the Institute of Electrical and Electronics Engineers. Published 2012 by John Wiley & Sons, Inc.

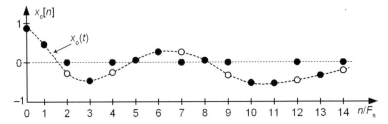

Figure 19–1   Corrupted $x_c[n]$ time sequence. ($A = 3$, $M = 2$, and $T = 5$.)

corrupted $x_c[n]$ signal. We can solve our missing sample recovery problem if the following two conditions are satisfied:

- $x_c[n]$ contains a T-length periodic pattern of A available samples and M missing samples.
- Frequency B is limited to:

$$B \le \frac{A}{T} \cdot \frac{F_s}{2}. \tag{19–1}$$

Next we explain how recover the missing samples using the above example of a periodic pattern where $A = 3$ and $M = 2$. (Our missing samples recovery scheme can be applied to any integer values of A and M.)

19.2 RECOVERY PROCESS

When the above two conditions are satisfied, our missing sample recovery technique begins by defining two sampling time vectors for the available and missing samples that reside within the $T = 5$ period. For our example in Figure 19–1, we define the available sample time vector as

$$t_A = [0, 1, 3]$$

whose length is $A = 3$, and define the missing sample time vector as

$$t_M = [2, 4]$$

whose length is $M = 2$.

Next we partition $x_c[n]$ into A subsequences of available samples based on the t_A time vector elements. The first subsequence of available samples is

$$x_o[t_A[1] + pT] = x_o[0], x_o[5], x_o[10], x_o[15], \text{etc.}$$

The second subsequence of available samples is

$$x_o[t_A[2] + pT] = x_o[1], x_o[6], x_o[11], x_o[16], \text{etc.}$$

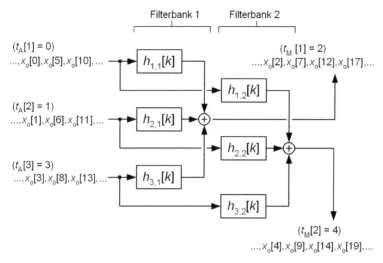

Figure 19–2   Reconstruction filters for missing sample recovery.

And finally, the third subsequence of available samples is

$$x_o[t_A[3] + pT] = x_o[3], x_o[8], x_o[13], x_o[18], \text{etc.}$$

where the integer p sequence is 0, 1, 2, 3, etc. (Because $x_o[n]$ can be infinite in extent, index p could range over both positive and negative integers. For convenience, we arbitrarily chose to define $x_o[n]$ as starting at index $p = 0$, with $x_o[n] = 0$ for $n < 0$.)

Using the above definitions, we can recover our missing samples in the form of $M = 2$ subsequences defined as

$$x_o[t_M[1] + pT] = x_o[2], x_o[7], x_o[12], x_o[17], \text{etc.}$$

and

$$x_o[t_M[2] + pT] = x_o[4], x_o[9], x_o[14], x_o[19], \text{etc.}$$

We implement our missing sample recovery technique by applying those available samples subsequences to the $M = 2$ reconstruction filterbanks shown in Figure 19–2. Each filterbank contains $A = 3$ tapped-delay line convolutional subfilters. The output sequences of the reconstruction filterbanks are the desired $x_o[n]$ samples that we set out to recover.

19.3 COMPUTING FILTER COEFFICIENTS

The subfilters used in the filterbanks are all variations of the well-known ideal fractional delay filter, which has an impulse response equal to

$$D_\Delta[k] = \text{sinc}(\Delta + k), \quad k = \dots, -1, 0, 1, \dots \tag{19–2}$$

where the delay parameter Δ is in the range $-1 < \Delta < 1$. Just like the sampled-sinc lowpass filter, the fractional delay filter is ideal in the sense that perfect implementation of this filter (for example, on a computer) is impossible. However, the approximation of the fractional delay filter has been studied extensively, and many approximation methods are known [1]. As always, the right choice of the fractional delay filter approximation depends strongly on the application.

We now proceed with the example in Figure 19–1 by using the simple but effective windowing approach to approximate the fractional delay. This results in a tapped-delay line subfilter with a finite number of coefficients (an FIR filter). For windowing the impulse response, the window must also be delayed in time by the fractional delay value Δ. We assume a definition of the window function $w(t)$ that is centered at 0, non-zero on the interval $[-1, 1]$, and zero elsewhere. Any window that is defined by sampling a function with a continuous time parameter "t" can be used (this includes, for example, Hanning or Hamming windows). Using the indexing variable $k = -K, -K + 1, \ldots, 1, 0, 1, \ldots, K - 1, K$, and (19–2) the impulse response of the windowed fractional delay filter is given by

$$d_\Delta[k] = D_\Delta[k] \cdot w\left(\frac{\Delta + k}{K}\right) \tag{19–3}$$

This makes computing the coefficients of the convolutional subfilters straightforward.

Next we define two indices that identify the individual subfilters. Because there are A subfilters in each filterbank, we will define a subfilter index of $a = 1, 2, \ldots, A$, and a filterbank index of $m = 1, 2, \ldots, M$. Using those indices, the coefficients of a single subfilter are represented by $h_{a,m}[k]$. For example, using this two-dimensional subscripted index notation, the coefficients of the third subfilter in the second filterbank are denoted as $h_{3,2}[k]$. Given this notation, the coefficients of the subfilters are defined by

$$h_{a,m}[k] = G_{a,m} \cdot d_{\Delta_{a,m}}[k] \cdot \cos(\pi k (A - 1)) \tag{19–4}$$

where $d_{\Delta a,m}[k]$ is given by (19–3) with Δ replaced by $\Delta_{a,m}$,

$$G_{a,m} = \prod_{q=1, q \neq a}^{A} \frac{\sin\left(\pi \cdot \dfrac{t_M[m] - t_A[q]}{T}\right)}{\sin\left(\pi \cdot \dfrac{t_A[a] - t_A[q]}{T}\right)} \tag{19–5}$$

$$\Delta_{a,m} = \frac{t_M[m] - t_A[a]}{T}. \tag{19–6}$$

Integer k ranges from $-K \leq k \leq K$.

19.4 DISCUSSION

We note that the process of removing the M missing samples from $x_o[n]$ is a form of decimation. Any noise components in $x_o[n]$ above the frequency B in (19–1) are aliased into the signal band by this decimation. Similarly, the recovery process using the filterbanks is a general form of upsampling and interpolation, and generates a signal that is bandlimited to B Hz, even if the original signal (which we are trying to recover) was not bandlimited. Some noise (images caused by upsampling) may appear above B Hz due to the finite attenuation of the FIR filters.

Although similar methods for recovering missing samples have been investigated before, to the best of our knowledge the powerful recovery technique presented here is new. The idea of using filterbanks to resample recurrent nonuniformly sampled signals is discussed in reference [2], however, instead of recovering missing samples in oversampled sequences those authors develop a filterbank that resamples the recurrent nonuniform samples to uniform samples spaced at a sampling period equal to $1/(2B)$. This can also be achieved using the method described in this chapter, by placing the missing samples at uniform intervals in one period.

The special case of recovery of an $M = 1$ missing sample per recurrence period was solved in [3] with a single symmetric FIR filter, also under the $x_o(t)$ signal bandwidth limitations in (19–1).

19.5 CONCLUSIONS

We presented a novel method to recover periodically missing samples from appropriately bandlimited signals. We used filterbanks with fractional delay filters as building blocks, and described a windowed-sinc method to approximate the fractional delays with FIR subfilters. Because the implementation of filterbanks and the approximation of fractional delay filters are well-studied problems, this makes the proposed recovery method flexible and easy to implement.

Space limitations do not allow us to present the complete derivation of the subfilter coefficient expression in (19–4). However, to aid in the reader's understanding and software modeling, a derivation (19–4), and MATLAB code to compute the reconstruction filterbanks' coefficients using the windowing method for the example in Figure 19–1 are made available at http://booksupport.wiley.com.

19.6 REFERENCES

[1] T. LAAKSO, V. VÄLIMÄKI, M. KARJALAINEN, and U. LAINE, "Splitting the Unit Delay," *IEEE Signal Processing Magazine*, vol. 13, no. 1, January 1996, pp. 30–60. [Software Online]: http://www.acoustics.hut.fi/software/fdtools.
[2] Y. ELDAR, and A. OPPENHEIM, "Filterbank *Reconstruction of Bandlimited Signals from Nonuniform and Generalized Samples*," *IEEE Transactions on Signal Processing*, vol. 48, no. 10, October 2000, pp. 2864–2875.

[3] R. ADAMS, "Nonuniform Sampling of Audio Signals," *J. Audio Eng. Soc.*, vol. 40, no. 11, November 1992, pp. 886–894.

EDITOR COMMENTS

The following material provides the sample values of the sequences in (19–2) through (19–6) for the example in Figure 19–1.

[I]   The parameters of the Figure 19–1 example are as follows:

$$A = 3$$
$$a = [1, 2, 3]$$
$$t_A = [0, 1, 3]$$
$$M = 2$$
$$m = [1, 2]$$
$$t_M = [2, 4]$$
$$T = 5$$

[II]   To keep the following tables manageable in size, a small value for K is used for this example.

$$K = 3 \text{ (number of subfilter taps is 7)}$$
$$k = [-3, -2, -1, 0, 1, 2, 3]$$

What follows is the intermediate computational results for Eqs. (19–2) through (19–6) for the Figure 19–1 missing samples example.

[III]   The Δ values in Eq. (19–2) are the $\Delta_{a,m}$ values computed using Eq. (19–6). They are:

$$\Delta_{1,1} = 0.40$$
$$\Delta_{2,1} = 0.20$$
$$\Delta_{3,1} = -0.20$$
$$\Delta_{1,2} = 0.80$$
$$\Delta_{2,2} = 0.60$$
$$\Delta_{3,2} = 0.20$$

[IV]   The $D_{\Delta_{a,m}}[k]$ sequences in Eq. (19–2), for each subfilter, are:

k	$D_{\Delta 1,1}[k]$	$D_{\Delta 2,1}[k]$	$D_{\Delta 3,1}[k]$	$D_{\Delta 1,2}[k]$	$D_{\Delta 2,2}[k]$	$D_{\Delta 3,2}[k]$
−3	0.1164	0.0668	−0.0585	0.0850	0.1261	0.0668
−2	−0.1892	−0.1039	0.0850	−0.1559	−0.2162	−0.1039
−1	0.5046	0.2339	−0.1559	0.9355	0.7568	0.2339
0	0.7568	0.9355	0.9355	0.2339	0.5046	0.9355
1	−0.2162	−0.1559	0.2339	−0.1039	−0.1892	−0.1559
2	0.1261	0.0850	−0.1039	0.0668	0.1164	0.0850
3	−0.0890	−0.0585	0.0668	−0.0492	−0.0841	−0.0585

[V]   The $w_{a,m}[k] = w((\Delta_{a,m} + k)/K)$ window sequences in Eq. (19–3), for each subfilter, were computed using the author's favored Knab* window. Those window sequences are:

k	$w_{1,1}[k]$	$w_{2,1}[k]$	$w_{3,1}[k]$	$w_{1,2}[k]$	$w_{2,2}[k]$	$w_{3,2}[k]$
−3	0.0011	0.0002	0.0000	0.0121	0.0041	0.0002
−2	0.1172	0.0622	0.0121	0.3119	0.1997	0.0622
−1	0.7538	0.6027	0.3119	0.9693	0.8825	0.6027
0	0.8825	0.9693	0.9693	0.6027	0.7538	0.9693
1	0.1997	0.3119	0.6027	0.0622	0.1172	0.3119
2	0.0041	0.0121	0.0622	0.0002	0.0011	0.0121
3	0.0000	0.0000	0.0002	0.0000	0.0000	0.0000

* KNAB, J: "An Alternate Kaiser Window," *IEEE Trans. on ASSP*, vol. ASSP–27, no. 5, October 1979.

[VI]   The $d_{\Delta_{a,m}}[k]$ sequences in Eq. (19–3), for each subfilter, are:

k	$d_{\Delta 1,1}[k]$	$d_{\Delta 2,1}[k]$	$d_{\Delta 3,1}[k]$	$d_{\Delta 1,2}[k]$	$d_{\Delta 2,2}[k]$	$d_{\Delta 3,2}[k]$
−3	0.0001	0.0000	0.0000	0.0010	0.0005	0.0000
−2	−0.0222	−0.0065	0.0010	−0.0486	−0.0432	−0.0065
−1	0.3803	0.1410	−0.0486	0.9068	0.6679	0.1410
0	0.6679	0.9068	0.9068	0.1410	0.3803	0.9068
1	−0.0432	−0.0486	0.1410	−0.0065	−0.0222	−0.0486
2	0.0005	0.0010	−0.0065	0.0000	0.0001	0.0010
3	0.0000	0.0000	0.0000	0.0000	0.0000	0.0000

[VII]   The $G_{a,m}$ gain factors in Eqs. (19–4) and (19–5) are:

$$G_{1,1} = -0.6180$$

$$G_{2,1} = 1.0000$$

$$G_{3,1} = 0.6180$$

$$G_{1,2} = 1.0000$$

$$G_{2,2} = -0.6180$$

$$G_{3,2} = 0.6180$$

[VIII]   The $\cos(\pi k(A-1))$ sequence in Eq. (19–4) is:

$$\cos(\pi k(A-1)) = \cos(\pi k(2)) = [1, 1, 1, 1, 1, 1, 1]$$

[IX]   Finally, the $h_{a,m}[k]$ coefficients in Eq. (19–4), for each subfilter, are:

k	$h_{1,1}[k]$	$h_{2,1}[k]$	$h_{3,1}[k]$	$h_{1,2}[k]$	$h_{2,2}[k]$	$h_{3,2}[k]$
−3	−0.0001	0.0000	0.0000	0.0010	−0.0003	0.0000
−2	0.0137	−0.0065	0.0006	−0.0486	0.0267	−0.0040
−1	−0.2351	0.1410	−0.0301	0.9068	−0.4128	0.0871
0	−0.4128	0.9068	0.5604	0.1410	−0.2351	0.5604
1	0.0267	−0.0486	0.0871	−0.0065	0.0137	−0.0301
	−0.0003	0.0010	−0.0040	0.0000	−0.0001	0.0006
3	0.0000	0.0000	0.0000	0.0000	−0.0000	0.0000

Chapter 20

Novel Adaptive IIR Filter for Frequency Estimation and Tracking

Li Tan and Jean Jiang
College of Engineering and Technology at Purdue University North Central

In many applications, a sinusoidal signal may be subjected to nonlinear effects in which possible harmonic frequency components are generated. In such an environment we may want to estimate (track) the signal's fundamental frequency as well as any harmonic frequencies. Using a second-order notch filter to estimate fundamental and harmonic frequencies is insufficient since it only accommodates one frequency component [1], [2]. On the other hand, applying a higher-order IIR notch filter may not be efficient due to adopting multiple adaptive filter coefficients.

In this chapter we present a novel adaptive harmonic IIR notch filter with a single adaptive coefficient to efficiently perform frequency estimation and tracking in a harmonic frequency environment. Furthermore, we devise a simple scheme to select the initial filter coefficient to insure algorithm convergence to its global minimum error.

20.1 HARMONIC NOTCH FILTER STRUCTURE

Consider frequency estimation of a measured signal $x(n)$ containing a fundamental frequency component and its harmonics up to Mth order as

$$x(n) = \sum_{m=1}^{M} A_m \sin[2\pi(mf)nT + \phi_m] + v(n) \qquad (20\text{--}1)$$

Streamlining Digital Signal Processing: A Tricks of the Trade Guidebook, Second Edition. Edited by Richard G. Lyons.
© 2012 the Institute of Electrical and Electronics Engineers. Published 2012 by John Wiley & Sons, Inc.

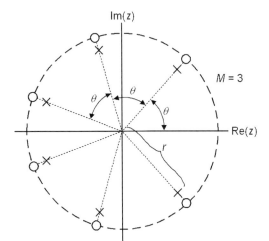

Figure 20–1   Pole-zero plot for the harmonic IIR notch filter for $M = 3$.

where A_m, mf, and ϕ_m are the magnitude, frequency (Hz), and phase angle of the mth harmonic component, respectively. $v(n)$ is a white Gaussian noise. Notice that n and T designate the time index and sampling period, respectively. To estimate frequency in such a harmonic frequency environment, we propose a harmonic IIR notch filter whose z-domain pole-zero plot is shown in Figure 20–1 for the case of $M = 3$ (the fundamental and two harmonics).

The idea is to place constrained pole-zero pairs with their angles equal to $\pm m\theta$ (multiples of the fundamental frequency angle θ) relative to the horizontal axis on the pole-zero plot for $m = 1, 2, \ldots, M$, respectively, to construct a multi-notch filter transfer function [1]–[3]. The zeros on the unit circle give us infinite-depth notches, and the parameter r controls the bandwidth of the notches. Parameter r is chosen to be close to, but less than, 1 to achieve narrowband notches and avoid any filter stability problems.

Hence, once θ is adapted to the angle corresponding to the fundamental frequency, each $m\theta$ ($m = 2, \ldots, M$) will automatically adapt to its harmonic frequency. We construct the filter transfer function in a cascaded form as

$$H(z) = \frac{Y(z)}{X(z)} = \prod_{m=1}^{M} H_m(z) \qquad (20\text{–}2)$$

where the transfer function $H_m(z)$ at the mth 2nd-order IIR section is defined as

$$H_m(z) = \frac{1 - 2z^{-1}\cos(m\theta) + z^{-2}}{1 - 2rz^{-1}\cos(m\theta) + r^2 z^{-2}}. \qquad (20\text{–}3)$$

From (20–2) and (20–3), the transfer function has only one adaptive coefficient θ. We determine the filter output $y_m(n)$ at the mth section as

$$y_m(n) = y_{m-1}(n) - 2\cos(m\theta)y_{m-1}(n-1) + y_{m-1}(n-2)$$
$$+ 2r\cos(m\theta)y_m(n-1) - r^2 y_m(n-2)$$
(20–4)

and $y_0(n) = x(n)$.

Once the adaptive parameter θ has converged to its fundamental frequency, the harmonic IIR notch filter will filter out all the $x(n)$ input signal's fundamental and harmonic frequency components. Thus the last second-order subfilter output $y_M(n)$ is expected to be $y_M(n) \approx 0$ for a noiseless $x(n)$ input signal. Our objective, then, is to minimize the power of the last subfilter output, $E[y_M^2(n)]$, at which time the converged parameter θ is our desired result. Here, we define the filter output $y_M(n)$ as the error signal $e(n)$, that is, $e(n) = y_M(n)$, in a sense that its power is to be minimized. Hence, we can express the power of the last subfilter output via the mean square error (MSE) function [1] as

$$E[e^2(n)] = E[y_M^2(n)]$$
$$= \frac{1}{2\pi j} \oint \left| \prod_{m=1}^{M} \frac{1 - 2z^{-1}\cos(m\theta) + z^{-2}}{1 - 2rz^{-1}\cos(m\theta) + r^2 z^{-2}} \right|^2 \Phi_{xx} \frac{dz}{z}$$
(20–5)

where Φ_{xx} is the power spectrum of the input signal. Equation (20–5) tells us that the MSE function is a nonlinear function of the adaptive parameter θ, and it may contain local minima. Selection of an initial value θ is critical for a global convergence of the adaptive algorithm. Fortunately, we need not evaluate (20–5) to obtain the MSE of $e(n)$. Instead, we estimate the MSE function for each given θ as follows:

$$MSE = E[e^2(n, \theta)] \approx \frac{1}{N} \sum_{n=1}^{N} y_M^2(n, \theta) \quad 0 \le \theta \le \pi/M$$
(20–6)

where N is the number of filter output samples averaged as θ varies over the range $0 \le \theta \le \pi/M$ radians. The cyclic frequency in Hz corresponding to θ in radians is $f = \theta f_s/(2M)$, where f_s is the sampling rate in Hz. To ensure the global minimum is at the fundamental frequency in case there are more than two global minima, we estimate another MSE function from the first subfilter corresponding to the fundamental frequency as follows:

$$MSE1 = E[e_1^2(n, \theta)] \approx \frac{1}{N} \sum_{n=1}^{N} y_1^2(n, \theta) \quad 0 \le \theta \le \pi/M$$
(20–7)

It is well known that the $MSE1 = E[e_1^2(n, \theta)]$ from this $m = 1$ subfilter always has a global minimum [1]. Hence, once the region of the global minimum is identified by verifying MSE functions from (20–6) and (20–7), we could determine a frequency capture range based on the plotted MSE function in (20–6) as described in the following example.

Figure 20–2 shows an example of the MSE and MSE1 functions for $M = 3$ and $r = 0.95$, where $f_s = 8000$ Hz, the fundamental frequency of the input signal is

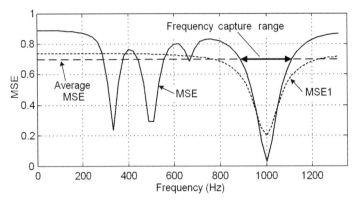

Figure 20–2   Notch filter MSE and MSE1 functions for $M = 3$ and $r = 0.95$.

1000 Hz, the input signal to noise power ratio (SNR) is 18 dB, and $N = 400$ filter output samples were averaged for each θ. Note that θ's range of $0 \leq \theta \leq \pi/M$ radians is based on the assumption that the $x(n)$ input's fundamental component and its $M - 1$ harmonic frequencies are less than $f_s/2$. Therefore, the corresponding cyclic frequency range in Figure 20–2 is from 0 Hz to $4000/M = 1333.33$ Hz for $M = 3$. For our Figure 20–2 example, there are four local minima of the MSE function from (20–6), where one global minimum is located at 1000 Hz. Clearly, the MSE1 function from (20–7) verifies the global minimum at the fundamental frequency. If we let the adaptive algorithm initially start from any point within the global minimum capture range in the MSE function from (20–6), the adaptive harmonic IIR notch filter parameter θ will converge to the global minimum of the MSE function, which corresponds to the fundamental frequency of the input signal.

20.2 ALGORITHM DEVELOPMENT

As discussed in the last section, we develop the adaptive algorithm to contain two major steps. The first step is to determine the initial value of the adaptive filter $\theta(0)$ coefficient, while the second step is to apply the least mean-square (LMS) algorithm to obtain the desired precise value for the filter's θ coefficient.

With a help of Figure 20–2, we can describe a strategy for selecting the initial value $\theta(0)$ as follows. After estimating MSE function values using (20–6) for $0 \leq \theta \leq \pi/M$, corresponding to $0 \leq f \leq f_s/(2M)$ Hz, we compute the average value of MSE (horizontal line in Figure 20–2). Next we use that average value to determine two intersection points in the global minimum valley of the MSE function to obtain a frequency capture range (the frequency range between two intersection points), as depicted in Figure 20–2. Notice that the best initial $\theta(0)$ value should be the point of f (Hz) on the horizontal axis, at which the MSE has the largest deviation below its average value, such as $f = 1000$ Hz. However, the LMS algorithm starting at any $\theta(0)$ point within the frequency capture range will converge to its global minimum.

For the second step, we apply the LMS algorithm to obtain the filter update equation to obtain the desired precise value for the filter's θ coefficient. Taking the derivative of $e^2(n) = [y_M(n)]^2$ and substituting the result to zero, we achieve

$$\theta(n+1) = \theta(n) - 2\mu y_M(n)\beta_M(n) \qquad (20\text{–}8)$$

where μ is a convergence step-size parameter and the gradient term $\beta_m(n)$ at subfilter section m is defined as

$$\beta_m(n) = \frac{\partial y_m(n)}{\partial \theta(n)} \qquad (20\text{–}9)$$

with

$$\beta_0(n) = \frac{\partial y_0(n)}{\partial \theta(n)} = \frac{\partial x(n)}{\partial \theta(n)} = 0. \qquad (20\text{–}10)$$

Using (4), $\beta_m(n)$ for $m = 1, 2, \ldots, M$ can be recursively computed using the following equation:

$$\beta_m(n) = \beta_{m-1}(n) - 2\cos[m\theta(n)]\beta_{m-1}(n-1) + 2m\sin[m\theta(n)]y_{m-1}(n-1) + \beta_{m-1}(n-2)$$
$$+ 2r\cos[m\theta(n)]\beta_m(n-1) - r^2\beta_m(n-2) - 2rm\sin[m\theta(n)]y_m(n-1). \qquad (20\text{–}11)$$

Again, for the first subfilter (when $m = 1$), $\beta_0(n) = \beta_0(n-1) = \beta_0(n-2) = 0$.

20.3 ALGORITHM STEPS

We summarize the adaptive harmonic notch filter algorithm below:

Step 1: Determine the initial adaptive coefficient $\theta(0)$:

- Calculate and plot $\mathrm{MSE} = E\big[y_M^2(n)\big]_\theta$ with (20–6) and $\mathrm{MSE1} = E\big[y_1^2(n)\big]_\theta$ with (20–7) using N filter outputs as θ is varied (scanned) over the range $0 \le \theta \le \pi/M$.
- Determine the MSE function's average and global minimum values.
- Based on the average and global minimum of the MSE, determine the frequency capture range.
- Choose the initial coefficient $\theta(0)$ to be within the frequency capture range.

Step 2: Apply the LMS algorithm for frequency estimation and tracking:

- For $m = 1, 2, \ldots, M$, apply (20–4) and (20–11).
- Apply the LMS process in (20–8) to obtain $\theta(n + 1)$. Note that μ should be a very small value ($\mu \ll 1$), and chosen empirically to guarantee the algorithm convergence.

• Convert $\theta(n)$ in radians to the desired estimated fundamental frequency in Hz using

$$f(n) = \frac{\theta(n)}{2\pi} \times f_s (\text{Hz}).\qquad(20\text{–}12)$$

20.4 COMPUTER SIMULATIONS

In our software simulations, the $x(n)$ input signal of 1000 Hz plus two harmonics, sampled at $f_s = 8000$ Hz, is given by

$$\begin{aligned} x(n) = &\sin[2\pi \times 1000 \times n/f_s] + 0.5\cos[2\pi \times (2 \times 1000) \times n/f_s] \\ &- 0.25\cos[2\pi \times (3 \times 1000) \times n/f_s] + v(n) \end{aligned}\qquad(20\text{–}13)$$

The input SNR is set to 18 dB by controlling the variance of the Gaussian $v(n)$ noise sequence in (20–13). We used $M = 3$ and $r = 0.95$ for constructing the harmonic notch filter. The MSE (20–6) and MSE1 (20–7) functions from Step 1 are plotted in Figure 20–2. The frequency capture range from 900 Hz to 1100 Hz (0.225π to 0.275π radians) was determined, and we chose $\theta(0) = 0.225\pi$ radians (900 Hz) and $\mu = 0.0001$ for the LMS algorithm.

To show the frequency tracking capabilities of the harmonic notch filter, Figure 20–3 shows the time-domain input and output at each filter subsection. The input fundamental frequency switches from 1000 Hz to 1075 Hz at time $n = 400$, while

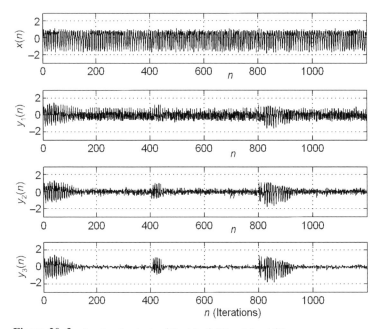

Figure 20–3 Input and outputs of the $M = 3$ IIR notch subfilters.

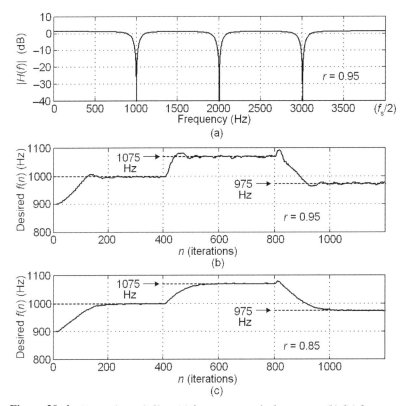

Figure 20–4   Harmonic notch filter: (a) frequency magnitude response; (b) $f(n)$ frequency tracking with $r = 0.95$; (c) reduced-noise tracking with $r = 0.85$.

at $n = 800$ the fundamental frequency switches to 975 Hz. As depicted in Figure 20–3, the algorithm converges to the global minimums after 150 LMS iterations.

Figure 20–4(a) shows our adaptive harmonic notch filter's frequency magnitude response, which has the null points located at the fundamental and harmonic frequencies. That figure indicates why we call our filter a *harmonic notch filter*. Figure 20–4(b) illustrates frequency tracking of the desired fundamental frequency $f(n)$ defined in (20–12), where the dashed curve indicates the true input fundamental frequencies. For this frequency tracking example, we can identify the fundamental frequencies after 150 LMS iterations.

As is well known, the LMS algorithm experiences reduced output fluctuations, but extended convergence time, when the value of μ is reduced. The harmonic notch filter exhibits similar behavior when the r bandwidth parameter is reduced to $r = 0.85$, as shown in Figure 20–4(c). (When parameter r is reduced, the global valley of the MSE function from (20–6) becomes less steep so that the LMS algorithm converges to its global minimum with less sensitivity.) For this $r = 0.85$ scenario, we can identify the fundamental frequencies after roughly 200 LMS iterations.

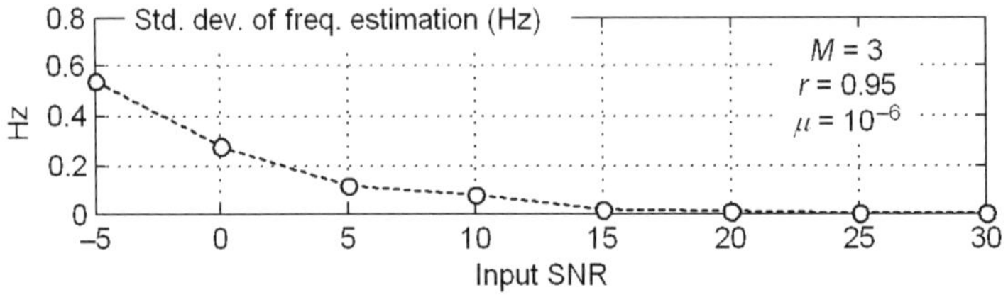

Figure 20–5   Standard deviation of frequency estimates versus input SNR.

20.5   TYPICAL ALGORITHM PERFORMANCE

As expected, the adaptive harmonic IIR notch filter algorithm's frequency estimation performance improves with increased input signal SNRs. Figure 20–5 shows our algorithm's performance for frequency estimation and tracking under various noise conditions, where the frequency deviation from the fundamental frequency of 1000 Hz was plotted versus the input signal's SNR. To obtain the results in Figure 20–5, we selected $\theta(0) = 0.225\pi$ radians (900 Hz), and used the M, r, and μ values shown in that figure. Next we generated $N = 20000$ input samples using (20–13) at various SNRs, and for each SNR data point we simply computed the standard deviation using the last 50 tracked $f(n)$ frequency values. As shown in Figure 20–5, for the worst case in our simulations (SNR = −5 dB) standard deviation of our frequency estimation was 0.55 Hz (less than 1 Hz). We observed that the algorithm performed well under high SNR conditions.

Our experiments also demonstrated that the harmonic notch filter possesses frequency tracking capability when the input signal contained fewer harmonics than we anticipated. On the other hand, if the number of harmonics in the input signal is greater than the number of second-order IIR subfilters, the harmonic notch filter still tracks the fundamental signal frequency but with some performance degradation.

20.6   CONCLUSIONS

In this chapter, we developed a novel adaptive harmonic IIR notch filter for frequency estimation and tracking in a multiharmonic frequency environment. This new frequency estimation algorithm, which has never before appeared in the literature, requires only a single adaptive coefficient and efficiently estimates fundamental and harmonic frequencies simultaneously.

20.7   REFERENCES

[1] J. CHICHARO and T. NG, "Gradient-Based Adaptive IIR Notch Filtering for Frequency Estimation," *IEEE Trans. Acoust., Speech, and Signal Processing*, vol. 38, no. 5, May 1990, pp. 769–777.
[2] J. ZHOU and G. LI, "Plain Gradient-Based Direct Frequency Estimation Using Second-Order Constrained Adaptive IIR Notch Filter," *Electronics Letters*, vol. 50, no. 5, 2004, pp. 351–352.
[3] L. TAN, *Digital Signal Processing: Fundamentals and Applications*. Elsevier/Academic Press, New York, 2007, pp. 358–362.

EDITOR COMMENTS

When trying to estimate, or track (periodically measure), the fundamental frequency of a signal, one might be inclined to merely perform the fast Fourier transform (FFT) of that time-domain signal and carry out some sort of "peak location" process in order to identify the frequency of the largest spectral component and its harmonics. When asked to compare this chapter's adaptive IIR notch filter frequency estimation/ tracking scheme to an FFT-based frequency estimation/tracking method, Professor Li Tan replied with the following.

- The proposed adaptive IIR notch filter method is superior to the FFT because the IIR notch filter method is adaptive, real-time, and a sample-based technique—meaning for every incoming sample, the frequency value is updated. The FFT is a frame-based technique and the frequency resolution depends on the frame size. For example, we may pick a frame of 8000 samples used in an FFT and obtain 1 Hz resolution at an 8 kHz sampling rate. The frequency resolution depends on the frame size.

- The proposed method has frequency tracking capability. It can track fundamental frequency changes during the adaptive processing. An FFT will have a large time latency depending on the frame size.

- The proposed method is a time-domain process, while FFT is a frequency-domain process.

- FFTs require more computation as well as peak location searching. The proposed method has a reduced computation requirement.

- The proposed method is much more flexible in terms of applications. For example, our scheme can capture the speed of a motor from its tachometer and the detected frequency can be sent to digital feedback control system for close-loop speed control. FFTs cannot do that for each incoming sample.

- The proposed method is an adaptive filter method. It can also be applied to other applications, such as frequency line enhancement, and interference cancellation, such as removing 60 Hz interference and its harmonics in ECG signal enhancement. FFTs cannot perform adaptive filtering.

- Since the proposed method adapts its fundamental frequency at the same time it captures its harmonics, the method works for frequency tracking for any periodic waveform such as a square wave, sawtooth, and so on. Even if the signal has aliased spectral components, the harmonic filter will also alias to catch those aliased harmonics.

Chapter 21

Accurate, Guaranteed-Stable, Sliding DFT

Krzysztof Duda
Department of Measurement and Instrumentation at the AGH University of Science and Technology

$\mathbf{T}$he sliding DFT (SDFT) is a recursive algorithm that computes a discrete Fourier transform (DFT) on a sample-by-sample basis. The SDFT is computationally efficient but it suffers from accumulated errors and potential instabilities. As we shall see, previous SDFT algorithms described in the literature compromise results accuracy for guaranteed stability [1]–[3].

This chapter describes a new SDFT algorithm that is guaranteed stable without sacrificing accuracy—an algorithm we call the *modulated SDFT* (mSDFT).

Our algorithm is based on the DFT modulation property; for the chosen DFT bin with index k we use this property to effectively shift that bin to the position $k = 0$. Next, we simply compute the running sum in the sliding window of length N. This approach allows us to exclude the complex twiddle factor from the feedback in the resonator and avoid accumulated errors and potential instabilities.

While the details of traditional SDFTs are found in [1], [2], the next section provides a brief summary of SDFT properties necessary to derive the mSDFT.

21.1 SLIDING DFT

Let us consider DFT computations in the time window of length M sliding along the signal x_n, such that $x_n = 0$ for $n < 0$:

$$X_n^k = \sum_{m=0}^{M-1} x_{q+m} W_M^{-km}, \qquad (21\text{–}1)$$

Streamlining Digital Signal Processing: A Tricks of the Trade Guidebook, Second Edition. Edited by Richard G. Lyons.
© 2012 the Institute of Electrical and Electronics Engineers. Published 2012 by John Wiley & Sons, Inc.

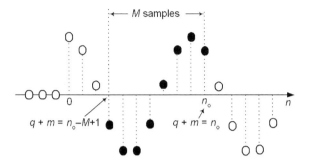

Figure 21–1   Solid dots represent the x_n samples used to compute X_n^k for $M = 8$.

where $q = n - M + 1$, $0 \le k \le M - 1$, and the complex twiddle factor equals $W_M = e^{j2\pi/M}$. To illustrate our time-domain indexing in (21–1), the solid dots in Figure 21–1 are the input x_n samples used to compute X_n^k when $n = n_o$.

In our notation the superscript k in X_n^k is not a traditional exponent but instead refers to the DFT's frequency (kth-bin) index, and the subscript n is a time index. As such, k is a constant and the n in X_n^k indicates that the DFT is computed from samples $x_{n-M+1}, x_{n-M+2}, \ldots, x_n$. In the case of sample-by-sample signal processing, consecutive kth-bin DFT output samples X_n^k, X_{n+1}^k, X_{n+2}^k, $\ldots$, are computed from the x_n time samples that differ only by the first and last samples. A recursive formula for computing (21–1) may be derived as follows:

$$
\begin{aligned}
X_n^k &= \sum_{m=0}^{M-1} x_{q+m} W_M^{-km} = \sum_{m=0}^{M-1} x_{q+m-1} W_M^{-k(m-1)} - x_{q-1} W_M^k + x_{q+M-1} W_M^{-k(M-1)} \\
&= W_M^k \sum_{m=0}^{M-1} x_{q+m-1} W_M^{-km} - x_{q-1} W_M^k + x_{q+M-1} W_M^k = W_M^k (X_{n-1}^k - x_{q-1} + x_{q+M-1}) \\
&= W_M^k (X_{n-1}^k - x_{n-M} + x_n)
\end{aligned}
\tag{21–2}
$$

The structure of difference equation (21–2) is depicted in Figure 21–2(a). This traditional SDFT filter is only marginally stable because it has a z-domain pole located at $z = W_M^k$ on the unit circle as shown in Figure 21–2(b).

In practice, finite precision representation of the twiddle factor is the cause of instabilities and accumulated errors in the traditional SDFT. Except for when a pole is located at $z = \pm 1$ or $z = \pm j$, our nonideal numerical precision of the W_M^k coefficient places that pole either slightly inside, or slightly outside, the unit circle. When the pole is just inside the unit circle, a small error is induced in an X_n^k output sample. Those errors accumulate with each new X_n^k output sample computation. On the other hand, if the W_M^k pole is just outside the unit circle, the network in Figure 21–2(a) becomes unstable.

It is easy to observe that for DFT bin $k = 0$ we have, from (21–2),

$$
X_n^0 = X_{n-1}^0 - x_{n-M} + x_n,
\tag{21–3}
$$

Computation of successive values of X_n^0 requires simple summations of the input samples in the sliding time window of length M.

Due to the absence of the typically imprecise W_M^k coefficient, the recurrence in (21–3) is unconditionally stable and does not accumulate errors. In the next section we use the DFT modulation property to effectively shift the DFT bin of interest to the position of $k = 0$ and then use (21–3) for computing that DFT bin output.

21.2 MODULATED SDFT

By the Fourier modulation property the X^k DFT bin may be shifted to the index $k = 0$ (zero Hz) by the multiplication of the input signal x_n by the modulation sequence W_M^{-km} [4]:

$$X_n^0 = X_{n-1}^0 - x_{n-M}W_M^{-k(m-M)} + x_n W_M^{-km}. \tag{21–4}$$

The W_M^{-km} term in (21–4) is the same complex twiddle factor used in the DFT definition in (21–1).

The structure of the mSDFT in (21–4) is depicted in Figure 21–3(a). Because there is no complex twiddle factor in the feedback of the resonator, the pole of the mSDFT resonator is located exactly (with no finite-precision numerical error) at $z = 1$ on the unit circle.

Unfortunately, if multiple DFT frequency bins are to be computed, which, for example, is the case for applying the time window in the frequency domain [1], one

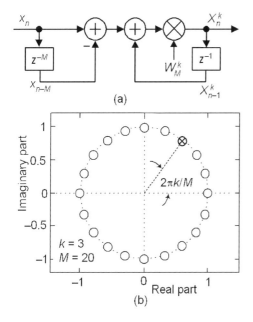

Figure 21–2   Traditional SDFT: (a) structure; (b) z-plane pole/zero locations for $k = 3$ and $M = 20$.

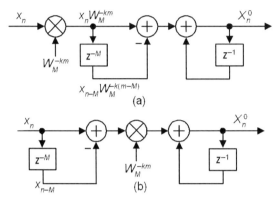

Figure 21–3   Modulated SDFT structures: (a) mSDFT in (21–4); (b) mSDFT in (21–5).

length-M delay buffer is needed in Figure 21–3(a) for each frequency bin. Note, however, due to the periodicity of W_M^{-km}, the difference equation (21–4) may be rewritten in the desired form of:

$$X_n^0 = X_{n-1}^0 + W_M^{-km}(-x_{n-M} + x_n). \qquad (21–5)$$

The structure of our mSDFT in (21–5) is depicted in Figure 21–3 (b). In the case of computing multiple DFT bins using (21–5), memory is saved because only one delay buffer is required.

21.3 MODULATING SEQUENCE

The phase of the modulating sequence is changing with index m. For m we have W_M^{-km}, but for the next sample $m + 1$ we have:

$$W_M^{-k(m+1)} = W_M^{-km}W_M^{-k}. \qquad (21–6)$$

Thus the phase of the modulating sequence equals 0 for $m = 0$, and increases by the W_M^{-k} in each iteration. From the definition in (21–1) it is seen that for $n = M$ the computations are started from $q = 1$, thus the phase of the analyzed signal refers to the time instant $q = 1$. On the other hand, the phase of the modulating sequence for $q = 1$ equals W_M^{-k} and not 0; thus we have a W_M^{-k} phase difference between modulating sequence and analyzed signal. Considering the above we have the following relation between the desired X_n^k, DFT of x_n, and the computed X_n^0 (21–4):

$$X_n^k = W_M^{k(m+1)}X_n^0. \qquad (21–7)$$

The twiddle factor in (21–7) corrects the phase of X_n^0. It is seen from (21–7) that the modulus of X_n^k and X_n^0 is always the same, and the phase is the same when $W_M^{km} = 1$.

The drawback of the structure from Figure 21–3(b), as compared with Figure 21–2(a), is that the modulating sequence is time varying (it depends on the index

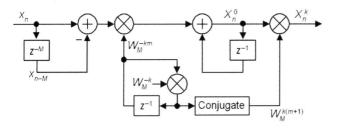

Figure 21–4   mSDFT structure with recursive computation of the modulating sequence in (21–8) and phase correction in (21–7).

m). That is, new sine and cosine factors must be computed, or stored in memory, for each new input x_n sample. To avoid that computational problem, in the following we derive a simple form of the complex oscillator that will replace the modulating sequence in Figure 21–3(b). By shifting time indices in (21–6) we obtain the recurrence for modulating sequence.

$$W_M^{-km} = W_M^{-k(m-1)}W_M^{-k}, \quad m = 0, 1, \ldots, M-1. \tag{21–8}$$

Note, that W_M^{-km} always has an integer number of periods in M samples as m cycles through the range $0 \le m \le M - 1$. Thus W_M^{-km} is automatically restarted every M samples with the value $W_M^{-km} = 1$ when $m = 0$. This periodic restarting of W_M^{-km} prevents our computations from accumulated errors. The mSDFT with recursive computation of the modulating sequence in (21–8) is depicted in Figure 21–4.

The phase of the frequency bin X_n^0 computed by mSDFT is related to the phase of the desired DFT bin X_n^k by (21–7). Figure 21–4 depicts mSDFT with the phase correction. The output of the mSDFT in Figure 21–4 is the same as the output of the SDFT in Figure 21–2(a).

It is worth noting that if our spectrum analysis application requires only DFT magnitude results, then the final complex multiplication in Figure 21–4 is not needed because $|X_n^k| = |X_n^0|$.

21.4 OTHER STABLE SDFT ALGORITHMS

In this section, for the purpose of comparison with the mSDFT, we describe two other stable SDFT algorithms.

The SDFT recurrence (21–2) would be stable if we had used rW_M^k, with $0 \ll r < 1$, as the multiplier instead of W_M^k as was done in [2]. Let us define following DFT:

$$\hat{X}_n^k = \text{DFT}\{x_{q+m}r^{M-m}\} = \sum_{m=0}^{M-1} x_{q+m}r^{M-m}W_M^{-km} \tag{21–9}$$

where $q = n - M + 1$, $0 \leq k \leq M - 1$, and $0 << r < 1$. A hat symbol in (21–9), and further equations, denotes an approximation of the DFT as defined by (21–1). From (21–9) we derive, similar to (21–2), the recurrence:

$$\hat{X}_n^k = rW_M^k(\hat{X}_{n-1}^k - x_{n-M}r^M + x_n).$$

(21–10)

From (21–9) it is seen that we compute the DFT of the signal multiplied by the sequence r^{M-m} and not the signal itself thus we trade of accuracy for stability. We will refer to (21–10) as the rSDFT algorithm.

Another stable SDFT was proposed in [3]. In the following derivation of that stable SDFT we assume, without the lost of generality, $M = 4$. Let us define following DFT:

$$
\begin{aligned}
\hat{X}_{n+0}^k &= x_{q+0}W_4^{-k0} + x_{q+1}W_4^{-k1} + x_{q+2}W_4^{-k2} + x_{q+3}W_4^{-k3} \\
\hat{X}_{n+1}^k &= x_{q+1}rW_4^{-k0} + x_{q+2}W_4^{-k1} + x_{q+3}rW_4^{-k2} + x_{q+4}W_4^{-k3} \\
\hat{X}_{n+2}^k &= x_{q+2}rW_4^{-k0} + x_{q+3}rW_4^{-k1} + x_{q+4}W_4^{-k2} + x_{q+5}W_4^{-k3} \\
\hat{X}_{n+3}^k &= x_{q+3}rW_4^{-k0} + x_{q+4}W_4^{-k1} + x_{q+5}W_4^{-k2} + x_{q+6}W_4^{-k3}
\end{aligned}
$$

(21–11)

In (21–11) there is no r in the upper equation; in the second equation we have r for $m = 0, 1, 2$; we have r for $m = 0, 1$ in the third equation; and one r for $m = 0$ in the last equation. The process of putting the r variable in the DFT definition repeats circularly. The definition in (21–11) may be computed recursively based on two difference equations [3]. From the first two lines of (21–11) and from the second and the third line of (21–11) we have a recurrence:

$$
\hat{X}_n^k = \begin{cases} rW_M^k(\hat{X}_{n-1}^k - x_{n-M}) + W_M^k x_n, & \text{if } n \text{ modulo } M = 0 \\ W_M^k(\hat{X}_{n-1}^k - x_{n-M}r + x_n), & \text{otherwise} \end{cases}
$$

(21–12)

We will refer to (21–12) as the Douglas and Soh (D&S) algorithm. It is seen from (21–10) and (21–12) that for $r = 1$ both algorithms become the SDFT in (21–2). In the next section we illustrate, by the example, the advantages of using the mSDFT rather than the rSDFT (21–10) or D&S (21–12) algorithms.

21.5 NUMERICAL SIMULATIONS

Figure 21–5 depicts the error of recursive DFT computations defined as:

$$\varepsilon_n = \left| X_{n\,DFT}^k - X_{n\,SDFT}^k \right|,$$

(21–13)

where $X_{n\,DFT}^k$ denotes the standard DFT bin result computed from the batch of data of length M, and $X_{n\,SDFT}^k$ stands for recursive computations of the mSDFT, rSDFT, and D&S algorithms. The test signal was zero-mean Gaussian noise with a standard deviation equal to 1, the length of the DFT was $M = 20$, and the $k = 3$ bin was computed. The simulation was carried out in a 64-bit double-precision MATLAB™

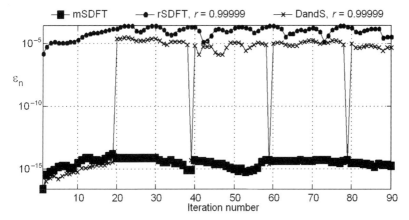

Figure 21–5   Error of single-bin DFT recursive computations for $M = 20$ and $k = 3$.

environment. We see from Figure 21–5 that, in the given example, the mSDFT is approximately 10 orders of magnitude more accurate than the rSDFT and D&S algorithms. The accuracy improvement depends on the value of r, for example, it is approximately 7 orders of magnitude for $r = 0.999999999$ and 15 orders for $r = 0.9$. The accuracy of the mSDFT, as presented in Figure 21–5, may be degraded by the limited precision of the twiddle factor representation in Figure 21–4. But for equal precision twiddle factors we observe better accuracy for the mSDFT than for the other two stable algorithms.

Finally, we make some remarks about preferable arithmetic formats. As an example consider $M = 4$, from (21–5) we have $X_5^0 = [W_4^{-k}x_1 + W_4^{-2k}x_2 + W_4^{-3k}x_3 + x_4 + W_4^{-k}x_5] - W_4^{-k}x_1$, where the values in brackets were previously summed. The mSDFT will run correctly for an unlimited time, without being reset, only if $X_5^0 = W_4^{-2k}x_2 + W_4^{-3k}x_3 + x_4 + W_4^{-k}x_5$ (and similarly for the next iterations) and this may be obtained only in integer arithmetic. Thus fixed point arithmetic is the best choice for implementing mSDFT.

In the case of floating point arithmetic we typically have $X_5^0 = W_4^{-2k}x_2 + W_4^{-3k}x_3 + x_4 + W_4^{-k}x_5 \pm err$, where err is a small roundoff error that may add or subtract. Let us assume that err is on the level of 10^{-10} and the worst case scenario that err only adds up, then it will take 10^{10} iterations to build up the overall error to a value of 1. For example, at a sampling frequency of 44 kHz it will take at least 63 hours of continual operation for the error to reach one. Thus, in practice, implementation in floating point arithmetic may be sufficiently accurate for a sufficiently long time, although it may not run correctly for an unlimited time.

21.6   COMPUTATIONAL WORKLOAD

Table 21–1 gives a computational workload comparison of the various sliding DFT algorithms, with the assumption that the input signal is real and one complex

Table 21–1   Computation Workload per Output Sample

Method:	Real multiplies:	Real adds:
SDFT (21–2) [Fig. 21–2a]	4	4
rSDFT algorithm, (21–10)	5	4
D & S algorithm, (21–12)	4, if n modulo $M = 0$, 5 otherwise	5, if n modulo $M = 0$, 4 otherwise
mSDFT to compute X_n^0 [Fig. 21–4]	6	5
mSDFT to compute X_n^k [Fig. 21–4]	10	7

multiplication requires four real multiplications and two additions, although it is also possible to compute one complex multiplication via three real multiplications and five additions [5].

21.7   CONCLUSIONS

This chapter presented a novel method of computing the SDFT that we call the modulated SDFT (mSDFT). The accumulated errors and potential instabilities inherent in traditional SDFT algorithms are drastically reduced in the mSDFT. We removed the twiddle factor from the feedback in a traditional SDFT resonator and thus the finite precision of its representation is no longer a problem.

We compared other algorithms for ensuring stability of the SDFT described in [1], [2], [3] to the stable mSDFT. Our comparison showed those algorithms trade accuracy for stability, whereas the mSDFT is always stable and more accurate than other stable algorithms in the same computing environment. As such, the mSDFT is an attractive option for recursive computation of DFT bins and is preferable to other SDFT algorithms.

A MATLAB™ program that implements our mSDFT in Figure 21–4, the rSDFT in (21–10), and the D&S in (21–12) algorithms is made available at http://booksupport.wiley.com.

21.8   REFERENCES

[1] E. JACOBSEN and R. LYONS, "The Sliding DFT," *IEEE Signal Processing Mag.*, vol. 20, no. 2, March 2003, pp. 74–80.

[2] E. JACOBSEN and R. LYONS, "An Update to the Sliding DFT," *IEEE Signal Processing Mag.*, vol. 21, no. 1, January 2004, pp. 110–111.

[3] S. DOUGLAS and J. SOH, "A Numerically Stable Sliding-Window Estimator and Its Application to Adaptive Filters," in *Proc. 31st Annual Asilomar Conf. on Signals, Systems, and Computers, Pacific Grove, CA*, vol. 1, November 1997, pp. 111–115.

[4] A. OPPENHEIM, R. SCHAFER, and J.R. BUCK, *Discrete-Time Signal Processing*, 2nd ed. Prentice-Hall, 1999, p. 576.

[5] R. LYONS, *Understanding Digital Signal Processing*, 3rd ed. Prentice-Hall, 2010, p. 686.

Chapter 22

Reducing FFT Scalloping Loss Errors without Multiplication

Richard Lyons
Besser Associates

This chapter discusses the estimation of time-domain sinewave peak amplitudes based on the fast Fourier transform (FFT) data. Such an operation sounds simple, but the scalloping loss characteristic of FFTs complicates the procedure. Here we present novel multiplier-free methods to accurately estimate sinewave amplitudes, based on FFT data, that greatly reduce scalloping loss problems.

22.1 FFT SCALLOPING LOSS REVISITED

There are many applications that require the estimation of a time-domain sinewave's peak amplitude based on FFT data. Such applications include oscillator and analog-to-digital converter performance measurements, as well as standard total harmonic distortion (THD) testing. However, the scalloping loss inherent in FFTs creates an uncertainty in such time-domain peak amplitude estimations. This section provides a brief review of FFT scalloping loss.

As most of you know, if we perform an N-point FFT on N real-valued time-domain samples of a discrete sinewave, whose frequency is an integer multiple of f_s/N (f_s is the sample rate in Hz), the peak magnitude of the sinewave's positive-frequency spectral component will be

$$M = \frac{A \cdot N}{2} \qquad (22\text{--}1)$$

where A is the peak amplitude of the time-domain sinewave. That phrase "whose frequency is an integer multiple of f_s/N" means that the sinewave's frequency is located exactly at one of the FFT's bin centers.

Streamlining Digital Signal Processing: A Tricks of the Trade Guidebook, Second Edition. Edited by Richard G. Lyons.
© 2012 the Institute of Electrical and Electronics Engineers. Published 2012 by John Wiley & Sons, Inc.

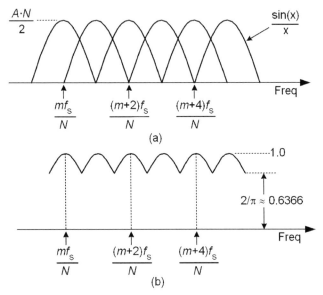

Figure 22–1   FFT frequency magnitude responses: (a) individual FFT bins; (b) overall FFT response.

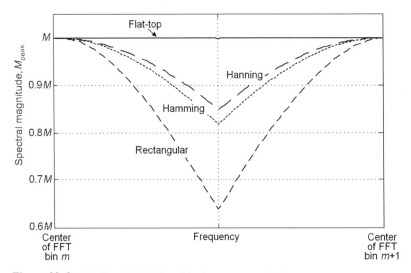

Figure 22–2   Windowed-FFT, bin-to-bin, frequency magnitude responses.

Now, if an FFT's input sinewave's frequency is between two FFT bin centers (equal to a non-integer multiple of f_s/N) the FFT magnitude of that spectral component will be less that the value of M in (22–1). Figure 22–1 illustrates this behavior. Figure 22–1(a) shows the frequency responses of individual FFT bins where, for simplicity, we show only the mainlobes (no sidelobes) of the FFT bins' responses.

What this means is that if we were to apply a sinewave to an FFT, and scan the frequency of that sinewave over multiple bins, the magnitude of the FFT's largest normalized magnitude sample value will follow the curve in Figure 22–1(b). That curve describes what is called the *scalloping loss* of an FFT [1].

(As an aside, the word *scallop* is not related to my favorite shellfish. As it turns out, some window drapery, and table cloths, do not have linear borders. Rather they have a series of circular segments, or loops, of fabric defining their decorative borders. Those loops of fabric are called scallops.)

What Figure 22–1(b) tells us is that if we examine the N-point FFT magnitude sample of an arbitrary-frequency, peak amplitude $= A$ sinewave, that spectral component's measured peak magnitude M_{peak} can be anywhere in the range of:

$$\frac{0.637 \cdot A \cdot N}{2} \le M_{peak} \le \frac{A \cdot N}{2} \qquad (22–2)$$

depending on the frequency of that sinewave. This is shown as the *rectangular window* curve in Figure 22–2, where the maximum scalloping error occurs at a frequency midpoint between two FFT bins. The variable M in the vertical axis label of Figure 22–2 is the M from (22–1). So if we want to determine a sinewave's time-domain peak amplitude A, by measuring its maximum FFT spectral peak magnitude M_{peak}, our estimated value of A, from (22–1), using

$$A = \frac{2M_{peak}}{N} \qquad (22–3)$$

can have an error as great as 36.3%. In many spectrum analysis applications such a large potential error, equivalent to 3.9 dB, is unacceptable. As shown by the magnitude-normalized curves in Figure 22–2, Hanning and Hamming windowing of the FFT input data reduce the unpleasant frequency-dependent fluctuations in a measured spectral M_{peak} value, but not nearly enough to satisfy many applications.

One solution to this frequency-dependent, FFT-based, measured amplitude uncertainty is to multiply the original N time-domain samples by an N-sample *flat-top* window function and then perform a new FFT on the windowed data. Flat-top window functions are designed to overcome the scallop loss inherent in rectangular-windowed FFTs. While such a flat-top-windowed FFT technique will work, there are more computationally efficient methods to solve our signal peak amplitude estimation uncertainty problem.

22.2 FREQUENCY-DOMAIN CONVOLUTION

Because multiplication in the time domain is equivalent to convolution in the frequency domain, we can convert rectangular-windowed (no windowing) FFT samples to windowed-FFT samples by way of convolution. For example, consider an N-point $w(n)$ window sequence whose time-domain samples are generated using:

$$w(n) = \sum_{k=0}^{K-1} (-1)^k h_k \cos(2\pi kn/N) \tag{22-4}$$

where the $w(n)$ sequence's generating polynomial has an integer K number of h_k coefficients.

Many window functions, including Hanning, Hamming, Blackman, and flat-top, are generated using (22–4). One popular flat-top window sequence, generated using (22–4), is MATLAB's *flattopwin(N)* routine where the h_k polynomial coefficients are [2]

$$h_0 = 0.2156, h_1 = 0.4160, h_2 = 0.2781, h_3 = 0.0836, h_4 = 0.0069. \tag{22-5}$$

(Very similar flat-top window generating coefficients are recommended in reference [3].) Thus, in implementing frequency-domain convolution, to compute a single flat-top windowed $X_{ft}(m)$ spectral sample from rectangular-windowed $X(m)$ spectral samples, we would compute

$$X_{ft}(m) = \frac{h_4}{2} X(m-4) - \frac{h_3}{2} X(m-3) + \frac{h_2}{2} X(m-2) - \frac{h_1}{2} X(m-1)$$
$$+ h_0 X(m) - \frac{h_1}{2} X(m+1) + \frac{h_2}{2} X(m+2) - \frac{h_3}{2} X(m+3) + \frac{h_4}{2} X(m+4) \tag{22-6}$$

where $X(m)$ is the rectangular-windowed FFT sample having the largest magnitude, and m is the FFT's frequency-domain sample index.

If we apply (22–6) to rectangular-windowed $X(m)$ FFT samples, and compute the flat-top windowed maximum FFT spectral peak magnitude $M_{peak} = |X_{ft}(m)|$, the estimated value of A from (22–3) will have an error of no more than 0.0166 dB. Such a small error is represented by the very flat, nearly ideal, solid curve labeled as flat-top in Figure 22–2.

That appealing flat-top curve in Figure 22–2 is the good news associated with the frequency-domain flat-top window convolution in (22–6). The bad news is that each computation of an $X_{ft}(m)$ sample requires, assuming we combine terms having identical coefficients, 18 real multiplies and 16 real additions. In what follows, we show how to drastically reduce the computational workload in computing an $X_{ft}(m)$ sample.

22.3   IMPROVED CONVOLUTION COEFFICIENTS

Reference [4], which discusses many different sets of window-generating polynomial coefficients, presents the following useful set of flat-top window coefficients

$$h_0 = 0.26526, h_1 = 0.5, h_2 = 0.23474 \tag{22-7}$$

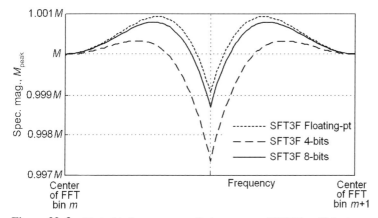

Figure 22–3   Bin-to-bin frequency magnitude response of SFT3F coefficients.

collectively called the *SFT3F coefficients*. Thus to obtain a single flat-top windowed $X_{ft}(m)$ spectral sample from rectangular windowed $X(m)$ samples, based on the SFT3F coefficients in (22–7), we compute

$$X_{ft}(m) = \frac{h_2}{2}X(m-2) - \frac{h_1}{2}X(m-1) + h_0 X(m) - \frac{h_1}{2}X(m+1)$$
$$+ \frac{h_2}{2}X(m+2). \tag{22-8}$$

The frequency-dependent scalloping loss of the floating-point SFT3F coefficients is shown as the black dotted curve in Figure 22–3. The variable M in Figure 22–3 is the M from (22–1). The computation of an $X_{ft}(m)$ sample using (22–8) results in an estimated value of A, from (22–3), having a scalloping error in the range of -0.0082 dB to $+0.0082$ dB. We call the coefficients in (22–7) "improved" because the computation in (22–8) requires only 10 real multiplies and 8 real additions.

Notice that flat-top window coefficients, such as those (22–7), have the interesting characteristic that they have both a scalloping loss and a scalloping gain versus frequency. (Compare the black dotted curve in Figure 22–3 with the lossy Hanning, Hamming, and rectangular curves in Figure 22–2, whose M_{peak} values are always less than M.)

22.4 FURTHER COMPUTATIONAL IMPROVEMENTS

We can take three steps to further reduce the computational workload of computing an $X_{ft}(m)$ sample using (22–8).

First Step

If we divide the coefficients in (22–7) by the first coefficient, h_0, we obtain the new coefficients

$$h_0 = 1.0, \, h_1 = 1.88494, \, h_2 = 0.88494. \qquad (22-9)$$

The coefficients in (22–9) eliminate the amplitude gain loss of the flat-top coefficients in (22–8) without changing their scalloping loss compensation performance. Given the flat-top window generating polynomial coefficients in (22–9), computing an $X_{ft}(m)$ sample proceeds as

$$
\begin{aligned}
X_{ft}(m) = X(m) - \frac{1.88494}{2}[X(m-1) + X(m+1)] \\
+ \frac{0.88494}{2}[X(m-2) + X(m+2)].
\end{aligned}
\qquad (22-10)
$$

The coefficients in the convolution expression in (22–10) are

$$
\begin{aligned}
g_0 &= 1.0, \\
g_1 &= -\frac{1.88494}{2} = -0.94247, \\
g_2 &= \frac{0.88494}{2} = 0.44247.
\end{aligned}
\qquad (22-11)
$$

Second Step

Next, we convert the coefficients in (22–11) to binary representation to simplify our processing by replacing the multiplications in (22–10) by arithmetic right-shifts. Doing so, the nonunity coefficients in (22–11) become

$$
\begin{aligned}
g_1 &= -0.94247 = -0.1111\,0001\,0100\,0101\ldots \\
g_2 &= 0.44247 = 0.0111\,0001\,0100\,0101\ldots
\end{aligned}
\qquad (22-12)
$$

The leftmost sequence of three consecutive zeros in coefficients g_1 and g_2 suggest that we can represent those coefficients with four fractional bits without inducing too much truncation error.

To simplify our equations, let's represent our five unwindowed frequency-domain samples in (22–10) with

$$
\begin{aligned}
c &= X(m) \\
p &= X(m-1) + X(m+1) \\
q &= X(m-2) + X(m+2) \\
r &= q - p.
\end{aligned}
$$

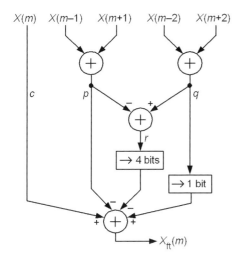

Figure 22–4   Multiplier-free scalloping loss compensation using 4-bit coefficients in (22–14).

Those assignments convert (22–10), using the first four fractional bits for g_1 and g_2 in (22–12), to

$$X_{\text{ft}}(m) = c - \frac{1}{2}p + \frac{1}{4}r + \frac{1}{8}r + \frac{1}{16}r \qquad (22\text{–}13)$$

allowing us to replace the multiplications in (10) with binary right-shifts. However, rather than implement the four separate binary right-shifts in (13), we can use canonical signed digit (CSD) notation to further streamline our computations. Using CSD, (22–13) becomes

$$X_{\text{ft}}(m) = c - p + \frac{1}{2}q - \frac{1}{16}r \qquad (22\text{–}14)$$

which is equivalent to, but simpler to compute than, (22–14). The signal flow implementation of (22–14) is given in Figure 22–4 and its performance is shown as the dashed "SFT3F 4-bits" curve in Figure 22–3.

Finally, we compute M_{peak}, using (22–14), as

$$M_{\text{peak}} = |X_{\text{ft}}(m)| \qquad (22\text{–}15)$$

and use that M_{peak} value in (22–3) to compute our desired A, the peak amplitude of the FFT's time-domain sinewave input. The computation of an M_{peak} value using (22–14) and (22–15) results in an estimated value of A, from (22–3), having a scalloping error in the range of −0.0229 dB to +0.003 dB.

We can achieve 8-bit accuracy of our binary coefficients in (22–12) by adding one more term to the approximation in (22–14) as

$$X_{\text{ft}}(m) = c - p + \frac{1}{2}q - \frac{1}{16}r + \frac{1}{256}r. \qquad (22\text{–}16)$$

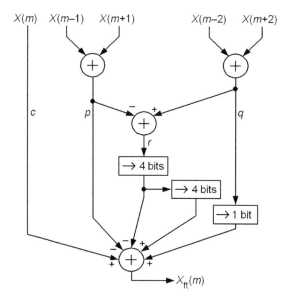

Figure 22–5 Scalloping loss compensation using 8-bit coefficients: initial implementation.

The signal flow implementation of (22–16) is given in Figure 22–5 and its performance is shown as the solid "SFT3F 8-bits" curve in Figure 22–3. The computation of an $X_{ft}(m)$ sample using (22–16) results in an estimated value of A, from (22–15) and (22–3), having a scalloping error in the range of −0.0113 dB to +0.0069 dB.

That's almost worth writing home about because the performance of the multiplier-free (22–16) is superior to the multiply-intensive computation in (22–6).

Third Step

In our relentless pursuit of accuracy, we employ one last binary arithmetic trick to reduce right-shift truncation error. Notice in Figure 22–5 that one of our complex data samples experiences a right-shift by eight bits. To reduce the truncation error of an 8-bit right shift, we use Horner's rule to convert (22–16) to

$$X_{ft}(m) = c - p + \frac{1}{2}\left(q - \frac{1}{8}\left(r - \frac{1}{16}r \right) \right). \tag{22–17}$$

This way, no data sample experiences a truncation error greater than a 4-bit right-shift. The signal flow implementation of (22–17) is given in Figure 22–6 and its performance is equal to that of (22–16).

To consolidate what we've covered so far, Table 22–1 shows the computational workload, and error performance in estimating a sinewave amplitude A, of the various scalloping loss compensation methods.

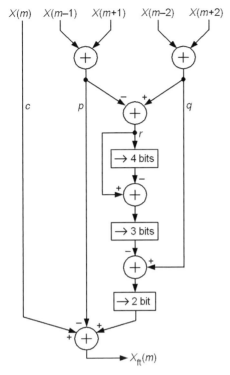

Figure 22–6   Scalloping loss compensation using 8-bit coefficients: reduced truncation error implementation.

Table 22–1   Computational Workload per $X_{ft}(m)$ Sample, and Performance

Single Computation	Real mults	Real adds	Binary right-shifts	Max. scalloping error (dB)
Eq. (22–6)	18	16	–	0.0166
Eq. (22–8)	10	8	–	0.0082
Eq. (22–10)	4	6	–	0.0082
Eq. (22–14)	–	12	4	0.0228
Eqs. (22–16) & (22–17)	–	14	6	0.0113

22.5 IMPLEMENTATION CONSIDERATIONS

There are two issues to keep in mind when using the above scalloping loss compensation methods.

- The flat-top window frequency-domain convolutions are most useful in accurately measuring the time-domain amplitude of a sinusoidal signal when that signal's spectral component is not contaminated by sidelobe leakage from a nearby spectral component. For example, if a positive-frequency spectral component is low in frequency, i.e., located in the first few FFT bins, leakage from the spectral component's corresponding negative-frequency spectral component will contaminate that positive-frequency spectral component. As such, empirical testing indicates that the convolutions in Figures 22–4, 22–5, and 22–6 should not be used for frequencies below the sixth FFT bin or above the ($N/2$–5)th FFT bin.

- The flat-top window frequency-domain convolutions discussed above are most useful when the FFT spectral component being measured is well above the background spectral noise floor.

22.6 CONCLUSION

We discussed the inherent scalloping loss uncertainty (potential error) of estimating sinewave peak amplitudes based on FFT spectral data. Then we briefly discussed the performance, and computational workload, of frequency-domain convolution using traditional five-term flat-top window coefficients to drastically reduce sinusoidal peak amplitude measurement uncertainty. Next we demonstrated a little-known three-term flat-top window polynomial that has very good scalloping loss compensation and a reduced computational workload. Finally, we presented a series of binary arithmetic tricks yielding a high-performance, efficient, multiplier-free implementation of scalloping loss compensation. MATLAB and C-code implementations of this material are available at http://booksupport.wiley.com.

22.7 REFERENCES

[1] F. HARRIS, "On the Use of Windows for Harmonic Analysis with the Discrete Fourier Transform," *Proceedings of the IEEE*, vol. 66, no. 1, January 1978, pp. 51–83.
[2] THE MATHWORKS INC., "Signal Processing Toolbox, Flat Top Weighted Window." [Online: http://www.mathworks.com/access/helpdesk/help/toolbox/signal/flattopwin.html.]
[3] S. GADE and H. HERLUFSEN, "Use of Weighting Functions in DFT/FFT Analysis (Part I)," *Brüel & Kjær Technical Review*, no. 3, 1987, pp. 19–21. [Online: http://www.bksv.com/doc/bv0031.pdf.]
[4] G. HEINZEL, A. RÜDIGER, and R. SCHILLING, "Spectrum and Spectral Density Estimation by the Discrete Fourier Transform (DFT), Including a Comprehensive List of Window Functions and Some New Flat-Top Windows," Max-Planck-Institut für Gravitationphysik Report, February 15, 2002. [Online: http://www.rssd.esa.int/SP/LISAPATHFINDER/docs/Data_Analysis/GH_FFT.pdf.]

EDITOR COMMENTS

That interesting constant of $2/\pi$ in Figure 22–1(b) is found as follows:

The frequency magnitude response of the $m = 3$ and $m = 4$ bins of an N-point discrete Fourier transform (DFT), $|X(m)|$, are shown in Figure 22–7. The solid curve

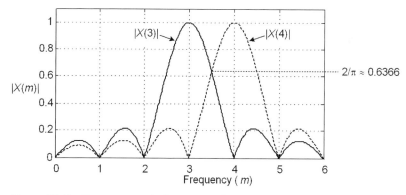

Figure 22–7   Frequency magnitude response of an N-point DFT.

is the $m = 3$ bin's reponse while the dotted curve is the $m = 4$ bin's magnitude reponse.

The solid $|X(3)|$ curve response is represented algebraically by

$$|X(3)| = \left| \frac{\sin[\pi(3-m)]}{\pi(3-m)} \right|. \tag{22–18}$$

The $|X(3)|$ curve interects the $|X(4)|$ curve at $m = 3.5$, so substituting 3.5 for m in (22–19) yields

$$|X(3)|_{=|X(4)|} = \left| \frac{\sin[\pi(3-3.5)]}{\pi(3-3.5)} \right| = \frac{\sin(\pi/2)}{\pi/2} = \frac{2}{\pi}. \tag{22–19}$$

Chapter 23

Slope Filtering: An FIR Approach to Linear Regression

Clay S. Turner
Pace-O-Matic, Inc.

While developing a communications receiver the author encountered a problem that was efficiently solved using a method called *slope filtering*, which is a novel application of the standard equations of linear regression. This chapter introduces the slope filtering process and its effective use in applications such as communications receiver carrier recovery, signal rate of change estimation, and signal transition and transition-polarity detection.

23.1 MOTIVATION FOR SLOPE FILTERING

The slope filtering technique was developed while attempting to design a carrier recovery process for a communications system that used quadrature (I/Q) modulation. The communications receiver had the typical problem that its local oscillator was not frequency locked to that of the received signal's original transmitter oscillator. Since both oscillators (transmitter and receiver) have nearly constant frequencies over short time intervals, when the instantaneous phase of the received signal was compared with the phase of its corresponding ideal equivalent (known from the protocol) the resulting phase offset error function was approximately linear with respect to time. To properly demodulate the received signal the receiver must estimate, and then eliminate, this phase error between the ideal and received signals.

Streamlining Digital Signal Processing: A Tricks of the Trade Guidebook, Second Edition. Edited by Richard G. Lyons.
© 2012 the Institute of Electrical and Electronics Engineers. Published 2012 by John Wiley & Sons, Inc.

Because the receiver must operate with low signal-to-noise ratio (SNR) input signals, the linear phase error function's excessive noise suggested that a statistical method was needed for the receiver to estimate the phase offset error. It was in solving this problem that the author developed the slope filtering scheme to manipulate classic linear regression equations into a form designed for efficient computation. We now look at how this manipulation is carried out, and then return to the communications receiver example.

23.2 LINEAR REGRESSION

The time-domain slope filtering process was developed by first borrowing the procedure of linear regression from the field of statistics [1], [2]. If we have a set of N ordered pairs $\{(x_0,y_0), (x_1,y_1), \ldots, (x_{N-1},y_{N-1})\}$ and we fit, via a least squares method, a straight line through the data then the regression line is $\hat{y} = \alpha + \beta\hat{x}$, where

$$\alpha = \frac{\left(\sum_k y_k\right)\left(\sum_k x_k^2\right) - \left(\sum_k x_k\right)\left(\sum_k x_k y_k\right)}{N\left(\sum_k x_k^2\right) - \left(\sum_k x_k\right)^2} \tag{23-1}$$

and

$$\beta = \frac{N\left(\sum_k x_k y_k\right) - \left(\sum_k x_k\right)\left(\sum_k y_k\right)}{N\left(\sum_k x_k^2\right) - \left(\sum_k x_k\right)^2} \tag{23-2}$$

where $k = 0, 1, \ldots, N - 1$. The slope of the best-fit line to an N-length segment of data is given by β. It is important to recall that (23–1) and (23–2) are valid for arbitrary sets of x_k values and there is no assumption of equal spacing for the x_k values.

As it is written, (23–2) doesn't appear amenable for efficient implementation in a signal processing application. But if we look at it from a DSP viewpoint, and apply some adroit logic and manipulations, we can make (23–2) acceptable for such use. Since programmable DSP chips themselves are designed to efficiently calculate vector dot products, we seek to manipulate (23–2) into such a form. This means we need to somehow decouple the y_k samples from the x_k samples, except for a final dot product type of formulation.

So starting with (23–2), we first change the index counter for y_k from k to i and then change the product of sums in (23–2) to a double summation. These changes result in

$$\beta = \frac{N\left(\sum_i x_i y_i\right) - \sum_i \left(\sum_k x_k\right) y_i}{N\left(\sum_k x_k^2\right) - \left(\sum_k x_k\right)^2} \qquad (23\text{--}3)$$

where index i, just like k, counts from 0 to $N - 1$. Next we factor out the y_k samples giving us

$$\beta = \sum_i \left(\frac{N \cdot x_i - \sum_k x_k}{N\left(\sum_k x_k^2\right) - \left(\sum_k x_k\right)^2}\right) y_i. \qquad (23\text{--}4)$$

Based on (23–4), we have our desired expression

$$\beta = \sum_i \beta_i y_i \qquad (23\text{--}5)$$

where β_i is defined as

$$\beta_i = \frac{N \cdot x_i - \sum_k x_k}{N\left(\sum_k x_k^2\right) - \left(\sum_k x_k\right)^2}. \qquad (23\text{--}6)$$

We have converted (23–2) into the vector dot product in (23–5), where the x_k and y_k samples are decoupled from each other. Fortunately, if all of the x_i in (23–6) are known a priori, then the β_i coefficients may be precomputed and stored in memory. So, if we have a block of data (N samples), we just use our pre-stored length-N β_i coefficients in a tapped-delay line filter structure implementing (23–5) to find the slope (rate of change) of the data. This is the process we call *slope filtering*.

23.3 APPLICATION: RECEIVER CARRIER RECOVERY

Now that we see how to transform the regression formula into a useful (for signal processing tasks) form, we return to its application to the communications receiver problem. Because the receiver uses an independent frequency reference for heterodyning, the baseband signal will have a residual frequency and phase offset when compared with the ideal. We need to estimate these errors to be able to correct them.

To find the error function both the received and ideal signals are represented in complex polar form and then the differences in the arguments of their corresponding samples become the samples of the error function. For relatively short time duration signals, the error function is well approximated as a linear function where its slope represents the frequency offset and the intercept represents the phase offset between the transmitter's and receiver's oscillators.

By applying (23–5) to the error function, the approximate rate of change of phase is found. And of course the rate of phase change is equal to the frequency, thus we can estimate the frequency offset of the receiver. The regression calculation may be used to find the frequency offset by applying it to the phase error function. Unwrapping the error function's phase may be needed to preserve the linearity of the error function but, since its underlying structure is linear, unwrapping is quite easy. If (23–1) is also manipulated into a dot product formulation, and it is applied to the error function, the initial (at the start of the data segment under analysis) phase error is found. A digital quadrature oscillator [3], initialized with the negatives of the frequency and phase offsets, can be used to correct the received signal's frequency and phase errors by simple modulation.

23.4 APPLICATION: SIGNAL RATE OF CHANGE ESTIMATION

Real-Time Rate of Change Estimation

For applications where continual signal rate of change estimations are needed, as opposed to the single dot product calculation of (23–5), when the x_i samples are equally spaced and ordered from low to high we can write $x_k = x_0 + k$ where $k = 0$, $1, 2, \ldots, N - 1$. Hence we are looking at a set of N consecutive equi-spaced x_k samples. Substituting our expression for x_k into (23–6) results in

$$\beta_i = \frac{N(x_0 + i) - \sum_k (x_0 + k)}{N\left(\sum_k (x_0 + k)^2\right) - \left(\sum_k (x_0 + k)\right)^2} \tag{23-7}$$

which after significant algebraic simplification becomes

$$\beta_i = \frac{12 \cdot i - 6(N - 1)}{N(N^2 - 1)} \tag{23-8}$$

giving us the β_i coefficients for use in an N-stage tapped-delay line FIR structure to compute our desired rate of change β in (23–5).

If the delay line structure used to implement (23–5) is designed for filter convolutions instead of correlations, then the β_i coefficients need to be reversed in order.

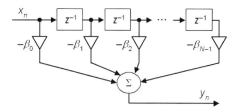

Figure 23–1   Real-time slope filtering structure.

Since the β_i coefficients have negative symmetry, a simple negation will be sufficient as shown in Figure 23–1.

The units for β in (23–5) are the units of y per sample interval. So, for example, if y is in volts and we are sampling at a 1 kHz rate, then the units of β are volts per msec.

Properties of Rate Change of Change Estimation

Equation (23–8) is remarkable for several reasons:

- It represents the coefficients of an odd symmetric linear-phase FIR filter, which means the delay is a constant of $(N-1)/2$ samples. In addition, half of the multiplications in (23–5) may be eliminated by using a "folded FIR" filter structure.

- Its result is origin independent! The β_i coefficients don't change as we traverse the data as evidenced by the fact the starting value x_0 is not a part of (23–8). So in terms of implementing (23–5) via a filter, this means the coefficients are time-invariant.

- The β_i coefficients define a linear ramp. This means we are in effect correlating our signal with a ramp to find the slope. This may be viewed as optimal in the sense of matched filtering allowing the user to easily pick an optimal N. That is, if our transitions are expected to span M samples, then we make the β_i coefficients M samples in length. And this brings us to the important observation that transitions usually don't exist in isolation, they are preceded and followed by generally level signals. This means that the span of the β_i coefficients may be longer than just the transition width. In fact, this *super-spanning* can increase the signal-to-noise ratio of our slope filter's output! Of course there are practical limits to the amount of super-spanning, but having a span 20% or 50% greater than the transition width is certainly acceptable. In the case of pulse detection with matched filtering in mind, the number of β_i coefficients may be set to the number of samples comprising the pulse. But if super-spanning is appropriate, a greater number of coefficients may yield the highest SNR.

23.5 APPLICATION: SIGNAL TRANSITION DETECTION

Another useful application of time-domain slope filtering is signal transition detection. Figure 23–2 illustrates this idea where a series of noisy input signals with transitions are the thin traces, and the outputs of the slope filter (delay compensated on the graphs) are shown by the bold traces. In that figure, the length of the input signal's transition is 80 samples. Examples of super-spanning are shown in Figures 23–2(b) and (c).

Why does time domain slope filtering work so well when compared with ideal differentiators? By inspecting the coefficients for each of these cases, we find that the regression method places the strongest weights at the ends of the data span and conventional differentiators weigh heaviest near the middle. Thus from a "torque" calculation point of view, conventional differentiators rely strongly on fewer terms (all near the middle) and therefore the result is more susceptible to noise. One can also note that the difference is that in time domain slope filtering a first-order polynomial is fitted to the data, whereas with a traditional differentiator a low-order trigonometric polynomial is fitted to the data.

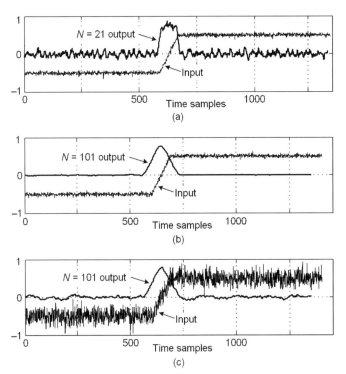

Figure 23–2   Time-domain slope filtering examples: (a) span $N = 21$; (b) $N = 101$; (c) $N = 101$ with a very noisy input signal.

Even if a differentiator is designed with the Parks-McClellan (PM) algorithm to closely match the performance shown in Figure 23–2(a), it will generally not fair well when the noise levels are increased. That is because a trigonometric polynomial fit admits polynomial terms higher than first-order. To see this, recall that the PM algorithm fits a weighted sum of sinusoids to the differentiator's desired frequency response. And, when transformed back to the time domain, its impulse response will contain polynomial terms higher than the first power.

23.6 APPLICATION: SIGNAL TRANSITION-POLARITY DETECTION

Our noise-tolerant time-domain slope filtering scheme also comes in handy in applications where one just wishes to know whether a signal is increasing or decreasing— for example, when detecting the polarity of signal transitions. In this case, we can multiply (23–8) by any positive number, since we only need the sign (positive or negative) of (23–5). For a start, just remove (multiply it out) the denominator of (23–8). This results in

$$\hat{\beta}_i = 12i - 6(N-1) \tag{23–9}$$

and also divide by 12 to yield

$$\tilde{\beta}_i = i - \frac{1}{2}(N-1). \tag{23–10}$$

Now (23–5) becomes (for polarity detection only)

$$\text{Polarity}_{\text{odd } N} = \text{sgn}\left[\sum_{i=0}^{N-1}\left(i - \frac{N-1}{2}\right)\cdot y_i\right]. \tag{23–11}$$

For example, with $N = 7$, we have

$$\text{Polarity}_{N=7} = \text{sgn}[-3y_0 - 2y_1 - y_2 + y_4 + 2y_5 + 3y_6]. \tag{23–12}$$

And for even N, we can rescale (23–10) by doubling all its coefficients so they are integers. For even N, this formulation results in

$$\text{Polarity}_{\text{even } N} = \text{sgn}\left[\sum_{i=0}^{N-1}(2i - (N-1))\cdot y_i\right]. \tag{23–13}$$

For example, when $N = 6$ we have

$$\text{Polarity}_{N=6} = \text{sgn}[-5y_0 - 3y_1 - y_2 + y_3 + 3y_4 + 5y_5]. \tag{23–14}$$

The polarity detection algorithms in (23–11) and (23–13) are expressed in forms amenable to fast calculation.

23.7 CONCLUSIONS

We presented a novel way of manipulating the standard equations of regression analysis into a form that is highly compatible with the methods of digital signal processing and, from them, derived a computationally efficient slope-filtering differentiator, as defined in (23–5) and (23–6), that is superior in its noise tolerance relative to Parks-McClellan designed differentiators. We extended the algorithm with a focus on both transition detection and transition-polarity detection in (23–11) and (23–13).

23.8 REFERENCES

[1] M. SPIEGEL, *Theory and Problems of Statistics*, Shaum's Outline Series, McGraw-Hill, New York, 1961, pp. 220.

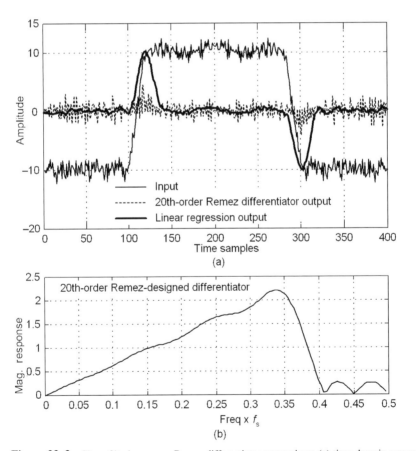

Figure 23–3   Slope filtering versus Remez differentiator comparison: (a) time-domain responses; (b) frequency-domain responses.

[2] P. MEYER, *Introductory Probability and Statistical Applications*, Addison-Wesley, Reading, MA, 1965, pp. 137–139.

[3] C. TURNER, "Recursive Discrete-Time Sinusoidal Oscillators," *IEEE Signal Processing Mag.*, vol. 20, no. 3, May 2003, pp. 103–111.

EDITOR COMMENTS

To emphasize the time-domain differentiating property of this slope filtering process, Figure 23–3(a) compares the time-domain responses of a linear regression process using 21 β_i coefficients (bold curve) and a traditional 21-tap, Remez-designed, finite impulse response (FIR) differentiator (dashed curve), whose frequency magnitude response is shown in Figure 23–3(b). Notice how the linear regression output, given the noisy input signal, in Figure 23–3(a) contains far less noise than the output of the 21-tap FIR differentiator.

Part Three

Fast Function Approximation Algorithms

Chapter 24

Another Contender in the Arctangent Race

Richard Lyons
Besser Associates

This chapter presents a computationally efficient algorithm for approximating the angle of a complex sample value. Estimating the angle θ of an $x = I + jQ$ complex sample has many applications in the field of signal processing. (The angle of x is defined as $\theta = \tan^{-1}[Q/I]$.)

24.1 COMPUTING ARCTANGENTS

Signal processing engineers interested in high-speed (minimum computations) arctangent computations typically use lookup tables where the values I and Q specify an address in read-only memory (ROM) containing an approximation of angle θ. The lookup table method is fast but requires much memory to provide acceptable arctangent accuracy. Those folk interested in high accuracy implement compute-intensive high-order algebraic polynomials to approximate angle θ such as the expression

$$\tan^{-1}(Q/I) \approx \theta' = \frac{(Q/I) + 0.372003(Q/I)^3}{1 + 0.703384(Q/I)^2 + 0.043562(Q/I)^4} \text{ radians} \qquad (24\text{--}1)$$

algorithm, whose maximum error is on the order of 0.003 degrees when $|\theta| \leq \pi/4$ radians [1].

Unfortunately, because it is such a nonlinear function, the arctangent is resistant to accurate small-length polynomial approximations. So we end up choosing the *least undesirable* method for computing arctangents; fast but inaccurate table lookup methods, or accurate but computationally intensive polynomial approximations.

Streamlining Digital Signal Processing: A Tricks of the Trade Guidebook, Second Edition. Edited by Richard G. Lyons.
© 2012 the Institute of Electrical and Electronics Engineers. Published 2012 by John Wiley & Sons, Inc.

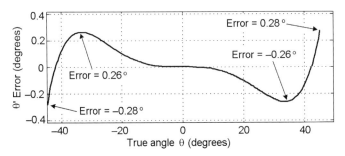

Figure 24–1   Error of estimated angle θ′.

Here we present another contender in the arctangent approximation race that uses neither lookup tables nor high-order polynomials. We can estimate the angle θ, in radians, of $x = I + jQ$ using the following approximation

$$\tan^{-1}(Q/I) \approx \theta' = \frac{Q/I}{1 + 0.28125(Q/I)^2} \text{ radians} \qquad (24\text{–}2)$$

where $-1 \le Q/I \le 1$. That is, θ is in the range $-45°$ to $+45°$ ($-\pi/4 \le \theta \le +\pi/4$ radians). Equation (24–2) has surprisingly good performance, particularly for a 90° ($\pi/2$ radians) angle range. Figure 24–1 shows the maximum error in θ′ is 0.28° using (24–2) when the true angle θ is within the angular range of $-45°$ to $+45°$.

A nice feature of this θ′ computation is it can be written as:

$$\theta' = \frac{IQ}{I^2 + 0.28125Q^2} \text{ radians} \qquad (24\text{–}3)$$

eliminating (24–2)'s Q/I division operation, at the expense of two additional multiplies. Another attribute of (24–3) is a single multiply can be eliminated with binary right shifts. The product $0.28125Q^2$ is equal to $(1/4 + 1/32)Q^2$, so we can implement the product by adding Q^2 shifted right by two bits to Q^2 shifted right by five bits.

We can extend the angle range over which our approximation operates. If we partition a circle into eight 45° octants, with the first octant being 0°-to-45°, we can compute the arctangent of a complex number residing in any octant. We do this by using the rotational symmetry properties of the arctangent:

$$\tan^{-1}(-Q/I) = -\tan^{-1}(Q/I) \qquad (24\text{–}4)$$

$$\tan^{-1}(Q/I) = \pi/2 - \tan^{-1}(I/Q). \qquad (24\text{–}4')$$

Those properties allow us to create Table 24–1, listing the appropriate arctan approximation based on the octant location of complex x. The maximum angle approximation error is 0.28° for all octants.

Table 24–1   Arctangent Expressions versus Octant Location

Octant	Arctan approximation (radians)
1st, or 8th	$\theta' = \dfrac{IQ}{I^2 + 0.28125Q^2}$
2nd, or 3rd	$\theta' = \dfrac{\pi}{2} - \dfrac{IQ}{Q^2 + 0.28125I^2}$
4th, or 5th	$\theta' = \text{sign}(Q)\pi + \dfrac{IQ}{I^2 + 0.28125Q^2}$
6th, or 7th	$\theta' = -\dfrac{\pi}{2} - \dfrac{IQ}{Q^2 + 0.28125I^2}$

Table 24–2   Identification of θ Octant

sign(I)	sign(Q)	$|Q| - |I|$	Octant
+	+	$|Q| - |I| < 0$	1st
+	+	$|Q| - |I| \geq 0$	2nd
−	+	$|Q| - |I| \geq 0$	3rd
−	+	$|Q| - |I| < 0$	4th
−	−	$|Q| - |I| < 0$	5th
−	−	$|Q| - |I| \geq 0$	6th
+	−	$|Q| - |I| \geq 0$	7th
+	−	$|Q - |I| < 0$	8th

So we have to check the signs of Q and I, and compare magnitudes $|Q|$ and $|I|$, to determine the original octant location of θ, and then use the appropriate approximation in Table 24–1.

With that thought in mind, Table 24–2 shows how to identify the original octant of θ based on the values of components I and Q.

24.2   CONCLUSIONS

This arctangent algorithm may be useful in a digital receiver application where I^2 and Q^2 have been previously computed in conjunction with an AM (amplitude modulation) demodulation process or envelope detection associated with automatic gain control (AGC). Although this chapter focused on estimating the angle of an $x = I + jQ$ complex sample value, we remind the reader that (24–2) can be viewed as a generic arctangent approximation represented by

$$\tan^{-1}(x) \approx \frac{x}{1+0.28125x^2} \text{ radians} \tag{24-5}$$

where $-1 \leq x \leq 1$. Such a generic viewpoint is discussed in Chapter 27.

24.3 REFERENCE

[1] C. ZAROWSKI, "Differential Evolution for a Better Approximation to the Arctangent Function," Nanodottek Report NDT3-04-2006, 26 April 2006.

Chapter 25

High-Speed Square Root Algorithms

Mark Allie
University of Wisconsin–Madison

Richard Lyons
Besser Associates

In this chapter we discuss several useful tricks for estimating the square root of a number. Our focus will be on high-speed (minimum computations) techniques for approximating the square root of a single value as well as the square root of a sum of squares for quadrature (I/Q) vector magnitude estimation.

In the world of DSP, computing a square root is found in many applications, such as calculating root mean squares, computing the magnitudes of fast Fourier transform (FFT) results, implementing automatic gain control (AGC) techniques, estimating the instantaneous envelope of a signal (AM demodulation) in digital receivers, and implementing three-dimensional graphics algorithms. The fundamental trade-off to be made in choosing a particular square root algorithm is execution speed versus algorithm accuracy. Here we disregard the advice of a legendary lawman ("Fast is important, but accuracy is everything"—Wyatt Earp [1848–1929]), and concentrate on various high-speed square root approximations. In particular, we focus on algorithms that can be implemented efficiently in fixed-point arithmetic. Other constraints we've set are: No divisions are allowed; only a small number of computational iterations are permitted; and only a small lookup table, if needed, is allowed.

The first two methods below describe ways to estimate the square root of a single value using iterative methods. The last two techniques describe methods for estimating the magnitude of a complex number.

25.1 NEWTON-RAPHSON INVERSE (NRI) METHOD

A venerable technique for computing the square root of x is to use the *Newton-Raphson square root* iterative technique to find $y(n)$, the approximation of

Streamlining Digital Signal Processing: A Tricks of the Trade Guidebook, Second Edition. Edited by Richard G. Lyons.
© 2012 the Institute of Electrical and Electronics Engineers. Published 2012 by John Wiley & Sons, Inc.

$$\text{sqrt}(x) \approx y(n+1) = [y(n) + x/y(n)]/2. \tag{25-1}$$

Variable n is the iteration index, and $y(0)$ is an initial guess of sqrt(x) used to compute $y(1)$. Successive iterations of (25–1) yield more accurate approximations to sqrt(x) because the number "bits of accuracy" doubles for each iteration.

However, to avoid that computationally expensive division by $y(n)$ in (25–1) we can use the iterative *Newton-Raphson inverse* (NRI) method; by first finding the approximation to the inverse square root of x using (25–2), where $p = 1/\text{sqrt}(x)$. Next we compute the square root of x by multiplying the final p by x,

$$p(n+1) = 0.5p(n)[3 - xp(n)^2] \tag{25-2}$$

Two iterations of (25–2) provide surprisingly good accuracy. As suggested by Cordesses and Araujo [1], the initial $p(0)$ value used in the first iteration is

$$p(0) = 1.63841001x^2 - 3.285040111x + 2.67057805. \tag{25-3}$$

Equation (25–3) can be evaluated using Horner's rule requiring only two multiplies and two adds as

$$p(0) = (1.63841001x - 3.28504011)x + 2.67057805. \tag{25-3'}$$

The square root function looks more linear when we restrict the range of our x input. As it turns out, it's convenient to limit the range of x to $0.25 \leq x < 1$. So if $x < 0.25$ then it must be *normalized*, and fortunately we have a slick way to do so. When x needs normalization, then x is multiplied by 4^n until $0.25 \leq x < 1$. A factor of 4 is 2 bits of arithmetic left-shift. The final square root result is *denormalized* by a factor of 2^n (the square root of 4^n). A denormalizing factor of 2 is a single arithmetic right-shift. After implementing this normalization, to ensure $0.25 \leq x < 1$, the error curve for this two-iteration NRI square root method is shown in Figure 25–2, where we see the maximum error is roughly 0.00011%. The curve is a plot of the error in our estimated square root divided by the true square root of x. That maximum error value is impressively small–almost worth writing home about.

This NRI method is not recommended for fixed-point format with its sign and fractional bits. The coefficients, $p(n)$, and intermediate values in (25–2) are greater than one. In fixed-point math, using bits to represent internal results greater than one increases the error by a factor of two per bit. (Certainly using more bits for the integer part of intermediate values without taking them away from the fractional part can be done, but only at the cost of more CPU cycles and memory.)

25.2 NONLINEAR IIR FILTER (NIIRF) METHOD

The next iterative technique, by Mikami et al. [2], specifically aimed at fixed-point implementation, is configured as a nonlinear IIR filter (NIIRF) as depicted in Figure 25–1, where again input x has been normalized to the range $0.25 \leq x < 1$. The output

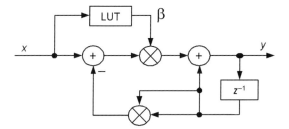

Figure 25–1 NIIRF implementation.

is given by (25–4) where the constant β, called the iteration *acceleration* factor, depends on the value of input x,

$$y(n+1) = \beta[x - y(n)^2] + y(n). \qquad (25\text{–}4)$$

The magnitude of the error of (25–4) is known in advance when $\beta = 1$. The β acceleration factor scales the square root update term, $x - y(n)^2$, so the new estimate of y has less error. The acceleration factor results in reduced error for a given number of iterations.

In its standard form, this technique uses a lookup table (LUT) to provide β and performs two iterations of (25–4). For the first iteration $y(0)$ is initialized using

$$y(0) = 2x/3 + 0.354167. \qquad (25\text{–}5)$$

The constants in (25–5) were determined empirically. The acceleration constant β, based on x, is stored in a LUT made up of 12 subintervals of x with an equal width of 1/16. This square root method is implemented with the range of x set between 4/16 and 16/16 (12 regions) as was done for the NRI method. This convenient interval directly gives the offset into the lookup table when using the four most significant mantissa bits of the normalized x value as a pointer, as shown in Table 25–1.

Figure 25–2 shows the *normalized* NIIRF error curve as a function of x. (Normalized, in this instance, means the error of the estimated square root divided by the true square root of x.) The NRI square root method is the most accurate, in floating-point math, followed by the NIIRF method. This is not surprising considering the greater number of operations needed to implement the NRI method.

The effect of the look-up table for β is clearly seen in Figure 25–2 where the dotted curve shows discontinuities (spikes) at the table lookup boundaries. These are, of course, missing from the NRI error curve.

One sensible thing to do, when presented with algorithms such as the NIIRF square root method, is to experiment with variations of the algorithm to investigate accuracy versus computational cost. (We did that for the above NRI method by determining the maximum error when only one iteration of (25–2) was performed.) With this *explore and investigate* thought in mind, we examined seven variations of this NIIRF algorithm. The first variation, a simplification, is the original NIIRF

Table 25–1   β versus Normalized-x LUT

4 MSBs of x	β fixed-point	β flt-point
0100	0x7b20	0.961914
0101	0x6b90	0.840332
0110	0x6430	0.782715
0111	0x5e10	0.734869
1000	0x5880	0.691406
1001	0x53c0	0.654297
1010	0x4fa0	0.622070
1011	0x4c30	0.595215
1100	0x4970	0.573731
1101	0x4730	0.556152
1110	0x4210	0.516113
1111	0x4060	0.502930

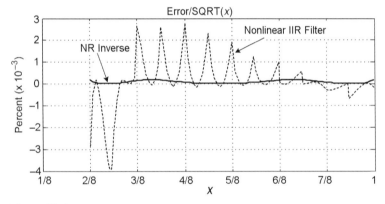

Figure 25–2   Normalized error for the NRI and NIIRF methods.

algorithm using only 1 iteration. We then studied the following quadratic function of x to find β

$$\beta = 0.763x^2 - 1.5688x + 1.314. \tag{25–6}$$

Next we used a linear function of x to find β, defined by

$$\beta = -0.61951x + 1.0688. \tag{25–7}$$

For the quadratic and linear variations to compute β, we investigated their performances when using both one and two iterations of (25–4).

For the final NIIRF method variation, we set β equal to the constants 0.633 and 0.64 for two and one iteration variations, respectively. Table 25–2 provides a

Table 25–2   Iterative Algorithm Normalized Absolute Error

Algorithm	Max. normalized error (%)	Mean normalized error (%)
NR inverse (NRI)		
NRI, 2 iters	1.1e–4	4.2e–5
NIIRF floating-pt		
LUT β, 2 iters	0.004	5.4e–4
LUT β, 1 iter	0.099	0.026
Quad. β, 2 iters	0.0013	2.8e–4
Quad. β, 1 iter	0.056	0.019
Linear β, 2 iters	0.024	0.0061
Linear β, 1 iter	0.28	0.088
β = 0.633, 2 iters	0.53	0.05
β = 0.64, 1 iter	1.44	0.23
NIIRF fixed-pt		
NIIRF, 2 iters	0.0035	5.1e–4
Quad. β, 2 iters	0.0019	4.1e–4
Linear β, 2 iters	0.011	0.0029

comparison of the error magnitude behavior of the original two-iteration (LUT) NIIRF algorithm and all the above variations.

For all of the variations listed in Table 25–2, the range of the input, x, was limited to $0.25 \le x < 1$. (The term *normalized* as used in Table 25–2 means the error of the estimated square root divided by the true square root of x.)

The error data in Table 25–2 was generated using double precision floating-point math. In comparing the accuracy and complexity of the NRI method to the NIIRF, we might ask why ever use the NIIRF method? Often, double-precision floating-point math is not available to us. Fixed-point data is one of the assumptions of this chapter. This fixed-point constraint requires that the algorithms must be reevaluated for the error when fixed-point math is used. It turns out this is why Mikami et al. developed the NIIRF square root method in the first place [2]. In fixed-point math most of the internal values are close to each other and less than one, so these values result in the lower error in the NIIRF method compared with the NRI method when both are implemented in fixed-point format. In fact, there is less error in the fixed-point compared with the floating-point implementations of NIIRF method. All of the NIIRF variations are well suited to fixed-point math. Next, we look at high-speed square root approximations used to estimate the magnitude of a complex number.

25.3   BINARY-SHIFT MAGNITUDE ESTIMATION

When we want to compute the magnitude M of the complex number $I + jQ$ the exact solution is, of course,

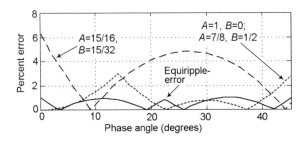

Figure 25–3   Error of *binary-shift* and *equiripple-error* methods.

$$M = \mathrm{sqrt}(I^2 + Q^2).\qquad(25\text{–}8)$$

To approximate the square root operation in (25–8), the following *binary-shift magnitude estimation* algorithm can be used to estimate M using

$$M \approx AM_{ax} + BM_{in}\qquad(25\text{–}9)$$

where A and B are constants, and M_{ax} is the maximum of either $|I|$ or $|Q|$ while M_{in} is the minimum of $|I|$ or $|Q|$.

Many combinations of A and B are provided in [3] yielding various accuracies for estimating M. (Chapter 16 discusses the use of $A = 1$, and $B = 4$ yielding estimates of magnitude M with a maximum error of 11.6% and a mean error of 3.2%.) However, of special interest is the combination $A = 15/16$, and $B = 15/32$ using

$$M \approx 15M_{ax}/16 + 15M_{in}/32.\qquad(25\text{–}10)$$

This algorithm's appeal is that it requires no explicit multiplies because the A and B values are binary fractions and (25–10) is implemented with simple binary right-shifts, additions, and subtractions. (For example, a 15/32 times z multiply can be performed by first shifting z right by four bits and subtracting that number from z to obtain 15/16 times z. That result is then shifted right one bit.) This algorithm estimates the magnitude M with a maximum error of 6.2% and a mean error of 3.1%.

The percent error of this binary-shift magnitude estimation scheme is shown as the dashed curve in Figure 25–3 as a function of the angle between I and Q. (The curves in Figure 25–3 repeat every 45°)

At the expense of a compare operation, we can improve on the accuracy of (25–10) [4]. If $M_{in} \leq M_{ax}/4$ we use the coefficients $A = 1$ and $B = 0$ to compute (25–9); otherwise if $M_{in} > M_{ax}/4$ we use $A = 7/8$ and $B = 1/2$. This dual-coefficients (and still multiplier-free) version of the binary-shift square root algorithm has a maximum error of 3.0% and a mean error of 0.95% as shown by the dotted curve in Figure 25–3.

25.4   EQUIRIPPLE-ERROR MAGNITUDE ESTIMATION

Another noteworthy scheme for computing the magnitude of the complex number $I + jQ$ is the *equiripple-error magnitude estimation* method by Filip [5]. This technique, whose maximum error is roughly 1%, is sweet in its simplicity: if $M_{in} \leq 0.4142135M_{ax}$, the complex number's magnitude is estimated using

$$M \approx 0.99M_{ax} + 0.197M_{in}. \qquad (25\text{--}11)$$

On the other hand, if $M_{in} > 0.4142135M_{ax}$, the following is used to estimate M:

$$M \approx 0.84M_{ax} + 0.561M_{in}. \qquad (25\text{--}12)$$

This algorithm is so named because its maximum error is 1.0% for both (25–11) and (25–12). Its mean error is 0.6%. Because its coefficients are not simple binary fractions, this method is best-suited for implementations on programmable hardware. The percent error of this equiripple-error magnitude estimation method is also shown in Figure 25–3 where we see the more computationally intensive method, equiripple-error, is more accurate. As usual, accuracy comes at the cost of computations. Although these methods are less accurate than the iterative square root techniques, they can be useful in those applications where high accuracy is not needed; such as when M is used to scale (control the amplitude of) a system's input signal.

One implementation issue to keep in mind when using integer arithmetic is that even though values $|I|$ and $|Q|$ may be within your binary word width range, the estimated magnitude value may be larger than can be contained within the numeric range. The practitioner must limit $|I|$ and $|Q|$, in some way, to ensure that the estimated M value does not cause overflows.

25.5 CONCLUSIONS

We discussed several square root and complex vector magnitude approximation algorithms with a focus on high-throughput (high-speed) algorithms as opposed to high-accuracy methods. We investigated the performance of several variations of iterative the NRI and NIIRF square root methods and found, not surprisingly, the number of iterations has the most profound effect on accuracy. The NRI method is not appropriate for implementation using fixed-point fractional binary arithmetic, while the NIIRF technique and the two magnitude estimation schemes lend themselves nicely to fixed-point implementation. The three magnitude estimation methods are less accurate but more computationally efficient than the NRI and NIIRF schemes. All algorithms described here are open to modification and experimentation, which will make them either more accurate and computationally expensive or less accurate and computationally cheap. Your accuracy and data throughput needs determine the path you take to a root of less evil.

A listing of a fixed-point DSP assembly code, simulated as an ADSP218x part, implementing this NIIRF is made available at http://booksupport.wiley.com. Various MATLAB square root modeling routines are also available.

25.6 REFERENCES

[1] L. CORDESSES and M. ARAUJO, Private communication, November 28, 2006.
[2] N. MIKAMI et al., "A new DSP-oriented algorithm for calculation of square root using a nonlinear digital filter," *IEEE Trans. on Signal Processing*, July 1992, pp. 1663–1669.
[3] R. LYONS, *Understanding Digital Signal Processing*, 2nd ed. Prentice Hall, Upper Saddle River, NJ, 2004, pp. 481–482.
[4] W. ADAMS and J. BRADY, "Magnitude approximations for microprocessor implementation," *IEEE Micro*, October 1983, pp. 27–31.
[5] A. FILIP, "Linear approximations to sqrt($x^2 + y^2$) having equiripple error characteristics", *IEEE Trans. on Audio and Electroacoustics*, December 1973, pp. 554–556.

EDITOR COMMENTS

The above material discusses two high-speed algorithms for approximating the square root function. While the first algorithm (the NRI method) is of interest when used in floating-point systems, the second algorithm (the NIIRF method) is most useful in fixed-point fractional binary arithmetic. After first learning of this second algorithm the authors did what all inquisitive DSP practitioners should do—they experimented with variations of the algorithm to compare computational workload versus algorithm accuracy.

To expand, a bit, on the NRI square root method, recall that the argument x was limited to the range $0.25 \leq x < 1$. That limitation can be relaxed, such that x need only be limited to be a positive number, if we store the appropriate $p(0)$ values in a lookup table (LUT). The $p(0)$ value retrieved from the LUT is based on the original value of x. An example of this implementation is described in Analog Devices Inc.'s "ADSP-21000 Family Application Handbook Volume 1." In that handbook an NRI square root algorithm is described where seven bits from the floating-point binary representation of x are extracted and used as address bits to access a read-only memory (ROM) containing initial inverse square root values, $p(0)$, accurate to 4 bits. In their NRI square root implementation, three iterations of (25–2) yields results accurate to 32 bits (32-bit mantissa).

Texas Instruments' document, "TMS320C4x General-Purpose Applications, User's Manual," describes their table lookup NRI square root method where the initial $p(0)$ value is accurate to 8 bits. So they achieve 32 bits of accuracy after only two iterations of (25–2).

Chapter 26

Function Approximation Using Polynomials

Jyri Ylöstalo
Nokia Siemens Networks

$\mathbf{D}$SP designers sometimes encounter a situation, where they have to use some of the elementary functions (logarithm, square root, exponential, trigonometric functions etc.) on a fixed-point processor. Although optimized math libraries are widely available for floating point processors, this is not necessarily true for the fixed-point case, which means that designers must implement the elementary functions using *home-grown* algorithms. This chapter describes a powerful technique for designing algorithms for such implementations, describes range-reduction methods to make the algorithms more accurate, and provides useful tips on using the proposed techniques in floating-point and fixed-point number systems.

In what follows we show how ordinary polynomials can be used for approximating the values of functions (logarithms, square root, trigonometric functions, etc.) of a real variable. The polynomial approximations discussed here are an attractive alternative for this purpose for several reasons:

- They require only multiplication and addition, which means that the calculation is fast if a hardware multiplier is available, as is the case with commercial DSP chips.

- The computation is noniterative, which makes efficient parallelization possible.

- They can be very accurate, and speed-versus-accuracy-versus-memory use trade-off can be tailored for the requirements of the application. For example, you can create a function approximation, which is fast, accurate, and uses a lot of memory, or one that is slow, accurate, and does not need much memory, or yet another one, which is fast, inaccurate, and uses very little memory.

- The same unified approach can be used for all continuous functions.

Streamlining Digital Signal Processing: A Tricks of the Trade Guidebook, Second Edition. Edited by Richard G. Lyons.
© 2012 the Institute of Electrical and Electronics Engineers. Published 2012 by John Wiley & Sons, Inc.

Approximation theory, in itself, is a rather broad subject and several extensive treatments have been written on it [1]–[3]. Here we provide a concise, practical presentation (no proofs for the theorems!) enabling the reader to create approximation polynomials for their own purposes and to utilize them sensibly. The approach presented here is a simple, basic one—those who want to delve really deep into elementary function implementation can study, for example, [4], [5].

26.1 USING LAGRANGE INTERPOLATION

Polynomial approximation is based on the classical Weierstrass theorem, according to which for any continuous function $f(x)$ defined within a closed range interval $[a,b]$, there exists a polynomial $p(x)$ for approximating the function in that range so that the maximum approximation error is kept below a given limit. The theorem doesn't say anything about how such a polynomial can be found. Usually the most straightforward method is to calculate the value of $f(x)$ for $n + 1$ distinct *fitting points* within the interval range $a \leq x \leq b$ and to satisfy the simultaneous equations $p(x_i) = f(x_i)$, $i = 0, 1, \ldots, n$. There is always a unique polynomial $p(x)$ of degree n, which satisfies these conditions and it is given by the so-called *Lagrange interpolation formula*:

$$p(x) = \sum_{i=0}^{n} p_i(x) \tag{26–1}$$

where each $p_i(x)$ term is defined by

$$p_i(x) = f(x_i) \prod_{j=0, j \neq i}^{n} \frac{x - x_j}{x_i - x_j}. \tag{26–1'}$$

For example, let's try to fit a second-degree polynomial to the fitting points $[x_i, f(x_i)] = (1,1), (2,3), (4,5)$. According to the above formulas, $n = 2$ and we get

$$p(x) = 1 \left[\frac{x-2}{1-2} \cdot \frac{x-4}{1-4} \right] + 3 \left[\frac{x-1}{2-1} \cdot \frac{x-4}{2-4} \right] + 5 \left[\frac{x-1}{4-1} \cdot \frac{x-2}{4-2} \right] \tag{26–2}$$
$$= -x^2/3 + 3x - 5/3.$$

26.2 FINDING THE OPTIMAL APPROXIMATION POLYNOMIAL

In principle, we could now improve the accuracy of the approximation as much as we want simply by adding new polynomial fitting points to the interval. For every new fitting point, the degree of the approximation polynomial would increase and calculation of the value of $p(x)$ would slow down correspondingly. Also, the amount of memory required for storing the parameters would increase. Instead of this kind of brute force approach, however, we could ask, "What is the optimal approximation,

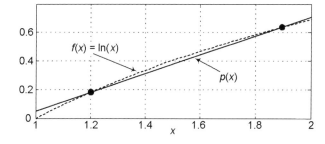

Figure 26–1 Nonoptimal, first-degree (linear) approximation of $\ln(x)$ with $x_0 = 1.2$ and $x_1 = 1.9$.

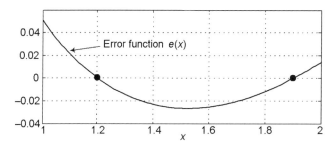

Figure 26–2 Error function of the nonoptimal, first-degree (linear) approximation of $\ln(x)$.

that is, the polynomial of a given degree that minimizes the maximum approximation error within the interval?" (Other definitions of *optimal approximation* are possible, but we won't deal with them here). It turns out that there is no straightforward mapping from the function being approximated to the optimal polynomial. We must, instead, use some kind of iterative algorithm to find our optimal approximating polynomial.

Consider the case of the first-degree polynomial $p(x)$ (the solid straight line in Figure 26–1), which attempts to approximate the function $f(x) = \ln(x)$ in the x interval $[a,b] = [1,2]$ using fitting points $[x_0 = 1.2, \ln(1.2)]$ and $[x_1 = 1.8, \ln(1.8)]$, which yield $p(x) = 0.6565x - 0.6054$. The approximation error $e(x) = f(x) - p(x)$ is depicted in Figure 26–2.

This approximation is in fact not optimal because, as we shall see, better approximations are available. We can perform an exhaustive search (e.g., using a 0.01 step size between new fitting points) using all possible pairs of fitting points on $f(x)$ in the interval. In doing so we find out that in the optimal case, the extrema of the error function have the same absolute value (0.03) and alternating signs as indicated in Figure 26–3. (It's because the maximum error in Figure 26–2 is greater than 0.03 that we said the first-degree polynomial approximation of $\ln(x)$ is not optimal.) *Equal absolute error extrema values* is an important and necessary property of *all* optimal (in the minimax sense) polynomial approximations of any degree for any function in any closed interval. This is called the *Chebyshev characterization theorem* and its first version was presented by Pafnuty Chebyshev in 1859.

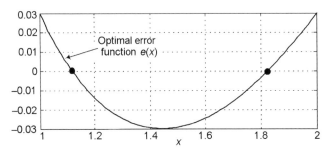

Figure 26–3   The optimal fit is reached with fitting points $x_0 = 1.12$ and $x_1 = 1.82$.

Exhaustive search is hardly a viable method for obtaining optimal polynomials of any degree higher than one. We could therefore try to use the above "equal absolute values of error extrema" property directly and move the fitting points gradually so that the absolute values of the extrema of the resulting error function would become more equal. Because the error is always zero at the fitting point, it seems clear that if we move a fitting point in the direction of an error peak, the height of the peak will decrease. Thus it might be a good idea to move each of the fitting points in the direction of the higher of its neighboring error peaks. Doing so leads us toward the optimal situation, where all of the error peaks have equal absolute values.

We now propose the following iterative algorithm for finding the optimal polynomial $p(x)$ for approximating the continuous function $f(x)$ in the interval $[a,b]$. This proposed process is not the fastest one available, but it is very simple and illustrative and in any case the iterations compute in mere seconds. Some hardcore mathematician might say to you that we should use the Remez exchange algorithm. It is admittedly a lot more efficient (2 to 6 iterations instead of 20 to 30), and there is very solid theory behind it as to why, when, and how it works. With our proposed technique you launch the routine using, say, MathCAD or MATLAB, go fetch yourself a cup of coffee, and upon returning you have parameters you can use for the rest of your life.

1. Choose approximation interval $[a,b]$, and the Degree of the approximating polynomial, D. The number of fitting points needed will be $D + 1$. The points are denoted as $x_0, f(x_0), \ldots, x_D, f(x_D)$.

2. Define initial iteration Step Size s, Decrease Factor d, and number of iterations n. Suitable values can be found by trial and error, however, $s = D/10$, $d = 0.9$, $n = 10D$ seem to work well.

3. Place a fitting point at both ends of the approximation interval (i.e. $x_0 = a$, $x_D = b$). These are the *terminal points* and they don't move during the iteration. Place the remaining $D - 1$ points (the *intermediate fitting points*) at suitable places in the interval (the initial positions are not very important, as long as two points don't have the same coordinates). Equidistantly scattered points work okay.

4. Fit a polynomial (Lagrange interpolation) to the $D + 1$ fitting points using (26–1) and (26–2) yielding the current $p(x)$.

5. For each of the intermediate fitting points $[x_i, f(x_i)]$, $i = 1 \ldots D - 1$, find the local error peak $|e(x_{ip})|$ in the interval defined by $[x_{i-1}, x_{i+1}]$, and compute the peak's x-position, x_{ip}.

6. Shift each x_i intermediate point by moving it in the direction of its interval's local error peak located at x_{ip}, using $x_{inew} = x_i + s(x_{ip} - x_i)$.

7. Define the next iteration Step Size: $s_{new} = sd$.

8. Return to step 4 until steps 4 through 8 have been performed n times.

(A feature of this algorithm is that you can easily fix points you don't want to move.)

The reason why we need two fixed terminal points at both ends of the interval is to assure continuity of the resulting approximated function. This becomes important when *piecewise polynomials* are used for approximation, as we will see later. Another good reason to fix the terminal points is to minimize error for critical values. For example, it might be embarrassing, or even devastating to your application, to have your logarithm approximation claim that $\log(1) = 0.0004$ instead of 0. But if neither of these concerns, continuity or critical values, applies to you, then by all means, move the terminal points in the same way as all the other points. In this way you will get tiny discontinuities, but slightly smaller error.

26.3 RANGE REDUCTION

Although in theory it is possible to approximate any continuous function over any closed interval with a polynomial, it hardly ever makes sense to try to fit one single polynomial to the entire possible input range. Such a polynomial might need to be of very high order to meet the desired accuracy specifications. In general, the shorter the approximation interval the closer to linear the function is in that interval, and the easier our job will be—the lower the degree of the polynomial required. We should therefore devise a suitable range reduction scheme, in order to be able to compute all possible output values from an approximation performed within a short, closed interval [4].

Consider the case of applying this range reduction notion to the logarithm (of any desired base k). From the *laws of exponents*, using $\log_k(y^n x) = n \cdot \log_k(y) + \log_k(x)$, we can write:

$$\log_k(x) = \log_k(2^n \cdot x) - n \cdot \log_k(2). \qquad (26–3)$$

Equation (26–3) tells us that to compute $\log_k(x)$, where x may cover a wide range of values, we can *normalize* x (reduce its range) to a smaller range of values for which a more accurate approximation polynomial can be found. Once we compute $\log_k(2^n \cdot x)$ we then *denormalize* that computation result by subtracting $n \cdot \log_k(2)$ to arrive at the desired $\log_k(x)$ value. In practice this means that we must decide upon some integer M and develop the approximation polynomial for the interval $[2^{M-1}, 2^M]$.

It is always possible to find a suitable n to normalize $2^n x$ between $[2^{M-1}, 2^M]$ and then use (26–3) to calculate $\log_k(x)$.

In the floating point case, the normalization can be done simply by adjusting the exponent part of the number x. In the fixed-point case let us define the operation HighestBit(x), for any integer $x > 0$, as the index of the most significant 1 bit of the number x. (The index of the MSB of an unsigned 16-bit number is 15, the index of the LSB is 0.) Luckily, some programmable DSP chips have a machine instruction for performing this kind of an operation in a single cycle, which of course speeds up calculation enormously. (For example, Texas Instruments DSP chips have the instruction LMBD, *Leftmost Bit Detection*.) Now, the required power of two for the normalization can be found with

$$n = M - 1 - \text{HighestBit}(x). \tag{26–4}$$

We can define $M = 0$, in which case the normalized logarithm, $\log_k(2^n x)$, is calculated for $2^n \cdot x$, which will be in the interval $[0.5, 1]$ and can be represented as a Q-15 number. For example, if we want to compute $\log(x)$ where $x = 1234$, we get HighestBit(x) = 10 and $n = -11$. The $x = 1234$ input to the approximation will thus be converted to $1234 \cdot 2^{-11} = 0.6025390625$, which is in the range $[0.5, 1]$.

A rather similar kind of range reduction scheme can be used for square roots as

$$\sqrt{x} = \sqrt{2^n \cdot x} \left(\sqrt{2}\right)^{-n} \tag{26–5}$$

where x is normalized to $x_{norm} = 2^n \cdot x$, with n being defined by (26–4). Note that the *denormalization* by $[\text{sqrt}(2)]^{-n}$ can be realized in the fixed-point case with shifting by $n/2$ in the case of n even. A possible approach is to perform the calculation in the interval $[0.25, 1]$. We can first calculate n just as in (26–4) and then if n is odd, decrement it by one. For example, say we want to compute the square root of $x = 1234$. First we determine $n = -11$, which is an odd number so we decrement it to $n = -12$. Next we normalize $x = 1234$ to $x_{norm} = 1234 \cdot 2^{-12} = 0.30126953125$, which is in the range $[0.25, 1]$, upon which we compute the square root. The final step is to *denormalize* the square root result by multiplying it by $[\text{sqrt}(2)]^{-n}$.

In the case of polynomial approximations of trigonometric functions, we need an entirely different type of range reduction: additive instead of multiplicative. A suitable reduced range is $[0, \pi/4]$. We first normalize the input x to be between $[-\pi/4, \pi/4]$: if x is outside this range, we add or subtract $\pi/2$ until the normalized input, x_{norm}, falls into this range. We get $x_{norm} = x - m\pi/2$. If x_{norm} is in the range $[-\pi/4, 0]$, we use the familiar relations $\cos(x) = \cos(-x)$, $\sin(x) = -\sin(-x)$. Now

$$\cos(x) = \cos(x_{norm}) \quad \text{if } m \bmod 4 = 0,$$

$$\cos(x) = -\sin(x_{norm}) \quad \text{if } m \bmod 4 = 1,$$

$$\cos(x) = -\cos(x_{norm}) \quad \text{if } m \bmod 4 = 2,$$

$$\cos(x) = \sin(x_{norm}) \quad \text{if } m \bmod 4 = 3.$$

For example, instead of calculating cos(9) (input in radians), we calculate $-\cos(6\pi/2{-}9)$.

26.4 SUBINTERVAL DIVISION

We don't need to stop our range reduction efforts here. One possibility is to further divide the approximation interval into S subintervals, which—let's assume for simplicity—are of equal width. For instance, we could divide the interval [0.5,1] into two subintervals, [0.5,0.75] and [0.75,1], or four subintervals, [0.5,0.625], [0.625,0.75], [0.75,0.875], [0.875,1]. Each of the subintervals has a separate approximation polynomial, which approximates a function $f(x)$ in that subinterval.

Subintervals are identified by their index i, which is defined in the following way:

$$i = \text{trunc}\left(\frac{S(x_{norm} - a)}{b - a}\right) \qquad (26\text{–}6)$$

where "trunc" stands for truncation of the decimal part and the input x has been previously normalized to the interval $a \le x_{norm} < b$. The index i starts from zero like in the C language. The speed of calculation of the approximation result is independent of the value of S, and if the parameters (S relative to $b - a$) are chosen smartly, division won't be needed and the correct subinterval for the normalized input x can be found with a simple shift operation. We can thus, at least in theory, increase accuracy as much as we want without any speed penalty simply by dividing the approximation interval to smaller subintervals—only the required amount of memory for storing the polynomial coefficients will increase. If the degree of approximation is D and the number of subintervals is S, we need $S(D + 1)$ polynomial coefficients.

When trying to improve the accuracy of a polynomial approximation, we have two alternatives: either increase D or S (i.e., use a higher degree of approximation or divide the approximation interval to smaller subintervals). Increasing D means increased computation time and a slight increase in memory use, whereas increasing S means a slightly faster increase in memory use *without any increase in computation time*. In modern systems, storage space is usually cheap while time isn't. Subinterval division is therefore an attractive option for improving accuracy. Another factor favoring increasing S instead of D is related to arithmetic problems using fixed-point calculations, as we shall soon see.

26.5 PRACTICAL CONSIDERATIONS

On a floating point system, calculating polynomial approximation results is trivial once a suitable range reduction scheme has been developed, the degree of approximation and subinterval division have been selected, and the polynomial coefficients have been calculated. Usually, in order to reach high accuracy it is preferable to use relatively high-degree polynomials, possibly taking advantage of available pipelining features and finding a suitable compromise between accuracy and speed.

With fixed-point systems things aren't necessarily as simple as this. The use of high-degree polynomials may easily lead to disappointing results for the following reason: Suppose that we have a 32-bit DSP with 16-bit hardware multiplier. In such a case it is usually advisable to represent the coefficients using 16 bits and accumulate the sum of the polynomial terms to a 32-bit variable. When we multiply the input x with itself, we have to discard half of the bits of the result in order to fit it into 16 bits. This introduces a quantization error, which becomes progressively larger in proportion to the result with higher powers of x. Therefore, on fixed-point systems the most viable alternative is usually to utilize low degree (e.g., quadratic or cubic) polynomials and increase the number of subintervals, if the accuracy must be improved.

26.6 ERROR STUDIES

Controlling approximation error is easy with a proper choice of S and D. On a fixed-point system the real problem is the quantization effects, which may spoil the outcome, no matter how excellent the results look of a floating point system like MATLAB. The most obvious solution to reduce quantization errors is to increase the multiplication wordlength. However, this will almost inevitably lead to increased execution time. It pays therefore to study the error generation a bit more precisely, in order to squeeze the best possible accuracy out of the least possible CPU cycles. Using the tricks described in this section, it should be possible to reduce the maximum approximation error to below 5 ppm.

For simplicity, let us concentrate on the case where we use 16×16-bit multiplication and 32-bit accumulation to calculate a second degree polynomial:

$$y = ax^2 + bx + c. \tag{26–7}$$

Because of quantization effects, this becomes

$$y = (a + \varepsilon_a)[(x + \varepsilon_x)(x + \varepsilon_x) + \varepsilon_{x2}] + (b + \varepsilon_b)(x + \varepsilon_x) + c + \varepsilon_c, \tag{26–8}$$

where

ε_a is the error made in the quantization of a,

ε_b is the error made in the quantization of b,

ε_c is the error made in the quantization of c,

ε_x is the error made in the quantization of x, and

ε_{x2} is the error made in the quantization of x^2.

We can first make an observation that has important practical consequences: the polynomial coefficient c is not multiplied by anything and therefore can (and should) be represented with 32 bits. In this way ε_c will be <1 ppm and can be considered insignificant. On the other hand, x, a, and b must be represented with 16 bits and therefore ε_x, ε_{x2}, ε_a, and ε_b are significant. However, if these 16-bit quantization error terms are multiplied with each other, the result is <1 ppm and can be considered insignificant. Therefore (26–8) reduces to

$$y \approx ax^2 + bx + c + 2ax\varepsilon_x + a\varepsilon_{x2} + x^2\varepsilon_a + b\varepsilon_x + x\varepsilon_b. \qquad (26\text{--}9)$$

We can see from the total error $2ax\varepsilon_x + a\varepsilon_{x2} + x^2\varepsilon_a + b\varepsilon_x + x\varepsilon_b$ that large values of a, b, and x have an amplifying effect on the quantization errors. Luckily, there is a simple cure for large values of x: always have x start from zero. Instead of approximating $f(x)$ with $p(x)$, approximate $f(x)$ with $p(x - A)$, where A is the start of the approximation interval. Note that this must be taken into account already when finding the coefficients for p.

It may also be possible to reduce the values of a and b by tuning range reduction suitably. In the case of log, sqrt, and similar type of functions, we notice that a and b tend to get smaller with large values of parameter M (see formula (26–4) above). On the other hand, a very large value of M may lead to mathematical problems in polynomial fitting, impractically small a, and so on. A value of $M = 4$ is usually a good compromise. The approximation interval will thus be [8,16] in the case of log and [4,16] in the case of sqrt.

26.7 FUNCTION APPROXIMATION EXAMPLE

To summarize all of the above, let's study a complete example of function approximation using polynomials. We will approximate the function $\ln(x)$ and use [8,16] as the approximation interval. We select $S = 4$ and $D = 3$ and thus have to find the coefficients for the polynomial $p(x - A) = a(x - A)^3 + b(x - A)^2 + c(x - A) + d$ in the subintervals [8,10], [10,12], [12,14], and [14,16]. Our optimal approximation polynomial search routine returns to us the polynomial coefficients shown in Table 26.1:

Let's now see what happens with the input $x = 987$. First x is normalized to the interval [8,16] by moving the binary point 6 places to the left. (In the fixed-point case we don't really move anything, we just "imagine" there is a point on the right side of the fourth bit from the left. Then we take this into account when doing the final denormalization.) The normalized input becomes $987.2^{-6} = 15.421875$. This

Table 26–1   Polynomial Coefficients versus Subinterval Range

Subinterval	a	b	c	d
$8 \leq x < 10$	0.0004627	−0.007606	0.124933	2.079442
$10 \leq x < 12$	0.0002524	−0.004910	0.099970	2.302585
$12 \leq x < 14$	0.0001526	−0.003427	0.083318	2.484907
$14 \leq x < 16$	0.0000992	−0.002526	0.071420	2.63906

value falls into the last subinterval [14,16]. With the values given in the table we get $p(x - 14) = 2.73579$. Denormalization yields $2.73579 + 6.\ln(2) = 6.89467$. The precise value for $\ln(987)$ is 6.89467004.

26.8 CONCLUSIONS

We have explained the use of polynomial approximations for calculating function values and demonstrated their viability in a fixed-point numerical environment. Because of their fast computation and capability of efficient parallelization, polynomial approximations are the algorithm of choice for the realization of elementary functions in modern, pipelined DSP architectures.

26.9 REFERENCES

[1] M. POWEL, *Approximation Theory and Methods*, Cambridge Univ. Press, Cambridge, UK, 1981.
[2] J. RICE, *The Approximation of Functions*, Addison Wesley, Reading, MA, 1964.
[3] A. TIMA, *Theory of Approximation of Functions of a Real Variable*, Dover Publications, New York, 1994.
[4] J. MULLER, *Elementary Functions, Algorithms and Implementation*, Birkhäuser, Boston, MA, 1997.
[5] P. TAN, "Table lookup algorithms for elementary functions and their error analysis," *Proceedings of the 10th IEEE Symposium on Computer Arithmetic*, Grenoble, France, June 1991, pp. 232–236.

EDITOR COMMENTS

Here we provide details of the derivation of (26–2) in order to clarify the indexing used in the Lagrange interpolation formula given in (26–1). Repeating (26–1) with $n = 2$, we have

$$p(x) = \sum_{i=0}^{2} p_i(x) = p_0(x) + p_1(x) + p_2(x).$$

Using Eq. (26–1′) we write polynomial $p(x)$ as

$$p(x) = f(x_0)\left[\frac{x - x_1}{x_0 - x_1} \cdot \frac{x - x_2}{x_0 - x_2}\right] + f(x_1)\left[\frac{x - x_0}{x_1 - x_0} \cdot \frac{x - x_2}{x_1 - x_2}\right] + f(x_2)\left[\frac{x - x_0}{x_2 - x_0} \cdot \frac{x - x_1}{x_2 - x_1}\right].$$

With

$$x_0 = 1, \quad f(x_0) = 1,$$
$$x_1 = 2, \quad f(x_1) = 3,$$
$$x_2 = 4, \quad f(x_2) = 5,$$

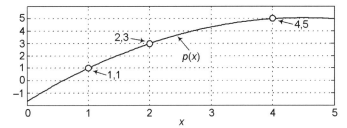

Figure 26–4   Second-degree *p*(*x*) polynomial.

polynomial $p(x)$ becomes

$$p(x) = 1\left[\frac{x-2}{1-2}\cdot\frac{x-4}{1-4}\right]+3\left[\frac{x-1}{2-1}\cdot\frac{x-4}{2-4}\right]+5\left[\frac{x-1}{4-1}\cdot\frac{x-2}{4-2}\right]$$

$$= \frac{(x-2)(x-4)}{3}+\frac{3(x-1)(x-4)}{-2}+\frac{5(x-1)(x-2)}{6}$$

$$= \frac{(-12)(x-2)(x-4)+(18)(3)(x-1)(x-4)+(-6)(5)(x-1)(x-2)}{-36}$$

$$= \frac{2x^2-18x+10}{-6}=\frac{-x^2}{3}+3x-\frac{5}{3}.$$

So the coefficients of the above polynomial $p(x)$ are –1/3, 3, and –5/3, agreeing with (26–2). Figure 26–4 depicts the continuous second-degree $p(x)$ polynomial defined by (26–2).

To expand on the explanation for the proposed function approximation algorithm, here is a detailed example of the method. Assume we are trying to approximate $\ln(x)$ between [1,2] using a third-degree polynomial, $D = 3$, and the $(x, f(x))$ coordinates of our initial four fitting points are (1,0), (1.4,ln(1.4)), (1.8,ln(1.8)), (2,ln(2)) as shown by the dots in Figure 26–5(a). Performing step 4 of the proposed algorithm, we use Lagrange interpolation to obtain our initial third-degree $p(x)$ polynomial of

$$p(x) = 0.09697899x^3 - 0.67342994x^2 + 2.03458402x - 1.45813308. \quad (26\text{–}10)$$

Evaluating $p(x)$ provides the curve in Figure 26–5(a). This $p(x)$ has the absolute approximation error $|e(x)| = |\ln(x) - p(x)|$ shown in Figure 26–5(b). The extrema (peaks) in the error function are not equal so our initial fitting points were not optimal. Step 5 of the proposed algorithm tells us how to move the intermediate fitting points 1.4 and 1.8. (The phrase *intermediate fitting points* means all the fitting points except the first and last points, so x_1 and x_2 are the intermediate fitting points in this example.)

Because we have two intermediate fitting points, x_1 and x_2, we must examine Interval 1 and Interval 2 to find the x-coordinates (x_{1p} and x_{2p}) of the error peaks in

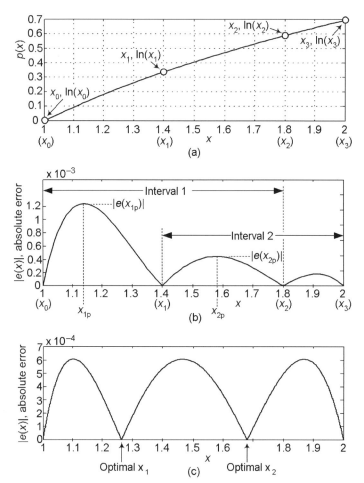

Figure 26–5   Approximating ln(x): (a) initial fitting points; (b) first-iteration error function; (c) final error function.

those two intervals. Knowing x_{1p} and x_{2p}, we perform step 6 by moving fitting point x_1 toward x_{1p} by the amount $s(x_{1p} - x_1)$. Next we move fitting point x_2 toward x_{2p} by the amount $s(x_{2p} - x_2)$. That completes the first iteration of the algorithm.

Continuing, we use the new fitting points (and Lagrange interpolation) to compute a new $p(x)$ polynomial, and begin analyzing its performance to find two new intermediate fitting points. We repeat this process and after $10D$ (30) iterations of steps 4 through 8 we arrive at the optimum fitting-points locations that yield the optimum $p(x)$ polynomial whose absolute approximation error is shown in Figure 26–5(c). Notice how the final extrema in that error function are now equal—that's what the author calls "optimal."

One last issue to consider, to reduce the computational workload in evaluating polynomials *Horner's rule* should be used. That rule reduces the necessary number of multiplies during polynomial computations. For example, computing

$$p(x) = ax^4 + bx^3 + cx^2 + dx + e$$

requires seven multiplications. Using Horner's rule to rewrite $p(x)$, we have the equivalent

$$p(x) = x(x(x(ax+b)+c)+d)+e$$

requiring only four multiplications. Horner's rule lends itself well to implementations using programmable chips having single-cycle *multiply and accumulate* (MAC) capabilities.

Chapter 27

Efficient Approximations for the Arctangent Function

Sreeraman Rajan, Sichun Wang, Robert Inkol, and Alain Joyal
Defence Research and Development

This chapter provides several efficient approximations for the arctangent function using Lagrange interpolation and minimax optimization techniques. These approximations are particularly useful for the implementation of the arctangent function when processing power, memory, and power consumption are important issues. In addition to comparing the errors and the computational workload of these approximations, we also give methods for extending arctangent approximations to all four quadrants.

27.1 ARCTANGENT APPROXIMATIONS

The evaluation of the arctangent function is commonly encountered in real-time applications of signal processing, such as biomedical engineering, instrumentation, communication, and electronic warfare systems. Numerous algorithms are available to implement the arctangent function when computational cost is unimportant. The most direct solution is based on the Taylor series expansion for $|x| < 1$, which is:

$$\arctan(x) = x - \frac{x^3}{3} + \frac{x^5}{5} - \frac{x^7}{7} + \dots \qquad (27\text{--}1)$$

Streamlining Digital Signal Processing: A Tricks of the Trade Guidebook, Second Edition. Edited by Richard G. Lyons.
© 2012 the Institute of Electrical and Electronics Engineers. Published 2012 by John Wiley & Sons, Inc.

Since $\arctan(x) = \pi/2 - \arctan(x^{-1})$, we can use (27–1) to calculate $\arctan(x)$ for all x. However, the series (27–1) converges slowly for values of x close to 1 and hence is inefficient. In order to expedite the convergence of the expansion, transformations such as the one discussed in [1] may be used. However, these transformations involve the computation of square roots. Another interesting transformation to handle slow convergence is the idea of series folding where the argument x is repeatedly transformed to a number close to zero by using the following transformation for various values of k [2]:

$$y = \frac{1 - kx}{k + x}, \quad k > 0. \tag{27–2}$$

The expressions, $\arctan(x)$ and $\arctan(y)$, are related by the following:

$$\arctan(y) = \arctan(k^{-1}) - \arctan(x). \tag{27–3}$$

Although this technique circumvents the convergence issues, it requires additional arithmetic operations. Hence this method may not lend easily to realization in hardware. It is also clear from the discussion that this process requires a lookup table for $\arctan(k^{-1})$.

Iterative algorithms, such as the CORDIC (*CO*rdinate *R*otation *DI*gital Computer) algorithm, have been successfully used to compute trigonometric functions [3]. These algorithms require only shifts and add operations and are thus multiplierless [4]. However, the sequential nature of these algorithms makes them less attractive when speed is a major concern while attempts to increase speed have been at the expense of additional hardware [5]. Lookup table-based approaches to the computation of inverse trigonometric functions are very fast, but require considerable memory [6], [7].

Polynomial and rational function approximations have been proposed in the literature [5] that are more suitable for numerical coprocessors. Most of these approximations are recursive in nature. Some of these approximations are linear combinations of orthogonal polynomials in a given closed interval. Given the precision requirements, a fixed formula for such approximations may be easily derived. Approximations using polynomials of large degrees are computationally expensive. Rational approximations are in principle more accurate than polynomial approximations for the same number of coefficients. However, the required division operations are relatively complex to implement; iterative techniques, such as those based on Newton's method are often used.

The aforementioned approaches are best suited for applications where processing power and/or memory are readily available. However, for many applications, simpler and more efficient ways of evaluating $\arctan(x)$ are desirable. Here we propose simple approximations for evaluating the arctangent function that may be easily implemented in hardware with limited memory and processing power.

27.2 PROPOSED METHODOLOGY AND APPROXIMATIONS

Approximations to the arctangent function can be obtained using second- and third-order polynomials, and simple rational functions. For these classes of approximation functions, Lagrange interpolation-based and minimax criterion-based approaches are used to obtain the polynomial coefficients. The following is our initial derivation of an arctangent approximation using Lagrange interpolation.

Consider the three points $x_0 = -1$, $x_1 = 0$, $x_2 = 1$. Let

$$\Phi(x) = \arctan\left(\frac{1+x}{1-x}\right), \quad -1 \leq x \leq 1.$$

According to the Lagrange interpolation formula, we have

$$\Phi(x) \approx \frac{(x-x_1)(x-x_2)}{(x_0-x_1)(x_0-x_2)}\Phi(x_0) + \frac{(x-x_0)(x-x_2)}{(x_1-x_0)(x_1-x_2)}\Phi(x_1) + \frac{(x-x_0)(x-x_1)}{(x_2-x_0)(x_2-x_1)}\Phi(x_2)$$

$$= \frac{\pi}{4}(x+1), \quad -1 \leq x \leq 1.$$

$$(27\text{–}4)$$

It can be verified that $\Phi(x) = \pi/4 + \arctan(x)$; hence, $\arctan(x)$ can be approximated by the first-order formula

$$\arctan(x) \approx \frac{\pi}{4}x, \quad -1 \leq x \leq 1. \qquad (27\text{–}5)$$

This linear approximation has been used in [8] for FM demodulation due to its minimal complexity. It requires only a scaling operation by a fixed constant and can be computed in one cycle in most processors. Figure 27–1 shows the deviation of this linear (first-order) approximation from the arctangent function on the interval $[-1,1]$.

This deviation is antisymmetric about $x = 0$ and attains a maximum at $x_{max} = \pm(4/\pi - 1)^{1/2}$, and is approximately quadratic in nature in $[0,1]$. The maximum error due to approximation (27–5) is about 0.07 radians (4°).

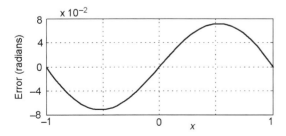

Figure 27–1 Approximation errors using linear approximation (27–5).

Here's a useful trick to improve the accuracy of the first-order approximation (27–5). We obtain an expression for the error in (27–5), we then subtract that error from (27–5) to yield a more accurate arctan approximation. Following this strategy, consider the error function given by

$$\Psi(x) = \arctan(x) - \frac{\pi}{4}x, \quad -1 \le x \le 1. \tag{27–6}$$

Applying Lagrange's interpolation formula to $\Psi(x)$ using $x_0 = 0$, $x_1 = x_{max}$, and $x_2 = 1$, $\Psi(x)$ can be approximated by

$$\Psi(x) \approx 0.285x(1 - |x|), \quad -1 \le x \le 1. \tag{27–7}$$

where the odd symmetry of $\Psi(x)$ has been applied. A second-order, and more accurate, approximation for $\arctan(x)$ thus becomes:

$$\arctan(x) \approx \frac{\pi}{4}x + 0.285x(1 - |x|), \quad -1 \le x \le 1 \tag{27–8}$$

with a maximum absolute error of 0.0053 radians (0.3°).

Exploring other second-order arctan algorithms, it is also of interest to approximate the error function, $\Psi(x)$, given by (27–6) by a second-order polynomial of the form $(\alpha x^2 - \alpha x)$ that passes through the endpoints of the interval $[0,1]$. The optimum parameter, $\alpha > 0$, may be obtained using the following minimax criterion:

$$J = \min_{\alpha} \{ \max_{0 \le x \le 1} \{ |\Psi(x) - \alpha x(x-1)| \} \} \tag{27–9}$$

The above criterion (27–9) does the following: For every α, the maximum absolute error between $\Psi(x)$ and the second-order polynomial $(\alpha x^2 - \alpha x)$ is determined. This error, as a function of α, is then minimized. This will yield a unique α that gives the least error J; hence the name minimax criterion and the approximation is called minimax approximation.

Using an extensive computer search, the optimum $\alpha \approx 0.273$ and the following approximation for the $\arctan(x)$ is obtained:

$$\arctan(x) \approx \frac{\pi}{4}x + 0.273x(1 - |x|), \quad -1 \le x \le 1 \tag{27–10}$$

with a maximum absolute error of 0.0038 radians (0.22°).

A third-order polynomial is also a good candidate to fit the error curve shown in Figure 27–1. When the third-order polynomial is constrained to be of the form $\alpha x^3 + \beta x$, the best minimax approximation for the arctangent function is given by

$$\arctan(x) \approx \frac{\pi}{4}x + x(0.186982 - 0.191942x^2), \quad -1 \le x \le 1. \tag{27–11}$$

The maximum absolute error for this approximation is 0.005 radians (0.29°) and is worse than that given by (27–10).

A better third-order polynomial for approximating the error is of the following form $x(x-1)(\alpha x - \beta)$ for x in the interval $[-1,1]$. Using the minimax criteria, the following polynomial is identified as the optimal approximation to arctan(x):

$$\arctan(x) \approx \frac{\pi}{4}x - x(|x|-1)(0.2447 + 0.0663|x|), \quad -1 \le x \le 1 \qquad (27\text{–}12)$$

with a maximum absolute error of 0.0015 radians (0.086°).

Another candidate for approximating the arctangent function is from the class of rational functions of the form $\tau(x) = x/(1 + \beta x^2)$ in the interval $[-1,1]$. For $0 \le \beta \le 1$, the first derivative of $\tau(x)$ is positive and the second derivative of $\tau(x)$ is negative. This implies that $\tau(x)$ has a shape very similar to that of the arctangent function in the same interval. Using minimax criteria, the following approximation is obtained:

$$\arctan(x) \approx \frac{x}{1+0.28086x^2}, \quad -1 \le x \le 1 \qquad (27\text{–}13)$$

with the maximum absolute error to be about 0.0047 radians (0.27°).

A similar idea was presented in [7] with a maximum absolute error of 0.0049 radians (0.28°). The approximation in [7] is as follows:

$$\arctan(x) \approx \frac{x}{1+0.28125x^2}, \quad -1 \le x \le 1. \qquad (27\text{–}14)$$

The scaling constant, 0.28125, was chosen to permit multiplication to be performed (as discussed later) with two arithmetic binary right shifts and a single addition. This yields a negligible increase in maximum error.

27.3 DISCUSSION

Now we compare the various arctangent approximations presented here. Table 27.1 contains the maximum error, number of adds, multiplies, and divides for various arctan approximations. The third-order approximation given by (27–12) has the least error among the proposed approximations, but has the highest computational cost. The second-order approximation given by (27–10) has the next lowest error and has fewer operations than (27–12). Hence (27–10) provides a favorable compromise between accuracy and computational cost. The linear approximation given by (27–5) has only one multiplication operation, but has the least accuracy. With a single cycle multiply and accumulate (MAC) processors, the evaluation of the arctangent using the (27–10) would take only two cycles, if the sign information of the argument is already available. However, one needs to check the sign of the argument, which may take an extra cycle.

Table 27–1   Maximum Approximation Error and Computational Workload

Equation	Error (rad.)	Adds	Multiplies	Divides
(27–5)	0.07	0	1	0
(27–8)	0.0053	1	2	0
(27–10)	0.0038	1	2	0
(27–11)	0.005	1	3	0
(27–12)	0.0015	2	3	0
(27–13)	0.0047	1	2	1
(27–14)	0.0049	2	1	1

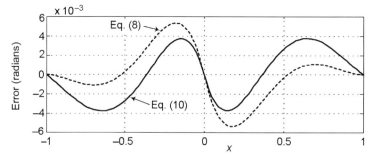

Figure 27–2   Approximation errors using second-order polynomials (27–8), (27–10).

It is interesting to note that all the second-order approximations given in this chapter have better accuracy than the third-order polynomial approximation given in reference [9]. The approximation in [9] was used in the computation of bearing and had an accuracy of only about 0.0175 radians (1°).

Comparison of the approximations to arctan(x) using the proposed two second-order approximations given by (27–8) and (27–10) are shown in Figure 27–2.

These approximations have maximum errors that are an order of magnitude better than that of the linear approximation (27–5). Furthermore, the second-order approximation given by (27–8) provides better accuracy for the subintervals $1 > x > 0.5$ and $-1 < x < -0.5$, where the maximum error is only about 0.001 radians (0.057°). No such observation can be made for (27–10).

Figure 27–3 shows that (27–12) provides the best accuracy among the third-order approximations considered in this chapter.

The error due to the rational approximations, (27–13) and (27–14), is presented in Figure 27–4. The behavior of the two rational approximations is almost identical, except near the maximum error.

The rational approximations given by (27–13) and (27–14) are computationally more expensive than the second-order ones given by (27–8) and (27–10), though this cost increase is partly offset by the elimination of the need for sign comparisons. Approximations using (27–13) or (27–14) have a division operation that may slow

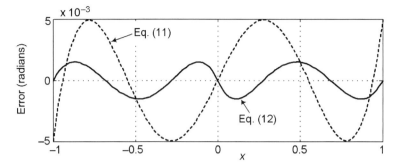

Figure 27–3   Approximation errors using cubic polynomials (27–11), (27–12).

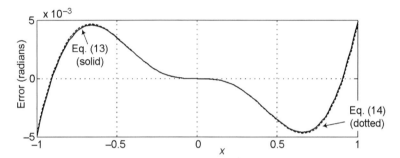

Figure 27–4   Approximation errors using rational functions (27–13), (27–14).

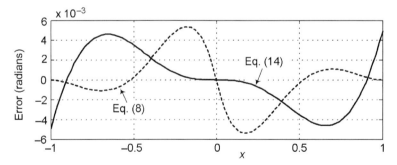

Figure 27–5   Approximation errors using second-order polynomial (27–8) and rational approximation (27–14).

down the processing. The approximation using (27–14) and the approximation given by (27–8) intersect at approximately $x_c = 0.3933$ in the interval $[0,1]$. For values of $|x| > x_c$, the proposed approximation given by (27–8) yields better accuracy as shown in the Figure 27–5 and for values of $|x| < x_c$, the approximation in (27–14) gives better accuracy.

An alternative approach to improving the accuracy, though at the expense of computational cost, would be to combine (27–8) and (27–14). This combined equation is given by

$$\arctan(x) \approx \gamma A + (1 - \gamma)B \tag{27–15}$$

where A is given by (27–14) and B is given by (27–8). The weighting variable γ is given by the following

$$\gamma = \begin{cases} 1, & 0 \leq |x| \leq x_c \\ 0, & x_c < |x| \leq 1. \end{cases} \tag{27–16}$$

Using this approach, the maximum error is less than 0.0025 radians (0.14°) as can be seen from Figure 27–5. Also the approximation given by (27–15) is better than (27–8) and (27–14) in the minimax sense.

27.4   FOUR-QUADRANT APPROXIMATIONS

The arctan algorithms presented thus far are applicable for angles in the range of $-\pi/4$ to $\pi/4$ radians. Here we provide the scheme for extending that angular range to $-\pi$ to π radians. Let $z = I + jQ$ be any complex number, and let $x = Q/I$ and $y = I/Q$. In signal processing terms, I may be considered as the in-phase component of a complex signal, while Q may be considered as the quadrature component. The four-quadrant arctangent function atan2(I,Q) can be evaluated over the four quadrants by substituting Q/I, or I/Q appropriately, for x in the above arctangent approximations. All the approximations given above can be extended in a straightforward manner to all the four quadrants. Due to space limitations, we only consider the extensions for approximations given by (27–10) and compare them to the four-quadrant arctangent expressions given for (27–13).

In order to obtain the four-quadrant arctangent approximations, the range over which the approximation operates is extended. The complex plane is divided into eight octants where octant I, for example, covers the angle range of 0 to $\pi/4$ radians. The four-quadrant calculations are thus reduced to the first-octant calculations and using the rotational symmetries of arctangent function, the approximations for the other octants can be easily obtained. Tables 27.2 and 27.3 provide the four quadrant approximations based on (27–10) and (27–13). It should be noted that Tables 27.2 and 27.3 use the following definition for determining the sign value of an argument:

$$\text{sign}(z) = \begin{cases} 1, & z \geq 0 \\ -1, & z < 0. \end{cases} \tag{27–17}$$

Many desirable features of the four-quadrant rational approximation based on (27–14) are given in [7]. The multiplication by 0.28125 in the denominator of (27–14) can be implemented as a sum of two arithmetic right shifts, $x^2/4$ and $x^2/32$.

Table 27–2   Second-Order Approximation (27–10) versus Octant Locations

Octant	Approximation
I, VIII	$\dfrac{Q}{I}[1.0584 - \mathrm{sign}(Q) \cdot 0.273(Q/I)]$
II, III	$\dfrac{\pi}{2} - \dfrac{I}{Q}[1.0584 - \mathrm{sign}(I) \cdot 0.273(I/Q)]$
IV, V	$\mathrm{sign}(Q) \cdot \pi + \dfrac{Q}{I}[1.0584 + \mathrm{sign}(Q) \cdot 0.273(Q/I)]$
VI, VII	$-\dfrac{\pi}{2} - \dfrac{I}{Q}[1.0584 + \mathrm{sign}(I) \cdot 0.273(I/Q)]$

Table 27–3   Rational Approximation (27–13) versus Octant Locations

Octant	Approximation
I, VIII	$\dfrac{Q/I}{1 + 0.28086(Q^2/I^2)}$
II, III	$\dfrac{\pi}{2} - \dfrac{I/Q}{1 + 0.28086(I^2/Q^2)}$
IV, V	$\mathrm{sign}(Q) \cdot \pi + \dfrac{Q/I}{1 + 0.28086(Q^2/I^2)}$
VI, VII	$-\dfrac{\pi}{2} - \dfrac{I/Q}{1 + 0.28086(I^2/Q^2)}$

(This fact accounts for the two adds and single multiply requirements for (27–14) in Table 27.1.) The constants appearing in the four-quadrant approximations using either (27–8) or (27–13) cannot be implemented with shifts. However, the complexity of the four-quadrant arctangent approximations using any of (27–8), (27–13), and (27–14) are the same. The four-quadrant arctangent approximation using (27–10) has better accuracy and deserves consideration.

27.5 CONCLUSIONS

Simple approximations to the arctangent function and four-quadrant arctangent functions have been introduced. The second-order polynomial in (27–10) provides a favorable compromise between accuracy and computational cost. Furthermore, it is well suited for implementation in hardware. The demodulators in digital receivers and software-defined radios are an important application where these approximations may be useful.

27.6 REFERENCES

[1] R. INKOL, A. JOYAL, and S. WANG, "Simple Approximations to the Four-Quadrant Arctangent Function," *Technical Note*, DRDC Ottawa TN 2003-180, November 2003.

[2] D. DAS, K. MUKHOPADHYAYA, and B. SINHA, "Implementation of Four Common Functions on an LNS Co-processor," *IEEE Transactions on Computers*, vol. *44*, no. 1, January 1995, pp. 155–161.

[3] W. WONG and E. GOTO, "Fast Hardware-Based Algorithms for Elementary Function Computations Using Rectangular Multipliers," *IEEE Transactions on Computers*, vol. *43*, no. 3, March 1994, pp. 278–294.

[4] D. HWANG, D. FU, and A. WILLSON, "A 400-mHz Processor for the Conversion of Rectangular to Polar Coordinates in 0.25-μm CMOS," *IEEE Journal of Solid-State Circuits*, vol. *38*, no. 10, October 2003.

[5] I. KOREN and O. ZINATY, "Evaluating Elementary Functions in a Numerical Coprocessor Based on Rational Approximations," *IEEE Transactions on Computers*, vol. *39*, no. 8, August 1990.

[6] M. RODRIGUEZ, J. ZURAWSKI, and G. GOSLING, "Hardware Evaluation of Mathematical Functions," *IEE Proc.*, vol. *128*, pt. E, no. 4, July 1981.

[7] R. LYONS, "Another Contender in the Arctangent Race," *IEEE Signal Processing Mag.*, vol. *20*, no. 1, January 2004, pp.109–111. [See Chapter 24 of this book.]

[8] J. SHIMA, "FM Demodulation Using a Digital Radio and Digital Signal Processing," *M.Sc. Thesis*, University of Florida, 1995.

[9] T. STOVER, "An Improved Arctangent Calculation Algorithm for DIFAR Bearing Computation in the ASP," *Report No. NADC-82013-30*, Naval Airsystems Command, U.S. Department of the Navy, March 1982.

EDITOR COMMENTS

Here we supply the details of how Lagrange interpolation was used to obtain the $\Psi(x)$ error expression in (27–7). We start by recalling that the value of x where (27–5) experiences its maximum error was given as

$$x_{\max} = \pm\sqrt{4/\pi - 1} = \pm 0.523.$$

That $x_{\max}$ value was obtained by setting the derivative of (27–6) to zero and solving for x.

In preparation for using the Lagrange interpolation method to find a second-order polynomial approximation to the positive-x portion of (27–5)'s error curve in Figure 27–1, we start by identifying three *fitting points* as those shown in Figure 27–6, with (27–5)'s error shown as the solid curve.

The three $[x_0,\Psi(x_0)],[x_1,\Psi(x_1)],[x_2,\Psi(x_2)]$ fitting points are $(0,0),(x_{\max},E_{\max}),(1,0)$ as shown in Figure 27–6. Using the principles of Lagrange interpolation we can approximate Figure 27–1's $\Psi(x)$ error curve as

$$\Psi(x) = (0)\frac{x - x_{\max}}{0 - x_{\max}}\frac{x - 1}{0 - 1} + (E_{\max})\frac{x - 0}{x_{\max} - 0}\frac{x - 1}{x_{\max} - 1} + (0)\frac{x - 0}{1 - 0}\frac{x - x_{\max}}{1 - x_{\max}}$$

$$= (E_{\max})\frac{x}{x_{\max}}\frac{x - 1}{x_{\max} - 1} = \frac{x(x-1)E_{\max}}{x_{\max}(x_{\max} - 1)}.$$

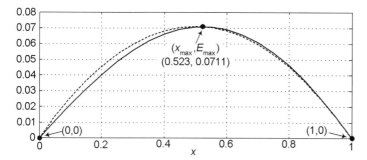

Figure 27–6   Error of the linear arctan approximation (27–5).

(For those readers unfamiliar with Lagrange interpolation, Chapter 26 provides tutorial material on that topic.) Knowing that $x_{max} = 0.523$, and from (27–6) that

$$E_{max} = \arctan(x_{max}) - \frac{\pi x_{max}}{4} = 0.0711,$$

we can write our desired $\Psi(x)$ error function's second-order polynomial approximation as

$$\Psi(x) = \frac{0.0711 \cdot x(x-1)}{0.523(0.523-1)} = -0.285 \cdot x(x-1)$$

for $0 \le x \le 1$. This $\Psi(x)$ error polynomial agrees with (27–7), which is what we set out to show. To indicate the accuracy of the Lagrange interpolation, $\Psi(x)$ is shown as the dashed curve in Figure 27–6 where it indeed passes through the three fitting points.

As a minor clarification, upon first glance it may not be obvious how the authors arrived at the number of multiplies (Mults) in Table 27.1. For example (27–12), over the range $0 \le x \le 1$, is repeated here as

$$\arctan(x) \approx \frac{\pi}{4}x - x(x-1)(0.2447 + 0.0663x).$$

Multiplying through by the factors in $\arctan(x)$ yields

$$\arctan(x) \approx -0.0663x^3 - 0.1784x^2 + 1.03x$$

which appears to require five multiplications. However, using Horner's rule reduces the number of multiplications from five to three. The above arctangent can be computed as

$$\arctan(x) \approx [(-0.0663x - 0.1784)x + 1.03]x.$$

One important aspect of the arctangent approximations described in this chapter is the source of the independent variable x. The computational requirements stated

Table 27–4 Max Error and Computational Workload When Only I and Q are Available

Equation	Error (rad.)	Adds	Multiplies	Divides
(27–5)	0.07	0	1	1
(27–8)	0.0053	1	2	1
(27–10)	0.0038	1	2	1
(27–11)	0.005	1	3	1
(27–12)	0.0015	2	3	1
(27–13)	0.0047	1	4	1
(27–14)	0.0049	2	3	1

in Table 27.1 assume that x has been previously computed and is available for use in the various arctan approximations. If however we are working with complex-valued samples, $x = Q/I$, and only I and Q are available, then a divide operation is required to compute x. In this situation the computational workloads of the various arctan approximations are quite similar as shown in Table 27.4. (We say "quite similar" because a divide operation typically requires more computational horse-power, processor cycles, than a multiply operation.) In this case (27–14) can be written as

$$\arctan(Q/I) \approx \frac{IQ}{I^2 + 0.28125Q^2}, \quad -1 \le Q/I \le 1 \qquad (27\text{–}18)$$

requiring three multiplies.

So the bottom line here is that the arctangent computational workloads are, of course, dependent on whether the values $x = Q/I$, I and Q only, or I^2 and Q^2 are available at the start of an arctan approximation computation. Again, the (0.28125) (Q^2) product in the denominator of (27–18) can be implemented as sum of two arithmetic right-shifts. That is,

$$0.28125Q^2 = Q^2/4 + Q^2/32.$$

Chapter 28

A Differentiator with
a Difference

Richard Lyons
Besser Associates

$\mathbf{I}$n this chapter we discuss a computationally efficient network that approximates the derivative of a low-frequency discrete-time sequence.

28.1 DISCRETE-TIME DIFFERENTIATION

Even though the concept of differentiation is not well defined for discrete signals, we can approximate the calculus of derivatives in our domain of discrete signals. To explain our approach, consider a continuous sinewave, whose frequency is ω_o radians/second, represented by

$$x(t) = \sin(2\pi f_o t) = \sin(\omega_o t). \qquad (28\text{–}1)$$

The derivative of that sinewave is

$$\frac{\mathrm{d}x(t)}{\mathrm{d}t} = \frac{\sin(\omega_o t)}{\mathrm{d}t} = \omega_o \cos(\omega_o t). \qquad (28\text{–}2)$$

Equation (28–2) tells us that the derivative of a sinewave is a cosine wave whose amplitude is proportional to the original $x(t)$ sinewave's ω_o frequency, and that an ideal differentiator's frequency magnitude response is a straight line increasing with frequency ω. A pair of discrete-time differentiators, a *first-difference* differentiator and a *central-difference* differentiator, are often used to approximate the time-domain derivative in (28–2).

Streamlining Digital Signal Processing: A Tricks of the Trade Guidebook, Second Edition. Edited by Richard G. Lyons.
© 2012 the Institute of Electrical and Electronics Engineers. Published 2012 by John Wiley & Sons, Inc.

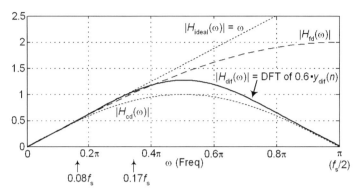

Figure 28–1   Differentiator performance responses.

The *first-difference* differentiator (what DSP purists call a *digital differencer*), the simple process of computing the difference between successive $x(n)$ signal samples, is defined in the time domain by

$$y_{\text{fd}}(n) = x(n) - x(n-1). \qquad (28\text{–}3)$$

The frequency magnitude response of that differentiator is the dashed $|H_{\text{fd}}(\omega)|$ curve in Figure 28–1. (For comparison, we also show an ideal differentiator's straight-line $|H_{\text{ideal}}(\omega)| = \omega$ magnitude response in Figure 28–1. The frequency axis in that figure covers the positive normalized frequency range $0 \le \omega \le \pi$ samples/radian, corresponding to a cyclic frequency range of 0 to $f_s/2$, where f_s is the $x(n)$ sample rate in Hz.)

Equation (28–3) is sweet in its simplicity but unfortunately its frequency response tends to amplify high-frequency noise that often contaminates real-world signals. For that reason the *central-difference* differentiator is often used in practice. The time-domain expression of the central-difference differentiator is

$$y_{\text{cd}}(n) = [x(n) - x(n-2)]/2 \qquad (28\text{–}4)$$

The central-difference differentiator's frequency magnitude response is the dotted $|H_{\text{cd}}(\omega)|$ curve in Figure 28–1. The price we pay for $|H_{\text{cd}}(\omega)|$'s desirable high-frequency (noise) attenuation is that its frequency range of linear operation is only from zero to roughly $\omega = 0.16\pi$ samples/radian ($0.08f_s$ Hz), which is, unfortunately, less than the frequency range of linear operation of the first-difference differentiator.

28.2   AN IMPROVED DIFFERENTIATOR

Here we propose a computationally efficient differentiator that maintains the central-difference differentiator's beneficial high-frequency attenuation behavior,

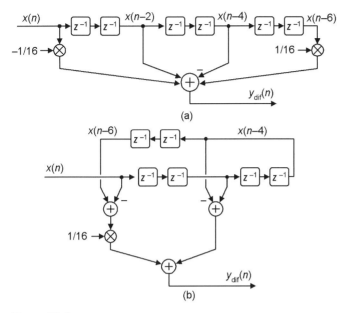

Figure 28–2   Efficient differentiator: (a) standard structure; (b) folded structure.

but extends its frequency range of linear operation. The improved differentiator is defined by

$$y_{\text{dif}}(n) = \frac{-x(n)}{16} + x(n-2) - x(n-4) + \frac{x(n-6)}{16} \qquad (28\text{–}5)$$

This novel differentiator's frequency magnitude response is the solid $|H_{\text{dif}}(\omega)|$ curve in Figure 28–1, where we see its frequency range of linear operation extends from zero to approximately $\omega = 0.34\pi$ samples/radian ($0.17f_s$ Hz). This is roughly double the frequency range of operation of the central-difference differentiator. The differentiator in (28–5) has a gain greater than that of the central-difference differentiator, so the solid $|H_{\text{dif}}(\omega)|$ curve in Figure 28–1 is the magnitude of the DFT of $0.6 \cdot y_{\text{dif}}(n)$ for easy comparison with $|H_{\text{cd}}(\omega)|$.

The implementation of the differentiator is shown in Figure 28–2(a). The *folded-FIR* structure for this differentiator is presented in Figure 28–2(b), where only a single multiply is needed to compute a $y_{\text{dif}}(n)$ output sample. The fortunate aspect of the $y_{\text{dif}}(n)$ differentiator is that its nonunity coefficients ($\pm 1/16$) are integer powers of two. This means that the multiplications in Figure 28–2 can be implemented with arithmetic right-shifts by four bits. Happily, using bit-shifting means the network is a multiplierless differentiator!

This differentiator has another useful property—due to its linear phase, and seven-sample impulse reponse length, its time delay (group delay) is exactly three sample periods ($3/f_s$). Such an integer delay makes the differentiator convenient for use when the $y_{\text{dif}}(n)$ output must be synchronized with other sequences in a system, such as with popular FM demodulation methods [1]–[3]. The only shortcoming of this very efficient differentiator is that its $x(n)$ input signals must be low frequency, less than $f_s/5$.

28.3 REFERENCES

[1] R. LYONS, *Understanding Digital Signal Processing*, 2nd ed. Prentice Hall, Upper Saddle River, NJ, 2004, pp. 549–552.
[2] A. BATEMAN, "Quadrature Frequency Discriminator," *GlobalDSP Magazine*, October 2002.
[3] T. HACK, "IQ Sampling Yields Flexible Demodulators," *RF Design Magazine*, April 1991.

Chapter 29

A Fast Binary Logarithm Algorithm

Clay S. Turner
Pace-O-Matic, Inc.

This chapter presents a computationally fast algorithm for computing logarithms. The algorithm is particularly well suited for implementation using fixed-point processors.

29.1 BACKGROUND

Long before the technological age, in 1427, the Persian mathematician and astronomer al Kashi presented an algorithm for efficient integral exponentiation via repeated squaring and multiplying [1]. Al Kashi's method now sees much use in modern cryptographic systems, where its efficiency with respect to very large numbers is paramount. Related but not as well known as al Kashi's method is an algorithm for finding binary logarithms via repeated squaring and dividing. Performing these operations in a radix two number format reduces the divisions to binary shifts, thus making the algorithm amenable to fixed-point implementations on low-complexity microprocessors and FPGAs.

29.2 THE LOGARITHM ALGORITHM

Now that we have revealed the two main types of computations for finding the binary logarithm, let's go through our algorithm's mathematical development. To start, we simply desire to find

$$y = \log_2(x). \tag{29–1}$$

Streamlining Digital Signal Processing: A Tricks of the Trade Guidebook, Second Edition. Edited by Richard G. Lyons.
© 2012 the Institute of Electrical and Electronics Engineers. Published 2012 by John Wiley & Sons, Inc.

Thus we can invert this and find

$$x = 2^y. \tag{29-2}$$

Since the iterative part of our log algorithm assumes x is in $1 \le x < 2$ we may need to "normalize" x. This normalization is performed by a simple succession of divides/multiplies by two. These divides/multiplies by two may be efficiently effected by binary shifts. The count of the divides/multiplies gives the characteristic of our logarithm result. This will be added to the mantissa (calculated below) to form the complete logarithm. The number of divides is taken as a positive number and the number of multiplies is taken as a negative number.

Now with x properly scaled, we know from (298–1) that $0 \le y < 1$. So let's expand y into a binary series, thus

$$y = y_1 \cdot 2^{-1} + y_2 \cdot 2^{-2} + y_3 \cdot 2^{-3} + y_4 \cdot 2^{-4} + \cdots \tag{29-3}$$

It will be efficacious to put this into nested parenthetical form:

$$y = 2^{-1}\left(y_1 + 2^{-1}\left(y_2 + 2^{-1}\left(y_3 + 2^{-1}\left(y_4 + \cdots\right)\right)\right)\right). \tag{29-4}$$

Putting (29–4) into (29–2) we have

$$x = 2^{2^{-1}\left(y_1 + 2^{-1}\left(y_2 + 2^{-1}\left(y_3 + 2^{-1}(y_4 + \cdots)\right)\right)\right)} \tag{29-5}$$

Now we are ready to see how to sequentially extract the bits comprising y.

Step 1: Square x as

$$x^2 = 2^{y_1} \cdot 2^{2^{-1}\left(y_2 + 2^{-1}\left(y_3 + 2^{-1}(y_4 + \cdots)\right)\right)} \tag{29-6}$$

We see the right-hand side of (29–6) is a product of two factors, where the left factor is equal to either 1 or 2 depending on the value of y_1. Thus we have the following two cases:

$$y_1 = 0: \quad x^2 = 2^{2^{-1}\left(y_2 + 2^{-1}\left(y_3 + 2^{-1}(y_4 + \cdots)\right)\right)} \tag{29-7}$$

or

$$y_1 = 1: \quad x^2 = 2 \cdot 2^{2^{-1}\left(y_2 + 2^{-1}\left(y_3 + 2^{-1}(y_4 + \cdots)\right)\right)} \tag{29-8}$$

Since both (29–7) and (29–8) contain the common factor $2^{2^{-1}\left(y_2 + 2^{-1}\left(y_3 + 2^{-1}(y_4 + \cdots)\right)\right)}$, and this factor will have a value that is greater than or equal to 1 and less than 2, we see the square of x will be greater than or equal to 2 if and only if $y_1 = 1$, otherwise $y_1 = 0$.

Step 2: Compare the result of squaring and if x^2 is greater than or equal to two, then set mantissa bit to "1" and divide x^2 by 2. Otherwise set mantissa bit to "0" and leave x^2 unchanged.

So now at the end of step 2, we have

$$x^2 = 2^{2^{-1}\left(y_2+2^{-1}\left(y_3+2^{-1}(y_4+\cdots)\right)\right)}. \tag{29–9}$$

Realizing that x^2 is just a number, we see that (29–9) has the exact same form as (29–5). Thus we can repeat steps 1 and 2 to extract the remaining bits in y.

After the prerequisite number of bits in the mantissa is obtained, then simply add the characteristic to the mantissa. If we require a logarithm having a radix other than two, then we may employ the following property of logarithms:

$$\log_a(x) = \frac{\log_2(x)}{\log_2(a)} \tag{29–10}$$

where the division by $\log_2(a)$ is implemented as a multiplication by the fixed value $1/\log_2(a)$.

We list the binary logarithm algorithm's steps as follows:

1. Initialize result to 0: $y = 0$
2. Initialize mantissa-bit decimal value to 0.5: $b = 1/2$
3. While $x < 1$, $x = 2x$, $y = y - 1$
4. While $x \geq 2$, $x = x/2$, $y = y + 1$
5. Go to step 3 and repeat until $1 \leq x < 2$
6. Square: $x = x \cdot x$
7. If $x \geq 2$, $x = x/2$, $y = y + b$
8. Scale for next bit: $b = b/2$
9. Go to step 6 and repeat until desired number of mantissa bits are found.
10. Final $\log(x)$ value: y

A "C" program, and MATLAB code, demonstrating the algorithm are made available at http://booksupport.wiley.com.

29.3 REFERENCE

[1] D. E. KNUTH, *The Art of Computer Programming: Seminumerical Algorithms*, vol. 2, 2nd ed. Addison-Wesley, 1981, pp. 441–466.

Chapter 30

Multiplier-Free Divide, Square Root, and Log Algorithms

François Auger, Bruno Feuvrie, and Feng Li
IREENA, University of Nantes

Zhen Luo
Guangdong University of Technology

$\mathbf{M}$any signal processing algorithms require the computation of the ratio of two numbers, the square root of a number, or a logarithm. These operations are difficult when using fixed-point hardware that lack dedicated multipliers, such as low-cost microcontrollers, application-specific integrated circuits (ASICs), and field-programmable gate arrays (FPGAs). This chapter presents straightforward, multiplier-free algorithms that implement both division and square roots, based on a technique known as *dichotomous coordinate descent*. (See [1] as a starting point, where several articles by the same authors are referenced.) We also make available a multiplier-free logarithm algorithm. All these algorithms are based on iterative methods, which compute the successive elements of a sequence of approximate solutions, just like the Gauss-Seidel, Jacobi, and conjugate gradient methods [2]. We begin our discussion with a scheme for computing the ratio of two real numbers.

30.1 COMPUTING THE RATIO OF TWO REAL NUMBERS

To compute the ratio of two real numbers, $x = b/a$, the dichotomous coordinate descent (DCD) technique performs the minimization of a criterion function $J(x) = ax^2 - 2bx$. As usual, $J(x)$ is minimum when its derivative $D(x) = dJ(x)/dx = 2ax - 2b$ equals zero. Since $D(x) = 0$ when $x = b/a$, finding the minimum of $J(x)$ is really equivalent to computing b/a. To this aim, the DCD algorithm defines two sequences x_n and $d_n = D(x_n)$, starting with $x_0 = 0$ and $d_0 = -2b$, and a positive real step-size

Streamlining Digital Signal Processing: A Tricks of the Trade Guidebook, Second Edition. Edited by Richard G. Lyons.
© 2012 the Institute of Electrical and Electronics Engineers. Published 2012 by John Wiley & Sons, Inc.

parameter h. At each iteration of the algorithm, x_{n+1} is chosen between $x_n + h$ and $x_n - h$ such that $J(x_{n+1}) < J(x_n)$. This way, $J(x_n)$ will be a decreasing sequence bounded by $-b^2/a$ that will converge, and x_n will go to b/a. Using straightforward computations, one can show that

$$\text{If } d_n < -a \cdot h \text{ then } J(x_n + h) < J(x_n) \tag{30–1}$$

$$\text{If } d_n > a \cdot h \text{ then } J(x_n - h) < J(x_n). \tag{30–2}$$

Expressions (30–1) and (30–2) mean that the correct value of x_{n+1} will be obtained by comparing d_n to $-a \cdot h$ and $a \cdot h$. If $d_n < -a \cdot h$, then $x_{n+1} = x_n + h$ and $d_{n+1} = d_n + 2ah$. If $d_n > a \cdot h$, then $x_{n+1} = x_n - h$ and $d_{n+1} = d_n - 2ah$. If none of these conditions are met, h is divided by two, and the d_n conditions in (30–1) and (30–2) are checked again. All these expressions require additions and only one multiplication for the computation of $a \cdot h$. But if h is initially chosen as a power of two, this multiplication will be simply implemented using binary right- or left-shifts. To present it more clearly, Figure 30–1 shows a computational flow diagram of this real-valued DCD divide algorithm.

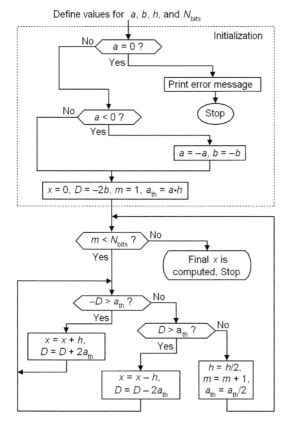

Figure 30–1   Computational flow diagram of the DCD real divide algorithm.

The initial value of h is really a key point of this algorithm. In theory, any non-zero value of the initial value of h can be chosen. But for this algorithm to be efficient, a power of two should be preferred. In this case, the product $a \cdot h$ becomes a binary left- or right-shift, and dividing h by two becomes a right-shift. This algorithm then becomes multiplication-free and well suited to fixed-point arithmetic on low-cost microcontrollers, FPGAs, or ASICs.

Moreover, for this algorithm to be as efficient as possible, the initial value of h should be chosen as the highest power of two that can be coded with the format of the output variable: suppose, for example, that a and b are coded using a fixed-point representation of 8 bits with 7 bits after the decimal point (all numbers are between $-128/128 = -1$ and $+127/128 \approx +1$). If $a = 1/128$, the smallest value of x will be -128 (obtained for $b = -128/128$) and the highest will be $+127$ (obtained for $b = +127/128$). As a consequence, the result of the division b/a must be coded with at least 8 bits before the decimal point, and the number of bits after the decimal point depends on the required accuracy. For example, one can choose to code the output variable of the algorithm, which is an approximation of the ratio b/a, using 9 bits before the decimal point and 7 bits after the decimal point. The DCD algorithm will then be as efficient as possible if the initial value of h is chosen as 128.

As an example, for $a = 1.357$, $b = 1.853$, an initial value of $h = 16$, and a result accurate to $N_{bits} = 28$ bits, the final result is reached after only 17 iterations. This means that for this example, the DCD real divide algorithm requires only 34 additions to compute our approximation of $b/a \approx x_{17} = 1.365512$. Figure 30–2(a) illustrates how fast the points $(x_n, J(x_n))$ reach the minimum of the criterion function, thanks to forward or backward "jumps" of decreasing size. Figure 30–2(b) shows how the approximation error $e_n = |x_n - b/a|$ decreases with n (the algorithm stops when e_n becomes smaller than $h/2^{N_{bits}-1} \approx 1.19 \times 10^{-7}$). A comparison of this

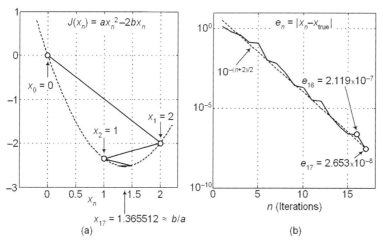

Figure 30–2   DCD real divide algorithm: (a) path of sequence $J(x_n)$ as a function of x_n; (b) error $e_n = |x_n - b/a|$ versus n (iterations).

approximation error e_n with an empirically derived exponential model, $10^{-(n-2)/2}$, shows that the common ratio nearly equals $10^{-1/2}$, which means that only two iterations are necessary to increase the accuracy by one decimal digit.

This divide algorithm clearly belongs to the family of *shift and add* algorithms [3], and is very close to the linear CORDIC (for *CO*ordinate *R*otation *DI*gital Computer) algorithm [4]. As the linear CORDIC algorithm, and as another very similar one [5], the DCD algorithm defines two sequences x_n and d_n, such as $2ax_n - d_n$ is kept constant ($2ax_{n+1} - d_{n+1} = 2ax_n - d_n = 2b$). But the DCD algorithm is deduced from a minimization problem, leading to particular definitions of the sequences x_n and d_n.

30.2 SOLVING A LINEAR SYSTEM $R_X = B$

The derivation of the DCD algorithm from a scalar minimization problem also allows extending it to a multidimensional one. This will allow one to find, by a similar algorithm, the solution of a linear system $Rx = b$, where R is a known symmetric, positive-definite $N * N$ matrix and b is an N-length vector of known real numbers. Many signal processing algorithms, such as parametric spectral analysis tools [6], Kalman state estimators [7], power amplifier linearization [8], and recursive least squares filtering [9], require us to find the solution such an equation.

The criterion function to be minimized, in this scenario, will now be defined as $J(x) = x^T Rx - 2x^T b$. This new quadratic criterion is minimum when its gradient $D(x) = dJ(x)/dx = 2Rx - 2b$ equals zero (i.e., when x is the solution of $Rx = b$). To reach this minimum, the DCD algorithm builds a sequence x_k initialized by $x_0 = 0$ and such that $J(x_k)$ decreases and $d_k = D(x_k)$ goes to zero. For this, all the components of the vector x_k are successively considered, checking if either $J(x_k + he_i)$ or $J(x_k - he_i)$ are lower than $J(x_k)$, where h is a step-size parameter and e_i is the ith element of the standard basis (i.e., the ith column of the $N * N$ identity matrix). The resulting final equations are:

$$\text{if } d_{i,k} < -hR_{ii}, \text{ then } d_{k+1} = d_k + 2hRe_i \text{ and } x_{k+1} = x_k + he_i \qquad (30\text{–}3)$$

$$\text{if } d_{i,k} > hR_{ii}, \text{ then } d_{k+1} = d_k - 2hRe_i \text{ and } x_{k+1} = x_k - he_i \qquad (30\text{–}4)$$

where $d_{i,k} = e_i^T d_k$ is the ith element of the vector d_k and $R_{ii} = e_i^T Re_i$ is the element of R at the ith row and the ith column. If none of these conditions are satisfied, the next component is considered. If none of these conditions are satisfied for the whole components, h is divided by two, and conditions (30–3) and (30–4) are checked again for all the successive components. Here again, the sequences x_k and d_k evolve such that $2Rx_k - d_k$ always remains equal to $2b$, with d_k going to zero.

30.3 COMPUTING THE RATIO OF TWO COMPLEX NUMBERS

As shown in [10], signal processing algorithms sometimes require one to compute the ratio of two complex numbers, $x = b/a$, with $b = b_r + jb_i$ and $a = a_r + ja_i$. For this, the previous algorithm could be used, since a two-dimensional linear system can be deduced from the equation $ax = b$. But this would require multiplications that we try to avoid.

To provide a multiplication-free algorithm, the DCD algorithm proposed in [1] uses another solution that we present here. In our discussion, a_r will be considered as a strictly positive number (a and b can always be simply modified to satisfy this condition without modifying their ratio). The basic principle is to define two complex sequences x_n and $b_n = b_{r,n} + jb_{i,n}$, with $x_0 = 0$ and $b_0 = b$, such that $ax_n + b_n$ is always kept constant:

$$ax_{n+1} + b_{n+1} = ax_n + b_n = ax_0 + b_0 = b. \tag{30-5}$$

If b_n goes to zero, x_n will, by necessity, go to b/a. For example, if b_n can be written as $b_{n+1} + ha$ (h being a step size parameter), such that b_{n+1} is "smaller" than b_n, then

$$ax_n + b_n = ax_n + b_{n+1} + ha = a(x_n + h) + b_{n+1} \tag{30-6}$$

allowing us to define x_{n+1} as $x_n + h$. If b_{n+1} is not "smaller" than b_n, then three other possibilities can be considered: $b_n = b_{n+1} - h$, $b_n = b_{n+1} + jha$, and $b_n = b_{n+1} - jha$. If any of those possibilities "decreases" b_n, the step size parameter h is divided by two and those four possibilities are checked again. Of course, the words *smaller* and *decreases* are questionable here, because two complex numbers cannot be compared. To be more precise, the real and imaginary parts of b_n are alternatively decreased. At even iterations, the real part $b_{r,2k}$ is decreased,

$$\text{if } b_{r,2k-1} > ha_r, \text{ then } b_{2k} = b_{2k-1} - ha \text{ and } x_{2k} = x_{2k-1} + h \tag{30-7}$$
$$\text{if } b_{r,2k-1} < -ha_r, \text{ then } b_{2k} = b_{2k-1} + ha \text{ and } x_{2k} = x_{2k-1} - h \tag{30-8}$$

whereas at odd iterations, the imaginary part $b_{i,2k}$ is decreased,

$$\text{if } b_{i,2k} > ha_r, \text{ then } b_{2k+1} = b_{2k} - jha \text{ and } x_{2k+1} = x_{2k} + jh \tag{30-9}$$
$$\text{if } b_{i,2k} < -ha_r, \text{ then } b_{2k+1} = b_{2k} + jha \text{ and } x_{2k+1} = x_{2k} - jh \tag{30-10}$$

As a conclusion, this algorithm adds or subtracts h to the real and imaginary parts of x_n as the real and imaginary parts of b_n decrease. Here also, the key point is that if the initial value of h is chosen as a power of two, this division of two complex numbers will require additions and left- or right-shifts, but no multiplications. Figure 30–3 provides a computational flow diagram describing this algorithm.

As an example of this complex divide algorithm, for $a = 3.6 + j3.2$, $b = 2.1 + j3.8$, an initial value of $h = 16$ and a result accurate to $N_{bits} = 18$ bits, the ratio b/a is approximated to $x = 0.8499756 + j\,0.3000488$ with the required accuracy after only 22 iterations (66 additions). Figure 30–4(a) illustrates how fast the points $(x_{r,n}, x_{i,n})$ spiral to their final values, performing horizontal or vertical "jumps" of decreasing size. A comparison of the approximation error $e_n = |x_n - b/a|$, given in Figure 30–4(b), with an empirically derived exponential model $0.751 \times 10^{-(n-2)/5}$, indicates that

Define values for a_r, a_i, b_r, b_i, h, and N_{bits}

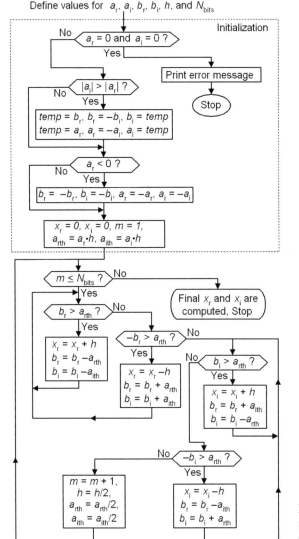

Figure 30–3 Computational flow diagram of the DCD complex divide algorithm.

only five iterations are necessary to increase the accuracy of the complex result x by one decimal digit.

30.4 COMPUTING SQUARE ROOTS

The DCD principle can have many other applications. For example, it can be applied to design a multiplication-free algorithm solving any quadratic equation written as $x^2 + bx = c$. To show this, we will consider here (without loss of generality) the

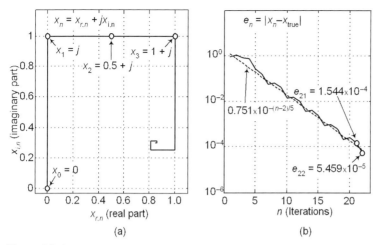

Figure 30–4 DCD complex divide algorithm: (a) path of sequence x_n; (b) error $e_n = |x_n - b/a|$ versus n (iterations).

particular case $b = 0$ and $c > 0$, leading to a square root computation algorithm. Here, again, the trick is to define two sequences x_n and c_n, with $x_0 = 0$ and $c_0 = c$, such that $x_n > 0$, $0 \leq c_{n+1} < c_n$ and $x_n^2 + c_n$ is kept constant:

$$x_{n+1}^2 + c_{n+1} = x_n^2 + c_n = x_0^2 + c_0 = c \qquad (30\text{–}11)$$

As c_n goes to zero, x_n will by necessity approach to the non-negative square root of c. As in all the previous algorithms presented here, we will check whether defining x_{n+1} as either $x_n + h$ or $x_n - h$ can lead to $0 \leq c_{n+1} < c_n$. The result is:

$$\text{if } c_n > h^2 + 2hx_n, \text{ then } c_{n+1} = c_n - 2hx_n - h^2 \text{ and } x_{n+1} = x_n + h \qquad (30\text{–}12)$$

$$\text{if } c_n > h^2 - 2hx_n \text{ and } x_n < h/2, \text{ then } c_{n+1} = c_n + 2hx_n - h^2 \text{ and } x_{n+1} = x_n - h \qquad (30\text{–}13)$$

As before, if the initial value of h is chosen as a power of two, this square root computation will be multiplication-free. Figure 30–5 shows a computational flow diagram of this DCD square root algorithm.

As an example, for $c = 2$, an initial value of $h = 2$ and a result accurate to $N_{bits} = 50$ bits, the result is reached after only 29 iterations. Figure 30–6(a) illustrates how fast the x_n result approaches its final value of 1.414214. Figure 30–6(b) shows how the approximation error $e_n = |x_n - c^{1/2}|$ decreases with n. The algorithms stops when e_n becomes lower than $h_{init}/2^{N_{bits}-1} \approx 3.55 \times 10^{-15}$. A comparison of this error plot with an exponential model $10^{-n/2}$ shows that the common ratio nearly equals $10^{-1/2}$, which means that around 2 iterations are necessary to increase the x_n result's accuracy by one decimal digit.

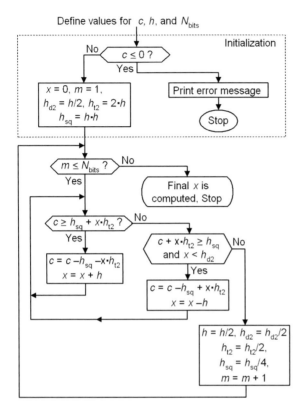

Figure 30–5 Computational flow diagram of the DCD square root algorithm.

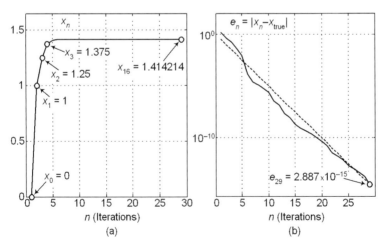

Figure 30–6   DCD square root algorithm: (a) path of sequence x_n; (b) error $e_n = |x_n - c^{1/2}|$ versus n (iterations).

30.5   COMPUTING LOGARITHMS

Recently, an algorithm to compute logarithms of positive real numbers was presented [11]. That algorithm required squaring operations, which require many elementary arithmetic operations. To this aim, the following fast multiplication-free algorithm can used.

To compute the logarithm of a positive number x, the principle of our method is to find the mathematical inverse of x, that is, the number y such that

$$x \cdot y = 1. \tag{30–14}$$

Taking the logarithm of both sides of (30–14) yields

$$\log(x) + \log(y) = 0$$

or

$$\log(x) = -\log(y). \tag{30–15}$$

The number y in (30–15) is expanded so that it is much easier to compute $\log(y)$ than it is to compute $\log(x)$. Once we compute $\log(y)$ we negate that value to obtain $\log(x)$. Based on this simple basic principle (which is of course not an algorithm) an algorithm can be designed. Our logarithm algorithm is multiplier-free and iterative, and therefore is a little more complicated than this basic principle.

30.6   THE LOG ALGORITHM

In our algorithm the value y in (30–14) takes the form:

$$y = (1+2^{-0})^{k_0}(1+2^{-1})^{k_1}(1+2^{-2})^{k_2}(1+2^{-3})^{k_3}\ldots(1+2^{-N})^{k_N} \tag{30–16}$$

where the k_i sequence of exponents are integers, defined such that

- $x \cdot 2^{k_0}$ is between 0.5 and 1,
- k_1 is the highest integer such that $x \cdot 2^{k_0}(1+2^{-1})^{k_1}$ is less than 1 (which means that $x \cdot 2^{k_0}(1+2^{-1})^{(k_1+1)}$ is greater than 1,
- k_2 is the highest integer such that $x \cdot 2^{k_0}(1+2^{-1})^{k_1}(1+2^{-2})^{k_2}$ is less than 1 (which means that $x \cdot 2^{k_0}(1+2^{-1})^{k_1}(1+2^{-2})^{(k_2+1)}$ is greater than 1,
- and so on . . .

As such, the product $x \cdot y$ becomes

$$x \cdot y = x \, [2^{k_0}(1+2^{-1})^{k_1}(1+2^{-2})^{k_2}(1+2^{-3})^{k_3}\ldots(1+2^{-N})^{k_N}] \approx 1. \tag{30–17}$$

Taking the logarithm, to the base b, of (30–17) gives us

$$
\begin{aligned}
\log_b(x \cdot y) = \log_b(x) &+ k_0 \log_b(2) + k_1 \log_b(1 + 2^{-1}) + k_2 \log_b(1 + 2^{-2}) \\
&+ k_3 \log_b(1 + 2^{-3})^{k_3} \dots + k_N \log_b(1 + 2^{-N}) \approx 0.
\end{aligned}
\tag{30–18}
$$

which implies

$$
\log_b(x) \approx -\sum_{i=0}^{i=N} k_i \log_b(1 + 2^{-i}).
\tag{30–19}
$$

Our final algorithm, then, will compute the series of products given in (30–19). However, while the absolute value of integer k_0 can be greater than one, the remaining k_1 through k_N integers in (30–19) can only be zeros or ones. For example, for $x = 5$, $k_0 = -3$ and all the following k_i are equal to zero except k_1, k_4, k_8, k_{16}, k_{32}, . . . , which are all equal to 1. This allows us to avoid performing multiplications when computing the right side of (30–19).

We completely eliminate multiplications by first precomputing the sequence $\log_b(1 + 2^{-i})$, for $0 \le i \le N$, and storing those constants in a lookup table (LUT). If, for example, $k_0 = -3$ we compute the first term in (30–19) by accumulating three of the $-\log_b(2)$ constants from the LUT. To that accumulated value, we add the remaining $\log_b(1 + 2^{-i})$ LUT constants based on whether k_i is a zero or a one. This way no multiplications are needed.

The value N in (30–19) is user-defined based on the desired precision of our $\log_b(x)$ result. The larger N, more terms will be accumulated in (30–19), and the more precise will be the computed $\log_b(x)$. It can hence be shown that

$$
\frac{2^N}{2^N + 1} < x \cdot y \le 1,
$$

which shows that for a large N, the product $x \cdot y$ is very close to 1.

30.7 COMPUTING THE K_I FACTORS

In an iterative implementation of our algorithm, shown in Figure 30–7, we don't actually compute the k_i factors explicitly. Here's what we do to compute $\log_b(x)$. Assuming that x is a positive number, we initialize an accumulator, $logx$ in Figure 30–7, to zero. Next, if x is greater than one, we perform binary right-shifts until the shifted x is less than one. For each right-shift we add $\log_b(2)$ to the $logx$ accumulator. If, on the other hand, the original x was less than 0.5 we perform left-shifts until the shifted x is greater than or equal to 0.5. For each left-shift we subtract $\log_b(2)$ from the $logx$ accumulator. Those additions (or subtractions) of $\log_b(2)$ complete the computation of the $-k_0 \log_b(2)$ term in (30–19).

Now that the shifted x is between 0.5 and 1, we iteratively subtract, where appropriate (i.e., when we can add $2^{-i} \cdot x$ to x without exceeding 1), the remaining

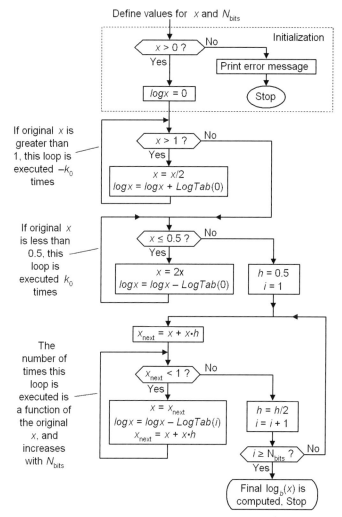

Define values for x and N_{bits}

Initialization

$x > 0$? — No → Print error message → Stop

Yes ↓

$logx = 0$

If original x is greater than 1, this loop is executed $-k_0$ times

$x > 1$? — No

Yes ↓

$x = x/2$
$logx = logx + LogTab(0)$

If original x is less than 0.5, this loop is executed k_0 times

$x \leq 0.5$? — No

Yes ↓

$x = 2x$
$logx = logx - LogTab(0)$

$h = 0.5$
$i = 1$

The number of times this loop is executed is a function of the original x, and increases with N_{bits}

$x_{next} = x + x \cdot h$

$x_{next} < 1$? — No

Yes ↓

$x = x_{next}$
$logx = logx - LogTab(i)$
$x_{next} = x + x \cdot h$

$h = h/2$
$i = i + 1$

$i \geq N_{bits}$? — No

Yes ↓

Final $log_b(x)$ is computed, Stop

Figure 30–7   Computational flow diagram of the logarithm algorithm.

LUT $log_b(1 + 2^{-i})$ terms from the *logx* accumulator until i reaches N and the accumulated *logx* reaches the desired accuracy. The value N is represented by variable N_{bits}, the desired number of bits in the $log_b(x)$ result, in Figure 30–7.

As an example of our log algorithm's performance, the natural logarithm of 5 is computed using only 20 additions, with an approximation error of 2.3×10^{-10}.

30.8   CONCLUSION

We have presented computationally simple multiplier-free algorithms for computing (1) the ratio of both real and complex numbers, (2) square roots, and (3) logarithms.

These algorithms are particularly useful when implemented on microcontrollers, FPGAs, or ASIC hardware using fixed-point arithmetic.

MATLAB programs that demonstrate the multiplier-free divide and square root algorithms are made available at http://booksupport.wiley.com.

30.9 REFERENCES

[1] J. Liu, Y. Zakharov, and B. Weaver, "Architecture and FPGA Design of Dichotomous Coordinate Descent Algorithms," *IEEE Trans. on Circuits and Systems I*, vol. *56*, no. 11, November 2009, pp. 2425–2438.

[2] G. Golub and C. Van Loan, *Matrix Computations*, 3rd ed. Baltimore, MD, The Johns Hopkins Univ. Press, 1996.

[3] J. Muller, *Elementary Functions, Algorithms and Implementation*, 2nd ed. Birkhäuser, Boston, MA, 1997.

[4] R. Andraka, "A survey of CORDIC algorithms for FPGA based computers," *Proc. 6th Int. Symp. on Field Programmable Arrays*, February 1998, pp. 191–200.

[5] A. Guyot, M. Renaudin, B. El Hassan, and V. Levering, "Self timed division and square root extraction," *Proc. 9th Int. Conf. on VLSI Design*, 1996, pp. 376–381.

[6] S. Kay and S. Marple, "Spectrum Analysis—A Modern Perspective," *Proc. of the IEEE*, vol. *69*, no. 11, 14–19 November 1981, pp. 1380.

[7] M.S. Grewal and A.P. Andrews, *Kalman Filtering: Theory and Practice Using MATLAB*, 3rd ed. Wiley-IEEE Press, September 2008.

[8] S.D. Muruganathan and A.B. Sesay, "A QRD-RLS-based predistortion scheme for high-power amplifier linearization," *IEEE Trans. Circuits Syst. II, Exp. Briefs*, vol. *53*, no. 10, pp. 1108–1112, October 2006.

[9] S. Haykin, *Adaptive Filter Theory*, 4th ed. Upper Saddle River, NJ, Prentice Hall, 2002.

[10] M. Ercegovac and J. Muller, "Design of a complex divider," *Proc. SPIE on Advanced Signal Processing Algorithms, Architectures, and Implementations XIV*, October 2004, pp. 51–59.

[11] C. Turner, "A fast binary logarithm algorithm," *IEEE Signal Processing Mag.*, vol. *27*, no 5, November 2010, pp 124–140. (Chapter 29 of this book.)

Chapter 31

A Simple Algorithm for Fitting a Gaussian Function

Hongwei Guo
Shanghai University

Gaussian functions are suitable for describing many processes in mathematics, science, and engineering, making them very useful in the fields of signal and image processing. For example, the random noise in a signal, induced by complicated physical factors, can be simply modeled with the Gaussian distribution according to the central limit theorem from the probability theory. Another typical example in image processing is the Airy disk resulting from the diffraction of a limited circular aperture as the point-spread function of an imaging system. Usually an Airy disk is approximately represented by a two-dimensional Gaussian function. As such, fitting Gaussian functions to experimental data is very important in many signal processing disciplines.

This chapter proposes a simple, and improved, algorithm for estimating the parameters of a Gaussian function fitted to observed data points.

31.1 GAUSSIAN CURVE FITTING

Recall that a Gaussian function is of the form

$$y = Ae^{-(x-\mu)^2/2\sigma^2}. \tag{31-1}$$

This function can be graphed with a symmetrical bell-shaped curve centered at the position $x = \mu$, with A being the height of the peak and σ controlling its width, and on both sides of the peak the tails (low-amplitude portions) of the curve quickly fall off and approach the x-axis. The focus of this chapter is on how we fit a Gaussian function to observed data points and determine the parameters, A, μ, and σ, exactly. The centroid method takes advantage of the symmetry of a Gaussian function, thus

Streamlining Digital Signal Processing: A Tricks of the Trade Guidebook, Second Edition. Edited by Richard G. Lyons.
© 2012 the Institute of Electrical and Electronics Engineers. Published 2012 by John Wiley & Sons, Inc.

allowing determining the Gaussian peak position very efficiently [1], [2]. Although this method is popularly used in image processing for the subpixel peak detection of a point or a line, it does not enable us to estimate the width or height of a peak. In practice, it is not easy to determine all the Gaussian parameters including A, μ, and σ because this problem is generally associated with the solution of an over-determined system of nonlinear equations, which is generated by substituting the observed data into (31–1).

The standard solution for such a nonlinear problem is to employ an iterative procedure like the Newton-Raphson algorithm, with which a sufficiently good initial guess is crucial for correctly solving for the unknowns, and it is possible that the procedure does not converge to the true solution [3]. By noting that a Gaussian function is the exponential of a quadratic function, a simpler method was proposed by Caruana et al. [4]. It calculates the natural logarithm of the data first and then fits the results to a parabola. Another method is to fit the straight line resulting from the differential of the quadratic function just mentioned [5], [6]. With these algorithms, however, noise in the observed data may induce relatively large errors in the estimated parameters, and the accuracies strongly depend on the y-amplitude range of the observed data points.

To overcome the aforementioned problems, this chapter analyzes the effects of noise on Caruana's algorithm and derives a simple and improved technique for estimating the Gaussian parameters. The technique uses a weighted least-squares method, deduced from a noise model, to fit the logarithm of Gaussian data. As such, the influences of the statistical fluctuations on the estimated Gaussian parameters are significantly reduced. Based on this principle, we also suggest an iterative procedure that is considerably less sensitive to the amplitude range of the observed data points. As a result, the initial guesses for the estimated parameters will no longer be critical. We proceed by reviewing Caruana's algorithm in the next section.

31.2   CARUANA'S ALGORITHM

Caruana's algorithm is based on the fact that a Gaussian function is the exponential of a quadratic function. Taking the natural logarithm of the Gaussian function in (31–1) yields

$$
\begin{aligned}
\ln(y) &= \ln(A) + \frac{-(x-\mu)^2}{2\sigma^2} \\
&= \ln(A) - \frac{\mu^2}{2\sigma^2} + \frac{2\mu x}{2\sigma^2} - \frac{x^2}{2\sigma^2} \\
&= a + bx + cx^2
\end{aligned}
\tag{31–2}
$$

where $a = \ln(A) - \mu^2/(2\sigma^2)$, $b = \mu/\sigma^2$, and $c = -1/(2\sigma^2)$. By doing this, the nonlinear equation with unknowns A, μ, and σ is transformed into a linear one with unknowns being a, b, and c, thus alleviating its computational complexity. The Gaussian parameters A, μ, and σ can be calculated from a, b, and c.

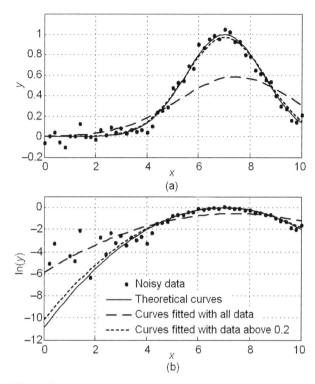

Figure 31–1   Results of Caruana's algorithm in the presence of noise.

Note that (31–2) denotes a parabola whose peak position is the same as that of the Gaussian function described in (31–1). We show an example of this in Figure 31–1, where the black solid curve in Figure 31–1(a) illustrates a Gaussian function with $A = 1$, $\mu = 7$, and $\sigma = 1.5$, and the black solid curve in Figure 31–1(b) plots its logarithm.

The fundamental principle of Caruana's algorithm is to fit this parabola in Figure 31–1(b) in the least squares sense so that its coefficients, a, b, and c, are determined, and then the Gaussian parameters, A, μ, and σ, are calculated as we shall show. In order to perform this task, an error function based on (31–2) is defined, namely

$$\delta = \ln(y) - (a + bx + cx^2). \tag{31–3}$$

Differentiating the sum of δ^2 with respect to a, b, and c and setting the resultant expressions to zero yields a linear system of equations

$$\begin{bmatrix} N & \Sigma x & \Sigma x^2 \\ \Sigma x & \Sigma x^2 & \Sigma x^3 \\ \Sigma x^2 & \Sigma x^3 & \Sigma x^4 \end{bmatrix} \begin{bmatrix} a \\ b \\ c \end{bmatrix} = \begin{bmatrix} \Sigma \ln(y) \\ \Sigma x \ln(y) \\ \Sigma x^2 \ln(y) \end{bmatrix}, \tag{31–4}$$

where N is the number of observed data points and Σ denotes $\sum_{n=1}^{N}$ for shortening the expression. After solving (31–4) for a, b, and c, the desired parameters of the Gaussian function are calculated using

$$\mu = \frac{-b}{2c}, \tag{31–5}$$

$$\sigma = \sqrt{\frac{-1}{2c}}, \tag{31–6}$$

and

$$A = e^{a - b^2/4c}. \tag{31–7}$$

31.3 EFFECTS OF NOISE

Caruana's algorithm is computationally efficient, since it is noniterative. In the presence of noise, however, its accuracy decreases dramatically. See Figure 31–1, for example, where the dots in Figure 31–1(a) denote the data obtained by sampling the black solid curve. The observed data is contaminated by zero-mean random noise having a standard deviation (SD) of 0.05. Excluding the negative-valued data points, their logarithms are plotted in Figure 31–1(b) also with dots. We see from them that the fluctuations of the data from their theoretical values, induced by the noise, may be magnified by the logarithmic operation, especially for the points far away from μ having small Gaussian values falling down. Using Caruana's algorithm, we fit the logarithmic data to a quadratic function. The resulting parabola is plotted in Figure 31–1(b) with the long-dashed curve, which noticeably deviates from the theoretical curve. The estimates of the Gaussian parameters are $A = 0.5946$, $\mu = 7.6933$, and $\sigma = 2.3768$, and the reconstructed Gaussian curve is shown in Figure 31–1(a) also with the long-dashed curve. From these results, the errors induced by noise are apparent. This phenomenon can be theoretically explained by considering an additive noise model.

If there is an additive random noise η, the data we observed is not the ideal value y but

$$\hat{y} = y + \eta. \tag{31–8}$$

Accordingly, the error function becomes

$$\begin{aligned}
\delta &= \ln(\hat{y}) - (a + bx + cx^2) \\
&= \ln(y + \eta) - (a + bx + cx^2).
\end{aligned} \tag{31–9}$$

Expanding it into a Taylor series and reasonably omitting the high-order terms, we have

$$\delta \approx \ln(y) - (a + bx + cx^2) + \frac{\eta}{y} \qquad (31\text{--}10)$$

so the expectation of δ^2 is

$$E\{\delta^2\} = [\ln(y) - (a + bx + cx^2)]^2 + \frac{\sigma_\eta^2}{y^2} \qquad (31\text{--}11)$$

where σ_η is the standard deviation of the noise.

Noting the second term of (31–11), if y is very small, the noise at this point will introduce very large errors in the estimates. This fact means that, when we use Caruana's algorithm, the observed data points used in our computations should be limited to those within a narrow range near the peak position of Gaussian curve (for example, within the x-interval of $\mu - 2\sigma \le x \le \mu + 2\sigma$), where the Gaussian function has relatively large values. In practice, because the parameters μ and σ are unknown, we usually set a threshold to exclude the data points having very small amplitude values.

If we use only the data points whose y-amplitude values are greater than 0.2, the fitting results are those illustrated with short-dashed curves in Figure 31–1. (The threshold value of 0.2 is determined empirically and must be several times greater than the noise SD value of 0.05.) In this scenario the estimated Gaussian parameters become $A = 0.9639$, $\mu = 6.9806$, and $\sigma = 1.5758$, which are close to the theoretical values.

Even so, the estimation using (31–11) remains more dependent on the observed points with small values than on those with large ones, and usually a manual intervention has to be performed for thresholding the data in advance. We shall solve this problem by employing our proposed weighted least squares algorithm in the next section.

31.4   WEIGHTED LEAST SQUARES ESTIMATION

The description of our proposed weighted least squares Gaussian curve fitting algorithm, which overcomes the noise sensitivity of Caruana's algorithm, begins by redefining the error function, using (31–3), as

$$\begin{aligned} \varepsilon &= y\delta \\ &= y[\ln(y) - (a + bx + cx^2)] \end{aligned} \qquad (31\text{--}12)$$

and in the presence of noise

$$\begin{aligned} \varepsilon &= y[\ln(y + \eta) - (a + bx + cx^2)] \\ &\approx y[\ln(y) - (a + bx + cx^2)] + \eta \end{aligned} \qquad (31\text{--}13)$$

so the expectation of ε^2 is

$$E\{\varepsilon^2\} = [y\ln(y) - y(a + bx + cx^2)]^2 + \sigma_\eta^2. \tag{31-14}$$

In (31–14), the influence of y on the second term is removed. Minimizing the sum of ε^2 implies an optimal weighted least squares estimation with the weights equaling y. Differentiating the sum of ε^2 with respect to a, b, and c and setting the resultant expressions to zero yields a linear system of equations of the form

$$\begin{bmatrix} \Sigma\hat{y}^2 & \Sigma x\hat{y}^2 & \Sigma x^2\hat{y}^2 \\ \Sigma x\hat{y}^2 & \Sigma x^2\hat{y}^2 & \Sigma x^3\hat{y}^2 \\ \Sigma x^2\hat{y}^2 & \Sigma x^3\hat{y}^2 & \Sigma x^4\hat{y}^2 \end{bmatrix}\begin{bmatrix} a \\ b \\ c \end{bmatrix} = \begin{bmatrix} \Sigma\hat{y}^2\ln(\hat{y}) \\ \Sigma x\hat{y}^2\ln(\hat{y}) \\ \Sigma x^2\hat{y}^2\ln(\hat{y}) \end{bmatrix}. \tag{31-15}$$

In this equation system, because the true values of y are unknown, we have to use $\hat{y}$ instead of y for the weights. Solving (31–15) for a, b, and c, the parameters of the Gaussian function are further calculated via (31–5) through (31–7).

Figure 31–2 illustrates the performance of our weighted least squares technique. The solid curve and dots in Figure 31–2(a) denote the theoretical Gaussian function

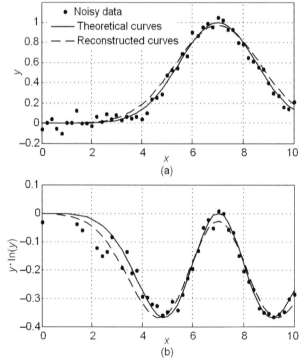

Figure 31–2 Results of the weighted least squares estimation in the presence of noise.

and its sampling data, respectively, which are the same as those in Figure 31–1(a). The curve reconstructed using our technique is plotted in Figure 31–2(a) with the dashed curve. The estimated Gaussian parameters are $A = 0.9689$, $\mu = 7.0184$, and $\sigma = 1.6251$, demonstrating that our technique, which does not exclude small-valued data, is more accurate than Caruana's method.

Of course if we limit our computations by using only the data points having amplitudes greater than 0.2, more accurate results may be obtained, say $A = 0.9807$, $\mu = 7.0114$, and $\sigma = 1.5683$.

The above results are obtained from one simulation, and a more convincing descriptor is the root-mean-square (RMS) error. This descriptor, as a statistic, is more suitable for describing the behavior of an algorithm in the presence of random noise. As such, we performed 5000 simulations repeatedly with varying noise (SD = 0.05) and the threshold held at 0.2. With Caruana's algorithm, the RMS errors for the parameters A, μ, and σ are 0.0340, 0.0431, and 0.0779, respectively; whereas when our proposed technique is used the RMS errors for the three parameters are reduced to 0.0179, 0.0315, and 0.0554, respectively. These results demonstrate the accuracy of our proposed technique over Caruana's algorithm when fitting a Gaussian peak.

The reason for the improvement of the proposed technique is graphically explained in Figure 31–2(b), where the vertical axis denotes the logarithms of the Gaussian data multiplied by the weights y. From it, we see that the random fluctuations of the data, induced by the noise, are much more uniform across the full data x-range. Therefore, data having small values, as indicated by (31–14), will not have an excessively detrimental impact on our estimation results.

31.5 ITERATIVE PROCEDURE

Our technique, compared with Caruana's algorithm, is less sensitive to the noise and more accurate in fitting a Gaussian peak. However, if a long tail of the Gaussian function curve is included in the observed data range, the large noise contamination in those data points far away from the peak position still inversely affect our curve-fitting accuracies, even lead to failure in the estimation, due to the following reasons. First, (31–10) cannot fully describe the behaviors of the noise, because the high-order terms of the Taylor series have been omitted in this equation. Second, the true values of y are unknown, so we have to use the noisy values $\hat{y}$ instead of y for the weights in (31–15). For the observed data points whose values of y are very small, the signal-to-noise ratios may decrease. In other words, the relative difference between $\hat{y}$ and y may be very large, thus inducing considerable error.

We solve this large noise contamination problem by iterating the estimating technique described by (31–15) with the weight values being updated in each iteration. The procedure is summarized by

$$
\begin{bmatrix}
\Sigma y^2_{(k-1)} & \Sigma x y^2_{(k-1)} & \Sigma x^2 y^2_{(k-1)} \\
\Sigma x y^2_{(k-1)} & \Sigma x^2 y^2_{(k-1)} & \Sigma x^3 y^2_{(k-1)} \\
\Sigma x^2 y^2_{(k-1)} & \Sigma x^3 y^2_{(k-1)} & \Sigma x^4 y^2_{(k-1)}
\end{bmatrix}
\begin{bmatrix}
a_{(k)} \\
b_{(k)} \\
c_{(k)}
\end{bmatrix}
=
\begin{bmatrix}
\Sigma y^2_{(k-1)} \ln(y) \\
\Sigma x y^2_{(k-1)} \ln(y) \\
\Sigma x^2 y^2_{(k-1)} \ln(y)
\end{bmatrix}
\qquad (31\text{–}16)
$$

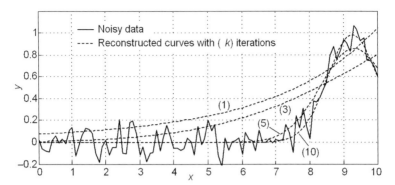

Figure 31–3   Results of the proposed iterative algorithm.

where

$$y_{(k)} = \begin{cases} \hat{y} & \text{for} \quad k = 0 \\ e^{a_{(k)}+b_{(k)}x+c_{(k)}x^2} & \text{for} \quad k > 0 \end{cases} \qquad (31\text{--}17)$$

with the parenthesized subscripts being the iteration indices. Compared with a standard iterative algorithm (e.g., Newton-Raphson) for solving a nonlinear system, our proposed iterative algorithm is computationally much simpler, and the initial guesses for the unknown a, b, and c are not required.

In order to verify the performance of this iterative procedure, we define a Gaussian function with the parameters $A = 1$, $\mu = 9.2$, and $\sigma = 0.75$. The x-range of observed data points is from 0 to 10, and the noisy data is illustrated with the black solid curve in Figure 31–3, where the SD of the additive noise is 0.1.

This noisy Gaussian function has a narrow peak located near the right edge of the data's x-range. On the left side of the peak there exists a long tail where Gaussian function has very small values and noise is relatively large (the signal-to-noise ratio here is very small). The weighted least squares technique in the previous section is not effective in this situation, but we can succeed by using the iterative procedure just introduced. Although the curves in Figure 31–3 show that the first several iterations cannot produce a satisfactory result, after ten iterations the reconstructed curve fits the theoretical noise-free Gaussian data quite well.

Table 31–1 lists the estimated Gaussian parameters versus the number of iterations. There we see the convergence of our iterative procedure. In the first two iterations, the large noise in the data makes the estimated σ value to be imaginary, but the final iteration results accurately approximate the theoretical values.

31.6 CONCLUSIONS

We proposed an improved technique for estimating the parameters of a Gaussian function from its observed data points. With it, the logarithms of Gaussian data are

Table 31–1 Estimated Gaussian Parameters ($A = 1$, $\mu = 9.2$, and $\sigma = 0.75$)

No. of iterations	A	μ	σ
1	0.0648	−6.8010	j7.0601
2	0.0270	0.4573	j3.4835
3	0.8175	11.0473	2.7769
4	0.8237	9.9600	1.5710
5	0.8890	9.1644	0.9004
6	0.9778	9.1482	0.7564
7	0.9966	9.1468	0.7272
8	0.9990	9.1473	0.7232
9	0.9994	9.1474	0.7226
10	0.9994	9.1474	0.7225

fitted by using a weighted least squares method derived from a noise model, so that the influences of random fluctuations on the estimation are effectively eliminated. Compared to Caruana's algorithm, our technique is much less sensitive to random noise. Based on our weighted least squares method we also suggested an iterative procedure suitable for reconstructing a Gaussian curve when a long tail of Gaussian curve is included in the observed data points. Because the iterative procedure starts directly from the original data, the initial guesses, which are generally crucial for guaranteeing the convergence of an iterative procedure, are not required, and the implementation of setting a threshold for excluding the small Gaussian data is also unnecessary. Although the techniques proposed in this material focused on fitting a one-dimensional Gaussian function, their principles are easy to extend to multi-dimensional Gaussian fitting.

31.7 REFERENCES

[1] Y. FENG, J. GOREE, and B. LIU, "Accurate Particle Position Measurement from Images," *Rev. Scient. Instrum.*, vol. 78, no. 5, May 2007, pp. 53–59.

[2] R. B. FISHER and D. K. NAIDU, "A Comparison of Algorithms for Subpixel Peak Detection," *Image technology: Advances in Image Processing, Multimedia and Machine Vision*, J. Sanz (Ed.), Springer-Verlag, Berlin and Heidelberg, 1996, pp. 385–404.

[3] W. PRESS, S. TEUKOLSKY, W. VETTERLING, and B. FLANNERY, *Numerical Recipes: The Art of Scientific Computing*, 3rd ed., Cambridge University Press, New York, 2007, pp. 733–836.

[4] R. CARUANA, R. SEARLE, T. HELLER, and S. SHUPACK, "Fast Algorithm for the Resolution of Spectra," *Anal. Chem.*, vol. 58, no. 6, May 1986, pp. 1162–1167.

[5] W. ZIMMERMANN, "Evaluation of Photopeaks in Scintillation Gamma-Ray Spectroscopy," *Rev. Scient. Instrum.*, vol. 32, no. 9, September 1961, pp 1063–1065.

[6] R. ABDEL-AAL, "Comparison of Algorithmic and Machine Learning Approaches for the Automatic Fitting of Gaussian Peaks," *Neural. Comput. Appl.*, vol. 11, no. 1, June 2002, pp. 17–29.

Chapter 32

Fixed-Point Square Roots Using *L*-Bit Truncation

Abhishek Seth   **Woon-Seng Gan**

Synopsys   **Nanyang Technological University**

In this chapter, we describe several techniques to reduce computational workload of the Newton-Raphson (NR)-based fixed-point square rooting method. Using the described techniques, the computational workload of NR methods can be reduced at the expense of memory. These new techniques outperform existing fixed-point square rooting methods both in terms of results accuracy and computational efficiency.

32.1 COMPUTING SQUARE ROOTS

Square root (SQRT) operations are used in a number of applications, like spectrum analysis, audio signal processing, digital communication, and 3-D graphics. Fixed-point digital signal processors (DSPs) are widely used in the aforementioned areas and many others. This makes fixed-point SQRT an important arithmetic operator. There are many methods described in the literature to find the SQRT of a number [1]. Based on their structures, some of them are more suitable for hardware implementation, while others are more suited for software implementation on DSPs with a hardware multiplier. In practice, only a limited amount of computational resources can be assigned to a square rooting routine. Choosing a fixed-point square rooting method requires a trade-off between its computational load and bit accuracy.

32.2 DIRECT NEWTON-RAPHSON (DNR) METHOD

The DNR method for computing square roots [2] is an iterative technique to find SQRT of a number x using:

Streamlining Digital Signal Processing: A Tricks of the Trade Guidebook, Second Edition. Edited by Richard G. Lyons.
© 2012 the Institute of Electrical and Electronics Engineers. Published 2012 by John Wiley & Sons, Inc.

$$y_{k+1} = y_k + \frac{1}{2\sqrt{x}}(x - y_k^2), \quad k = 0,1,2\ldots, \tag{32-1}$$

where y_{k+1} is the estimated value of SQRT of x obtained after $(k + 1)$ iterations.

Here we will use the following notation:

$$SQRT(x) = \sqrt{x}, \text{ and } ISQRT(x)/2 = \frac{1}{2\sqrt{x}}. \tag{32-2}$$

DNR requires an initial approximation for SQRT(x) and ISQRT(x)/2. The initial approximation for SQRT(x) is represented by y_0 and this approximation is improved after each iteration. Initial approximations of SQRT(x) and ISQRT(x)/2 can be obtained using either polynomial expansions (PE) or lookup tables (LUTs) [1]. The block diagram of a general DNR using two iterations to find the SQRT of a 16-bit fixed-point number is shown in Figure 32–1. In our notation the subscripted number 16 in x_{16} and $y_{0,16}$, for example, in Figure 32–1 means that those samples are 16 bits in width.

As shown in Figure 32–1, a single iteration of DNR requires two multiplications and two additions. For the SQRT(x) and ISQRT(x)/2 initialization in (32–1), either PE or LUT can be used. PE increases the computational load, whereas LUT increases memory consumption. A variation of DNR, known as the nonlinear infinite impulse response filter (NLIIRF) method, is described in [3]. The NLIIRF method, which uses PE for SQRT(x) initialization and an LUT for ISQRT(x)/2 initialization, is described next.

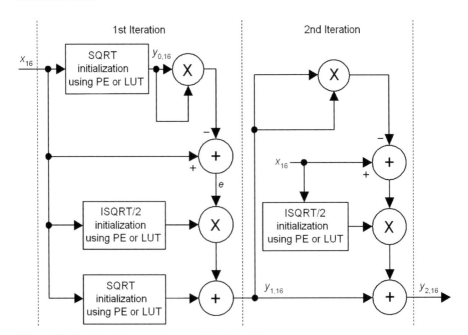

Figure 32–1   Block diagram of a general DNR algorithm.

32.3 NLIIRF METHOD

This NLIIRF method is an iterative technique to find the square root of a number [3]. The input range of the NLIIRF method, described in [3], is stated as $0.25 \leq x < 1$. Any value of x lying outside this range needs to be normalized to this range. The iterative computation used to determine the SQRT of x is given by:

$$y_{k+1} = y_k + \beta(x - y_k^2), \quad k = 0,1,2,\dots \tag{32-3}$$

The nonlinear difference expression in (32–3) can be implemented by an IIR filter and hence the name NLIIRF method. In (32–3), y_k is the approximation for SQRT(x) at the kth iteration. However, during the first iteration of (32–3), the starting approximation for SQRT(x) (stated as y_0) needs to be found using a linear PE. In addition, β can be considered as an approximation of ISQRT(x)/2 and determined by different methods. In [3], β is stored in a LUT. In [4], the authors suggested that a linear PE of the form of

$$\beta = -0.61951x + 1.0688, \tag{32-4}$$

or a quadratic PE of the form of

$$\beta = 0.763x^2 - 1.5688x + 1.314 \tag{32-5}$$

can be used. The NLIIRF method can also be depicted by Figure 32–1.

While the NLIIRF square root method is viable, we next propose a specialized DNR method that reduces the computational workload below that of the NLIIRF method.

32.4 PROPOSED DNR$_T(N)$ METHOD

For a 16-bit fixed-point processor, the NLIIRF method can produce highly accurate results with two iterations (using five multiplications and seven additions). We therefore aim to formulate a method, which can generate equal, or more accurate, results on a 16-bit fixed-point processor with fewer than five multiplications and seven additions. We call our proposed square root method "DNR$_T(n)$," where the symbol "n" in DNR$_T(n)$ denotes the number of multiplications used by the algorithm.

The proposed DNR$_T(n)$ method uses LUTs to store initial approximate values of both SQRT(x) and ISQRT(x)/2 . A direct advantage of using LUTs over PE is that it saves a number of additions and multiplications at the expense of memory needed for the LUTs.

The input range for the DNR$_T(n)$ spans from $0.25 \leq x < 1$ as described in [4]. To generate the SQRT(x) and ISQRT(x)/2 LUTs, the interval $0.25 \leq x < 1$ is divided into subintervals with equal width of $2^{-(L-1)}$ ($L < 16$). This gives $0.75 \cdot 2^{(L-1)}$ different values, denoted as x_L, obtained by rounding x_{16} to L bits (1 bit for sign and $(L-1)$ bits for magnitude). The SQRT(x_L) and ISQRT(x_L)/2 values are computed for each

x_L, rounded to 16 bits, and stored in the SQRT(x) and ISQRT(x)/2 LUTs, respectively. So the SQRT(x) LUT and ISQRT(x)/2 LUTs each have $0.75 \cdot 2^{(L-1)}$ entries, where each entry is 16 bits in width. The index of each LUT memory location, integer I_{DX}, is in the range $0 \le I_{DX} \le 0.75 \cdot 2^{(L-1)} - 1$ and is computed using

$$I_{DX} = (x_L - 0.25) \cdot 2^{(L-1)} \qquad (32\text{--}6)$$

requiring one subtraction operation and a multiply implemented by an arithmetic left-shift. The same I_{DX} value is used for SQRT(x) and ISQRT(x)/2 LUT. Sample x_{16} is rounded to x_L by adding 2^{-L} to x_{16} and truncating the lower $(16 - L)$ bits of the result. Therefore, including (32–6), one addition and one subtraction are required to generate index for LUTs.

In the first iteration of Figure 32–1, a SQRT(x) LUT value is squared to produce an x_L data sample. That x_L is subsequently subtracted from x_{16} to generate error term e as

$$e = x_{16} - x_L. \qquad (32\text{--}7)$$

We can compute e without using (32–7). If x_{16} is truncated to L bits, the truncation error term, denoted by e_T, is obtained by extracting the least significant $(16\text{-}L)$ bits of x_{16}. Our desired error term e is then calculated from e_T as

$$e = (e_T \leftarrow L) \rightarrow L \qquad (32\text{--}8)$$

where "$\leftarrow L$" ("$\rightarrow L$") means an arithmetic left (right) shift by L bits. Therefore, our first trick is to replace the squaring and subtraction in the first iteration of Figure 32–1 with L-bit truncation and arithmetic shifts as shown in Figure 32–2.

Next we show a two-multiply $DNR_T(2)$ scheme that approximates the $y_{2,16}$ output in Figure 32–2.

32.5   A $DNR_T(2)$ METHOD

Due to the range of data values at various nodes in Figure 32–2, we can make a few reasonable assumptions about that data and approximate the $y_{2,16}$ output using only one processing iteration. We call that single-iteration scheme the "$DNR_T(2)$ method" and depict it in Figure 32–3.

The $y_{2,16}$ output in Figure 32–2 is approximated by the $y_{1,16}$ output in Figure 32–3, requiring only two multiplications, using

$$y_{1,16} \approx y_{2,16} \approx LUT_{SQRT}(i) + LUTI_{SQRT/2}(i) \cdot (e - e^2/2). \qquad (32\text{--}9)$$

The derivation of (32–9) is provided in the appendix to this chapter.

If the $e^2/2$ term in (32–9) is neglected, we see that (32–9) represents the $DNR_T(1)$ method in Figure 32–2. Next we show an improved $DNR_T(n)$ scheme that uses enhanced precision LUTs.

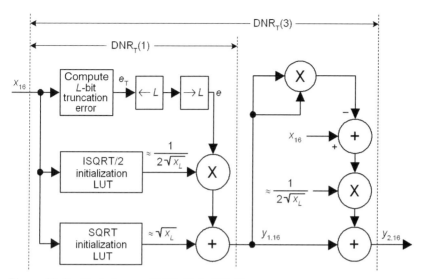

Figure 32–2   Block diagram of $DNR_T(1)$ & $DNR_T(3)$.

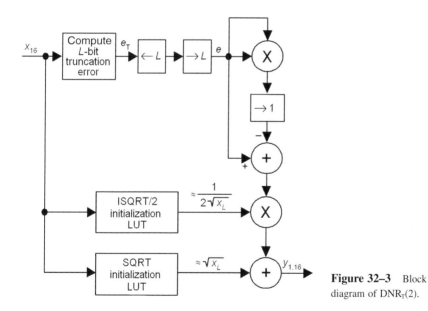

Figure 32–3   Block diagram of $DNR_T(2)$.

32.6 ENHANCED PRECISION LUTs

Because the SQRT(x) and ISQRT(x)/2 values are always positive, the sign bit of the binary representations of those values is always "0". Now, for the x_{16} input interval $0.25 \leq x_{16} < 1$, the contents of both LUTs are in the range of $0.5 \leq$ table value < 1. So the most significant bit (bit next to the sign bit) is always "1", for each entry of

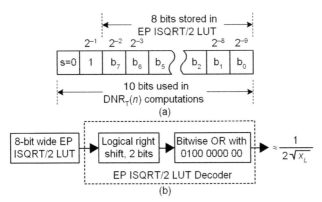

Figure 32–4   Generating EP ISQRT(*x*)/2 LUT.

Table 32–1   Normal and EP LUT Size and Performance Comparison

		Total EP LUT size (bytes)		% of output samples with bit precision > 15	
$DNR_T(n)$	Subinterval width, *L*	Normal LUT	EP LUT	Normal LUT	EP LUT
$DNR_T(1)$	2^{-7}, 8	384	288	99.74	100
$DNR_T(1)$	2^{-6}, 7	192	144	93.81	98.65
$DNR_T(2)$	2^{-7}, 8	384	288	99.91	100
$DNR_T(2)$	2^{-6}, 7	192	144	99.41	99.67
$DNR_T(3)$	2^{-5}, 6	96	72	99.99	99.98
$DNR_T(3)$	2^{-4}, 5	48	36	98.10	98.22

the LUTs. Our next trick is to make use of these facts to generate enhanced precision (EP) SQRT(*x*) and ISQRT(*x*)/2 LUTs. Figure 32–4(a) shows how an EP ISQRT(*x*)/2 LUT value having 10-bit precision can be stored in an 8-bit LUT memory location. When that ISQRT(*x*)/2 LUT value is accessed for a computation, the 8-bit value is converted to an enhanced precision 10-bit value as shown in Figure 32–4(b).

Similarly, we can generate a 16-bit EP SQRT(x) LUT. Our simulation results show that we do not need to generate a 16-bit wide EP ISQRT(x)/2 LUT because an 8-bit wide EP ISQRT(x)/2 LUT can generate sufficiently accurate results. This saves 25% of LUT memory.

Table 32–1 compares the performance of 16-bit-wide normal LUTs and EP LUTs. In Table 32–1, a listed value of *L* is the *L* used for *L*-bit rounding. Those *L* values are a consequence of the chosen SQRT(*x*) and ISQRT(*x*)/2 LUT subinterval widths. For example, if the SQRT(*x*) and ISQRT(*x*)/2 LUT subinterval width is 2^{-4} then $L = (1 + 4) = 5$. We interpret $(1 + 4)$ to be one sign bit and four magnitude bits (because processing is in Q1.15 binary arithmetic), so in this case x_{16} is rounded to five bits.

We see from Table 32–1 that using EP LUTs not only reduces memory requirements but also generates more accurate results than using normal LUTs.

32.7 PROCESSING RESULTS COMPARISON

The above NLIIRF and $DNR_T(n)$ methods are simulated for 8-bit and 16-bit fixed-point processors using MATLAB. The fixed-point SQRT output is compared with the reference of a double-precision (DP) floating-point (FP) SQRT output, y generated from MATLAB, to evaluate the accuracy error. The negative $\log_2$ of the magnitude of error values is used to calculate bit precision for each output sample, as

$$Bit\ precision\ of\ y_{k,N} = -\log_2\left(|y - y_{k,N}|\right) \qquad (32\text{–}10)$$

For 8-bit and 16-bit fixed-point processors, error values are generated for all values of x that can be represented by 8 and 16 bits, respectively, lying in the interval [.25, 1). For N-bit (8 or 16) fixed-point processors, number of additions, multiplications, bytes required, and percentage of output with bit precision $>N - 1$ bits are shown in Table 32–2. The L values for the $DNR_T(n)$ configurations in Table 32–2 are same as those in Table 32–1.

For both 8-bit and 16-bit fixed-point precision, the NLIIRF (LUT) method performs better than NLIIRF (linear PE) and NLIIRF (quadratic PE) methods. Even though it uses fewer multiplications and fewer LUT bytes, the 8-bit fixed-point $DNR_T(1)$ achieves the same degree of accuracy as the NLIIRF (LUT) method. Since bit accuracy >7 bits can be achieved with $DNR_T(1)$ for all output samples, $DNR_T(2)$,

Table 32–2 Comparison of Various Fixed-Point Square Root Methods on 8-Bit and 16-Bit Fixed-Point Processors

Method	Multiplications		Additions		Total LUT size (bytes)		% of output samples with bit precision $>N - 1$	
	8 bits	16 bits	8 bits	16 bits	8 bits	16 bits	$N = 8$ bits	$N = 16$ bits
$DNR_T(1)$	1	1	3	3	12	288/144	100	100/98.65
$DNR_T(2)$	N.A.	2	N.A.	4	N.A.	288/144	N.A.	100/99.67
$DNR_T(3)$	N.A.	3	N.A.	5	N.A.	72/36	N.A.	99.98/98.22
NLIIRF (LUT)	3	5	5	7	15	30	100	99.3
NLIIRF (Quadratic PE)	6	8	5	7	7	14	100	97.3
NLIIRF (Linear PE)	4	6	6	6	6	12	87.50	40.19

N.A.—Not applicable.

and $DNR_T(3)$ are not required for 8-bit fixed-point precision. For 16-bit fixed-point numbers, $DNR_T(1)$, $DNR_T(2)$, and $DNR_T(3)$ are used.

As shown in Table 32–2, for 16-bit fixed-point precision, $DNR_T(1)$ with a total LUT ($SQRT(x)$ LUT plus $ISQRT(x)/2$ LUT) size of 144 bytes ($L = 7$) produces 98.65% of output samples with bit accuracy >15 bits. This percentage value increases to 100% when the total LUT size is increased to 288 bytes ($L = 8$). Similar observations can be made for $DNR_T(2)$ and $DNR_T(3)$. For 16-bit fixed-point precision, $DNR_T(n)$ for $n = 1$, 2, and 3, either outperforms or generate comparable results as compared with the NLIIRF(LUT). For 16-bit fixed-point processor, $DNR_T(n)$ requires fewer number of multiplications and additions as compared with the NLIIRF(LUT). But on the downside, it needs longer LUTs. However, $DNR_T(n)$ provides the flexibility of decreasing the computational load at the expense of memory without compromising the performance. This flexibility is not available in the NLIIRF method.

32.8 CONCLUSION

This chapter presented a square rooting method suitable for 8-bit and 16-bit fixed-point implementation. $DNR_T(n)$ outperforms the NLIIRF method and its variants in terms of the number of operations (additions and multiplications) and accuracy. But $DNR_T(n)$ needs more memory to store LUTs. Because current fixed-point processors come with internal memories ranging up to megabytes, the memory requirements of $DNR_T(n)$ can be considered negligible for both 8-bit and 16-bit platforms. In this regard, $DNR_T(n)$ turns out to be a highly accurate and computationally cheap square rooting method for both 8-bit and 16-bit fixed-point processors. Moreover, the various $DNR_T(n)$ methods provide options for reducing computational load at the expense of memory. This scalability can be exploited in various situations making $DNR_T(n)$ a versatile algorithm. A website providing MATLAB m-files for NLIIRF and $DNR_T(n)$ is provided for the reader's reference [5].

32.9 REFERENCES

[1] P. MONTUSCHI and M. MEZZALAMA, "Survey of Square Rooting Algorithms," *IEE Proceedings*, vol. *137*, January 1990, pp. 31–40.
[2] C. RAMAMOORTHY, J. GOODMAN, and K. KIM, "Some Properties of Iterative Square-Rooting Methods Using High-Speed Multiplication," *IEEE Trans. Computers*, vol. *21*, August 1972, pp. 837–847.
[3] N. MIKAMI, M. KOBAYASHI, and Y. YOKOYAMA, "A New DSP-Oriented Algorithm for Calculation of Square Root Using a Nonlinear Digital Filter," *IEEE Trans. Signal Processing*, vol. *40*, July 1992, pp. 1663–1669.
[4] M. ALLIE and R. LYONS, "A Root of Less evil," *IEEE Signal Processing Magazine*, vol. *22*, March 2005, pp. 93–96.
[5] A. SETH and W. Seng GAN, "Matlab Square Root Files," 2011, [Online: http://www3.ntu.edu.sg/home/aseth/.]

APPENDIX

The derivation of the $DNR_T(2)$ processing expression in (32–9) proceeds as follows: to keep the notation simple we refer to the *i*th entry of the $ISQRT(x)/2$ LUT,

$LUT_{ISQRT/2}(i)$, as I and refer to the ith entry of the SQRT(x) LUT, $LUT_{SQRT}(i)$, as S. Referring to Figure 32–2, denoting the difference between x_{16} and x_L by e, we write

$$e = x_{16} - x_L, \tag{32–A1}$$

where the output of first iteration is given by

$$y_{1,16} = S + e \cdot I. \tag{32–A2}$$

Subsequently, the $y_{2,16}$ output of the second iteration can be written as

$$\begin{aligned} y_{2,16} &= [x_{16} - (S + e \cdot I)^2] \cdot I + S + e \cdot I \\ &= S + I \cdot (x_{16} - S^2 - e^2 \cdot I^2) + e \cdot I - 2 \cdot S \cdot I \cdot e \cdot I. \end{aligned} \tag{32–A3}$$

Because $S \cdot I \approx 1/2$, the last two terms in (32–A3) cancel each other, and the $x_{16} - S^2$ terms can be replaced by e, giving us

$$y_{2,16} \approx S + I \cdot (e - e^2 \cdot I^2), \tag{32–A4}$$

or

$$y_{2,16} \approx LUT_{SQRT}(i) + LUTI_{SQRT/2}(i) \cdot [e - e^2 \cdot (LUTI_{SQRT/2}(i))^2]. \tag{32–A5}$$

Our empirical studies have shown that $(LUT_{ISQRT/2}(i))^2$ can be replaced by the factor 1/2 without inducing significant error in the desired results. As such, this gives a single-iteration approximation of

$$y_{1,16} \approx y_{2,16} \approx LUT_{SQRT}(i) + LUTI_{SQRT/2}(i) \cdot (e - e^2/2)$$

as presented in (32–9).

Part Four

Signal Generation Techniques

Chapter 33

Recursive Discrete-Time Sinusoidal Oscillators

Clay S. Turner
Pace-O-Matic, Inc.

Every few years a DSP article emerges that presents a method for generating sinusoidal functions with a DSP. While each oscillator structure has been developed pretty much on its own, there has not been presented a simple overlying theory that unifies all of the various oscillator structures and can easily allow one to look for other potential oscillator structures. We can find some guidance for our quest by first examining classical oscillators.

The German physicist Heinrich Barkhausen, during the early 1900s, formulated the necessary requirements for oscillation. He modeled an oscillator as an amplifier with its output fed back to its input via a phase shifting network. From this model, he deduced and stated two necessary conditions for oscillation. The Barkhausen criteria, as they are now known, require the total loop gain to be equal to one and the total loop phase shift needs to be a multiple of 2π radians. So, if we are to unify discrete time oscillator designs into a single theory, we need to find the discrete-time equivalent of the Barkhausen criteria and use them to develop our theory.

But first, we will look at how some very old trigonometric formulae, viewed in a not-so-usual way, can be utilized for sinusoidal generation. Several oscillators have been designed via this approach. A sum and difference of angles formula written explicitly in recursive form is:

$$\cos(\varphi + \theta) = 2\cos(\theta)\cos(\varphi) - \cos(\varphi - \theta) \qquad (33–1)$$

We will refer to this as the *biquad* form. This is also called the *direct form* implementation. If the value θ is viewed as a step angle, then we can immediately see how this formula can be used to calculate the next sample of a sinusoid given two

Streamlining Digital Signal Processing: A Tricks of the Trade Guidebook, Second Edition. Edited by Richard G. Lyons.
© 2012 the Institute of Electrical and Electronics Engineers. Published 2012 by John Wiley & Sons, Inc.

319

known samples spaced θ apart and a step factor that is just 2cos(θ). To see how this can be used for iterative generation of a sinusoid, just view the formula as follows:

$$NextCos = 2\cos(\theta) \times CurrentCos - LastCos \tag{33–2}$$

Instead of using a single equation for a recurrence relation, a pair of trigonometric formulae may be used. An example that can be used to recursively generate sinusoids is:

$$\cos(\varphi + \theta) = \cos(\varphi)\cos(\theta) - \sin(\varphi)\sin(\theta)$$
$$\sin(\varphi + \theta) = \cos(\varphi)\sin(\theta) + \sin(\varphi)\cos(\theta) \tag{33–3}$$

We will refer to this as the *coupled* form. The coupling is evident in that each equation uses not only its past value, but also the past value produced by the other equation. Again θ is used as the step angle per iteration, and this leads to an oscillator output frequency of $\theta f_s/2\pi$ where f_s is the sample rate. Just like the biquad form, the coupled form requires the use of two past values. This turns out to be the case for all sinusoidal oscillators that are limited to the use of real numbers.

33.1 MATRIX DERIVATION OF OSCILLATOR PROPERTIES

Now we desire to find a general enough process that can be used to express both the biquad and the coupled form equations. First we will denote the two past values (these are actually called state variables) as x_1 and x_2, and their "hatted" versions as the new values. Plus we notice that the update equations are linear for both forms, so we can write them in terms of matrix multiplication.

The matrix form of the update iteration for the biquad is:

$$\begin{bmatrix} \hat{x}_1 \\ \hat{x}_2 \end{bmatrix} = \begin{bmatrix} 2\cos(\theta) & -1 \\ 1 & 0 \end{bmatrix} \begin{bmatrix} x_1 \\ x_2 \end{bmatrix} \tag{33–4}$$

Likewise the coupled form's update iteration is written as:

$$\begin{bmatrix} \hat{x}_1 \\ \hat{x}_2 \end{bmatrix} = \begin{bmatrix} \cos(\theta) & \sin(\theta) \\ -\sin(\theta) & \cos(\theta) \end{bmatrix} \begin{bmatrix} x_1 \\ x_2 \end{bmatrix} \tag{33–5}$$

The interpretation of the matrix iteration is that the column vector on the right-hand side contains the old state values, and when they are multiplied by the rotation matrix, you get a new set of state values. Then for the next iteration the *new values* from the last iteration are used as the *old values* for the next iteration. Thus each iteration is just performed by a matrix multiplication times the state variables. While the idea of matrix math may seem to unnecessarily complicate things, it actually allows us to go and find new types of oscillators.

The term *rotation* is used since the matrix multiplication can be viewed from a special vantage point as doing a rotation. This special vantage point will be explained in more detail later. A general form that fits both of these aforementioned oscillators is the use of a 2-by-2 *rotation matrix* and two state variables. So a general oscillator iteration is written as:

$$\begin{bmatrix} \hat{x}_1 \\ \hat{x}_2 \end{bmatrix} = \begin{bmatrix} a & b \\ c & d \end{bmatrix} \begin{bmatrix} x_1 \\ x_2 \end{bmatrix} \tag{33-6}$$

We will now look at a numerical example of an oscillator iteration. (It is neither the biquad nor the coupled form mentioned earlier.) We will use

$$\begin{bmatrix} 0.95 & -1 \\ 0.0975 & 0.95 \end{bmatrix} \tag{33-7}$$

for the rotation matrix, and use

$$\begin{bmatrix} 1 \\ 0 \end{bmatrix} \tag{33-8}$$

for the initial state values. A graph of both state variables' values for the first 80 iterations is shown in Figure 33–1.

Now that we've seen a numerical example and two analytical examples, naturally the question becomes, "What are the constraints on the values of the four elements in the rotation matrix that will still allow the matrix to function for an oscillator iteration?" It turns out that there are two constraints and these are the discrete-time equivalent of the Barkhausen criteria. They are:

$$ad - bc = 1$$
$$|a + d| < 2 \tag{33-9}$$

The first constraint says the determinant of the rotation matrix must be one. This is analogous to saying the loop gain is unity. The sectiond constraint (assuming the first constraint is met) says the matrix has complex eigenvalues. This means the

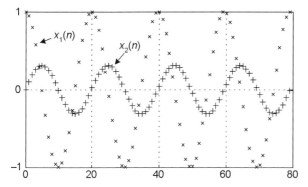

Figure 33–1   Output from example numeric oscillator.

oscillator output will eventually repeat. This is a discrete-time equivalent to Barkhausen's criterion for periodicity. While not obvious, it can be shown from these two constraints that both matrix elements, b and c, must be nonzero – thus the rotation matrix can't be triangular. What we have done here is basically come up with a set of rules that can be easily applied to any 2-by-2 matrix to see whether it can be used to make an oscillator. And soon we will come up with more rules that will allow you to identify the type of oscillator just by looking at its matrix.

Now we will make a brief sojourn into eigenvalues and eigenvectors. *Eigen* comes from German, meaning *characteristic*. The reason we need to make this side trip is this theory will allow us to perform an adroit factoring of the rotation matrix. And this will allow us to ascertain the oscillator's properties and determine the initial values for the state variables.

We are used to the idea of identity operations such as adding zero and multiplying by one. An analogous question in matrix theory is, "Is there a vector x that when multiplied by a matrix A results in a scalar multiple, λ, of the original vector?" Mathematically this is written as:

$$A x = \lambda x \tag{33–10}$$

When this is satisfied, λ is an eigenvalue and x is its corresponding eigenvector of the matrix A. For our 2-by-2 matrices that obey the Barkhausen criteria, there will be two eigenvalues–in fact they are complex conjugates of each other and each has a magnitude of one. For the rest of this chapter, the rotation matrix, A, is assumed to be a 2-by-2 matrix consisting of four real-valued elements and it obeys the aforementioned Barkhausen criteria. Explicitly matrix A is:

$$A = \begin{bmatrix} a & b \\ c & d \end{bmatrix} \tag{33–11}$$

If we let the vector x contain the initial state values, the nth output, $y(n)$, of the oscillator can be written as:

$$y(n) = A^n x \tag{33–12}$$

Now we will use a wonderful result of eigenvalue theory and factor the general matrix into a product of three matrices. This actually makes raising a matrix to a power much easier to perform.

$$A = S D S^{-1} \tag{33–13}$$

Thus A^n can be written as $(SDS^{-1})(SDS^{-1})(SDS^{-1}) \ldots (SDS^{-1})$, and after canceling out paired $S^{-1}S$ terms, we get the following simplification for our oscillator output:

$$y(n) = S D^n S^{-1} x \tag{33–14}$$

While at first blush this doesn't seem to help, let's talk about the contents of the S and D matrices. First the D matrix is a diagonal matrix whose only non-zero elements are on the main diagonal (upper left and lower right for our case). These elements are the eigenvalues. Also raising a diagonal matrix to a power is simply the raising of the diagonal elements to the same power. In terms of the original rotation matrix elements, the diagonal matrix is found thus:

$$D = \begin{bmatrix} e^{j\theta} & 0 \\ 0 & e^{-j\theta} \end{bmatrix} \qquad (33\text{--}15)$$

where

$$\theta = \cos^{-1}\left(\frac{\Delta}{2}\right) \qquad (33\text{--}16)$$

and

$$\Delta = a + d \qquad (33\text{--}17)$$

It is at this point that we can see how raising the matrix D to a power effects a rotation. In fact, θ, is the step angle of the oscillator per iteration. Now the *change of basis* matrix, S, contains the eigenvectors that correspond to the eigenvalues used in matrix D. Again in terms of the original rotation matrix elements, the matrix S is found thus:

$$S = \begin{bmatrix} 1 & 1 \\ \psi e^{j\phi} & \psi e^{-j\phi} \end{bmatrix} \qquad (33\text{--}18)$$

where

$$\psi = \sqrt{\frac{-c}{b}} \qquad (33\text{--}19)$$

and

$$\phi = \arg(\eta) \qquad (33\text{--}20)$$

where

$$\eta = \frac{(d-a) + j\sqrt{4 - \Delta^2}}{2b} \qquad (33\text{--}21)$$

It is interesting to comment on the fact that a real-valued rotation matrix is factored into a product of complex-valued matrices. However, the implemented oscillators will only use real-valued numbers. This brings to mind a saying attributed to the French mathematician Jacques Hadamard: "The shortest path between truths in the real domain passes through the complex domain."

Now the term *change of basis* was mentioned in connection with matrix S. This is used since the matrices S and S^{-1} map between external and internal space. This should be viewed as the state variables undergoing three processes. The first is a transformation to internal space representation. The second process is a pair of rotations performed on them, and the third is a transformation back to external space. In internal space, the two rotations are in opposite directions—this allows us to combine complex numbers so as to result in only real numbers. Now let the variable z be an internal representation as follows:

$$z = S^{-1} x \qquad (33\text{–}22)$$

Next let x have an initial value so that

$$z = \begin{bmatrix} 0.5 \\ 0.5 \end{bmatrix} \qquad (33\text{–}23)$$

Then our oscillator output is simply written as

$$y(n) = S D^n z \qquad (33\text{–}24)$$

This after simplification yields

$$y(n) = \begin{bmatrix} \cos(n\theta) \\ \psi \cos(n\theta + \phi) \end{bmatrix} \qquad (33\text{–}25)$$

Likewise, given our initial choice for z, then the initial state values are

$$x = S z \qquad (33\text{–}26)$$

This after simplification yields

$$x = \begin{bmatrix} 1 \\ \psi \cos(\phi) \end{bmatrix} \qquad (33\text{–}27)$$

which is of course just $y(0)$.

33.2   INTERPRETING THE ROTATION MATRIX

The attractive aspect of this analysis method is that we can now evaluate an oscillator's behavior by merely looking at its rotation matrix! We see that our analysis includes two angles. The angle θ is the step angle per iteration and the angle ϕ is the phase shift between the two state variables. So, we can see that if we desire an oscillator to have quadrature outputs (the two state variables), then ϕ must be $\pm 90°$, which in terms of the matrix elements means that the two values on the main diagonal must be the same! So if $a = d$, we have a quadrature oscillator. Likewise the scaling factor ψ tells us the amplitude of the second state variable relative to the first one. If we desire the two outputs to have equal amplitudes, then the off-diagonal elements must be negatives of each other! That is, $b = -c$. So looking back at the

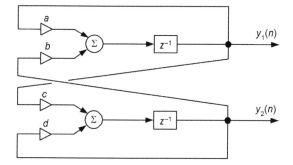

Figure 33–2   Generic oscillator iteration network diagram.

matrix used in our numerical example, we can quickly ascertain the outputs are in quadrature but will have unequal amplitudes as its graph confirms.

Now when it comes to programming the oscillator in a DSP chip, we would not normally elect to implement the iteration as a matrix multiply–something simpler may be possible! So we will derive the network diagram that represents the general rotation matrix multiplication. In the world of signal processing, it is customary to try to only use addition, multiplication, and delays, and that is all we will use. The centerpiece of the network diagram in Figure 33–2 is the pair of delay elements that are interpreted to hold the state variables. For each iteration, the outputs of the delay elements are used as the past values and the inputs to the delay elements are the new values. For example, the input to the upper delay element in Figure 33–2 is calculated thus: $\hat{y}_1 = ay_1 + by_2$.

This generic form requires four multiplies and two additions for each iteration. If the rotation matrix has some values in common or some are simply zero or one, then the iteration computations may become simpler.

33.3 CATALOG OF OSCILLATORS

Now that we have gone through the theory of discrete-time recursive oscillators, we will list some common oscillators along with their attributes. I have chosen five oscillators that span the gamut based on the number of multiplies per iteration and their type of outputs. All oscillators represented by 2-by-2 matrices will produce two sine waves simultaneously. These two sine waves will always have the same frequency, be out of phase with each other, and may have differing amplitudes. If the outputs are 90° out of phase, then we have a quadrature oscillator. Likewise, if the two sine wave outputs have the same amplitude, then we have an equi-amplitude oscillator. Four of the five oscillators in this catalog are in use in industry as they are mentioned in various application notes and trade journals. The quadrature oscillator with staggered update is one I put together using this matrix theory. Often you can make a new oscillator by permuting the matrix elements of a known oscillator. Just apply the Barkhausen criteria to the resulting matrix to see if they still hold. Also you can multiply an oscillator matrix by another matrix and often create another oscillator. So as you can gather, this catalog hardly exhausts the possible list of oscillators. In order to show the structure of an oscillator's matrix in its simplest form, the concept of tuning parameter is introduced. This parameter, k, is related to the step angle, θ. The exact relation, which depends on the particular oscillator used, will be provided.

33.4   BIQUAD

The biquad oscillator was one of the first discrete oscillators to see use in signal processing applications. I recall an application patent issued in the 1980s that used this oscillator for generating call progress tones used in telephony. I found this interesting since François Viète discovered the trigonometric recurrence relation (33–1) long before. His result was published in the year 1571! The biquad oscillator has equi-amplitude outputs, which turn out to have a relative phase shift of θ.

$$k = 2\cos(\theta) \tag{33–28}$$

$$\begin{bmatrix} k & -1 \\ 1 & 0 \end{bmatrix} \tag{33–29}$$

When the rotation matrix (33–29) elements are substituted in Figure 33–2, the generic oscillator network becomes the biquad oscillator shown in Figure 33–3.

33.5   DIGITAL WAVEGUIDE

The *digital waveguide oscillator* is the simplest (in terms of the number of multiplies) oscillator with quadrature outputs. For k near zero, the outputs become nearly equal in amplitude. This means this oscillator can be effectively used to phase lock a signal near 1/4 the sampling rate. More on dynamic tuning (i.e., changing frequency while in operation) of oscillators later. Figure 33–4 shows the network form for the digital waveguide oscillator.

$$k = \cos(\theta) \tag{33–30}$$

$$\begin{bmatrix} k & k-1 \\ k+1 & k \end{bmatrix} \tag{33–31}$$

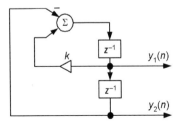

Figure 33–3   Biquad oscillator.

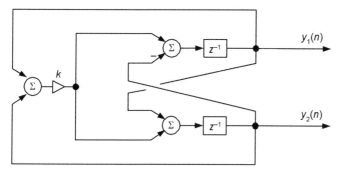

Figure 33–4 Digital waveguide oscillator.

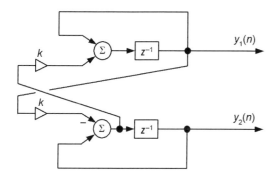

Figure 33–5 Equi-amplitude staggered update oscillator.

Note: The digital waveguide oscillator is a claimed item under U.S. patent #5701393, so consult the patent's owner before using this oscillator in a product.

33.6 EQUI-AMPLITUDE STAGGERED UPDATE

The staggered update oscillators take their name from the fact that one state variable is first updated and then that new value is used in the update of the remaining variable. This oscillator's outputs are equi-amplitude and quasi-quadrature, with the quadrature relation being reached in the limit of small k. To explicitly show the oscillator iteration as being staggered, its matrix along with a factoring of the matrix is shown. Notice how the matrix factors into a product of two triangular matrices, neither of which can function as an oscillator alone. This oscillator is shown in Figure 33–5.

$$k = 2\sin\left(\frac{\theta}{2}\right) \tag{33–32}$$

$$\begin{bmatrix} 1-k^2 & k \\ -k & 1 \end{bmatrix} = \begin{bmatrix} 1 & k \\ 0 & 1 \end{bmatrix}\begin{bmatrix} 1 & 0 \\ -k & 1 \end{bmatrix} \tag{33–33}$$

33.7   QUADRATURE STAGGERED UPDATE

This quadrature oscillator has nearly equi-amplitude outputs when k is small. Again as before, we show the factoring of its matrix, but here we notice that the right-hand factor is effectively a biquad oscillator. So the left-hand factor is used to change the configuration of the right-hand oscillator. This oscillator is shown in Figure 33–6.

$$k = \cos(\theta) \tag{33-34}$$

$$\begin{bmatrix} k & 1-k^2 \\ -1 & k \end{bmatrix} = \begin{bmatrix} 1 & -k \\ 0 & 1 \end{bmatrix} \begin{bmatrix} 0 & 1 \\ -1 & k \end{bmatrix} \tag{33-35}$$

33.8   COUPLED STANDARD QUADRATURE

The coupled standard quadrature oscillator features both quadrature and equi-amplitude outputs. However, there is a cost–this oscillator requires four multiplies per iteration. This oscillator (like the biquad) may be derived directly from trigonometric formulas as shown in (33–3). Here the two trigonometric functions are written in terms of the single parameter, k. This oscillator is shown in Figure 33–7.

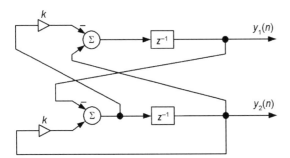

Figure 33–6   Quadrature staggered update oscillator.

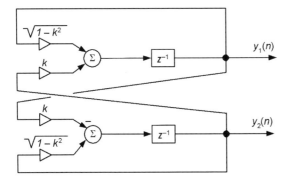

Figure 33–7   Coupled standard quadrature oscillator.

Table 33–1   Recursive Oscillator Properties

Oscillator	Multiplies/ iteration	Equi- amplitude	Quadrature output	$k =$	Rotation matrix
Biquad	1	Yes	No	$2\cos(\theta)$	$\begin{bmatrix} k & -1 \\ 1 & 0 \end{bmatrix}$
Digital waveguide	1	No	Yes	$\cos(\theta)$	$\begin{bmatrix} k & k-1 \\ k+1 & k \end{bmatrix}$
Equi-amplitude staggered update	2	Yes	No	$2\sin(\theta/2)$	$\begin{bmatrix} 1-k^2 & k \\ -k & 1 \end{bmatrix}$
Quadrature staggered update	2	No	Yes	$\cos(\theta)$	$\begin{bmatrix} k & 1-k^2 \\ -1 & k \end{bmatrix}$
Coupled standard quadrature	4	Yes	Yes	$\sin(\theta)$	$\begin{bmatrix} \sqrt{1-k^2} & k \\ -k & \sqrt{1-k^2} \end{bmatrix}$

$$k = \sin(\theta) \tag{33–36}$$

$$\begin{bmatrix} \sqrt{1-k^2} & k \\ -k & \sqrt{1-k^2} \end{bmatrix} \tag{33–37}$$

Gathering our catalog of oscillator properties yields the data in Table 33–1.

33.9 DYNAMIC AMPLITUDE CONTROL

So far the oscillators we have described are ballistic in the sense that they are loaded with some values and allowed to free run. If the run time is short, then this may be adequate. But errors can and may accumulate to the point where the output no longer meets your requirements. Thus a way of controlling the oscillator's amplitude is needed. A standard approach uses automatic gain control (AGC). Since we are iterating the oscillator, why don't we just measure the strength of the output and correct it after each iteration. Also, if the errors are small, the corrections only need to be approximate.

Now when we are measuring the oscillator's output we will use powers (squares of the amplitude) instead of the amplitudes to avoid square roots. Square root calculations tend to be costly in a DSP. The instantaneous power, P, is given by the following formula (in terms of state variables, matrix elements, and phase shift):

$$P = \frac{x_1^2 - \dfrac{b}{c}x_2^2 - 2\sqrt{\dfrac{-b}{c}}x_1 x_2 \cos(\phi)}{\sin^2(\phi)} \tag{33–38}$$

This formula looks painful to implement, but for fixed frequencies, the calculation only involves six multiplies and two subtracts. However, looking ahead to being able to change frequency, we see that if we restrict ourselves to the quadrature oscillators ($\phi = \pm 90°$), then the formula for the power becomes much simpler! It is just:

$$P = x_1^2 - \frac{b}{c} x_2^2 \qquad (33\text{--}39)$$

Along with the power measurement, we need to find the gain needed to properly scale the state variables. A general gain formula is:

$$G = \frac{P_0^q}{P^q} \qquad (33\text{--}40)$$

In this formula, P is the measured power, P_0 is the set point power, and q is a convergence factor. Since we'd rather not perform division, we will use the first order Taylor's approximation—it has the neat property of turning division into subtraction. It is:

$$G \approx 1 + q - q \frac{P}{P_0} \qquad (33\text{--}41)$$

Since we are using G to scale the amplitudes (the state variables), it is best to let $q = 1/2$. Also, when using fixed-point DSPs it becomes convenient to let the set point power also be $1/2$ (amplitude ≈ 0.707). Thus our correction formula becomes

$$G = \frac{3}{2} - P \qquad (33\text{--}42)$$

But we still have one more potential problem, and that is G will nominally be 1, and sometimes we will need to multiply the state variables by a number slightly greater than 1. A trick that can be used is to multiply the state variables by $G/2$ and then double the results. So in summary, the AGC approach consists of the following steps per iteration:

1. Perform one oscillator iteration to update the state variables.
2. Measure the oscillator's output power, P.
3. Calculate a gain factor, G.
4. Scale the state variables by this gain factor.

33.10 DYNAMIC FREQUENCY CONTROL

Finally we get to the subject of dynamic frequency control. Since there are some applications where would like to have a numerically controlled oscillator, we will briefly look at what it takes to do this with these oscillators. Changing an oscillator's frequency merely requires modifying the rotation matrix's k value for any new θ frequency value. The difficult part, however, is maintaining amplitude control during dynamic frequency changes. Since our general power formula (33–38) shows both a dependence on the frequency and amplitude ratio, this can prove computationally inefficient in the general case. Thus the problem is updating the coefficients in the power formula as the frequency is changed. Plus some oscillators will also change output amplitudes.

If we apply some restrictions to the type of oscillator we can simplify the situation. If we look at just using a coupled–standard quadrature oscillator, all of these problems go away. In this case, the formula for the power becomes independent of the matrix elements altogether! However, the difficulty in this case lies in the matrix coefficients where two of them involve radicals. One can use a first-order binomial expansion for these two terms, but then the determinant is not quite one, so the AGC must make up for the error. The coupled–standard quadrature rotation matrix that uses first-order binomial expansions for the two terms with radicals is:

$$
\begin{bmatrix}
1 - \dfrac{k^2}{2} & k \\[2ex]
-k & 1 - \dfrac{k^2}{2}
\end{bmatrix}
\tag{33–43}
$$

Equation (33–43), which can now be used to approximate (33–37), has a determinant that's nearly one (specifically $1 + k^4/4$), so the gain compensation works well for a wide range of k. But even if we use a non-equi-amplitude quadrature oscillator, we can find a simple solution for its amplitude control. In fact, we move the approximation in the process from the rotation matrix to the power measurement. An example using the digital waveguide oscillator will now be shown. The power formula for the digital waveguide oscillator in terms of the tuning parameter, k, is

$$
P = x_1^2 - \frac{k-1}{k+1}x_2^2
\tag{33–44}
$$

But knowing that we will be calculating the power on every iteration, and preferring to avoid division, we will use a first-order series expansion of the denominator, which gives the following approximation for P:

$$
P \approx x_1^2 + (1-k)^2 x_2^2
\tag{33–45}
$$

This approximation works well when k is small, and k is small for frequencies near 1/4 the sampling rate. If the tuning range needs to be enlarged, a higher-order expansion may be used.

So far we have been making the oscillators easily controllable by using low-order series expansions for the difficult to calculate terms. But these approximations carry a price tag: that is, we either must use the oscillator in a narrow tuning range where k is small, or we must use a higher-order approximation to accommodate a larger range of operation. This results from the limitation that the low-order approximation is only good over a finite range. Sometimes our problem won't really be one of needing a large range of operation, but rather we need to operate over a narrow range where k is centered around some non-zero value. This case is easily handled by the use of a frequency translation matrix. This matrix in effect, shifts the center frequency of our oscillator to the point of interest. The oscillator's center frequency is taken to be the frequency where k is zero and thus the approximations are exact at this frequency. The steps for one iteration of a dynamic frequency–amplitude controlled oscillator that uses a frequency translation matrix are as follows:

1. Perform one oscillator iteration using a tunable oscillator to update the state variables. This can be one of the aforementioned quadrature oscillators.

2. Update the state variables using the frequency translation matrix. This matrix is simply the same as shown in (33–37), where k is fixed to represent the translation(shift) frequency. Since the values in this matrix are interpreted to be constant, these values can be calculated prior to run time.

3. Measure the oscillator's output power, P. This is just applying (33–38) or one of its simpler incarnations to the state variables.

4. Calculate a gain factor, G.

5. And finally, scale the state variables by this gain factor.

It should be understood that there is only one pair of state variables involved in the previous set of steps. Steps 1, 2, and 5 operate on the one pair of values. And step 3 just uses the same pair to calculate the power. This can be viewed as two oscillators operating on the same state variables. It just happens that one oscillator allows for dynamic frequency control, and the other just shifts the frequency. The networks representing the oscillator iterations are still the same as previously cataloged. Explicitly we can write the translation (shift) matrix as:

$$\begin{bmatrix} \cos(\omega) & \sin(\omega) \\ -\sin(\omega) & \cos(\omega) \end{bmatrix} \tag{33–46}$$

In a practical implementation, you will set $\omega = 2\pi f_o/f_s$, where f_o is the shift frequency in Hertz, and f_s is the sample rate. The shift frequency may be different from the new desired center frequency, since the natural center frequency for some of the oscillators is one-fourth of the sampling rate.

So we've learned a recipe for recursive oscillator design:

1. Pick your step angle based on the desired oscillator frequency by using $\theta = 2\pi f/f_s$.

2. Select the desired oscillator network based on the properties in Table 33–1.

3. Define k using its relation to θ in Table 33–1.

4. Determine network coefficients from the appropriate rotation matrix in Table 33–1.

5. Establish the initial conditions from (33–27).

6. Implement the oscillator using the target hardware environment to determine whether dynamic amplitude control is necessary.

7. If dynamic frequency control is being used, determine whether a translation matrix is needed.

33.11 FSK MODULATOR DESIGN EXAMPLE

For our example, we will highlight the design of a simple modulator for a 1200-baud FSK modem. The modem generates either 1300 or 2100 Hz depending on whether it is sending a zero or a one bit. We will let our sample rate be 8 kHz, which is common in telephony. Since the data rate does not divide evenly into the sample rate, a sample rate conversion will be needed. This also means the oscillator will generate intermediate frequencies other than 1300 and 2100 Hz. This modulator example will benefit from the use of a translation matrix as well.

Looking at our two frequencies, namely 1300 and 2100 Hz, we see that we can use a fixed frequency oscillator at 1700 Hz and let the variable frequency oscillator range between +400 and −400 Hz. So we will use two oscillator matrices in tandem where one has a fixed frequency and the other varies depending on whether we are sending a zero or a one. The first oscillator will function as the frequency translation oscillator and hence uses (33–46) for its work. This oscillator will not change frequency during the modem's operation, so the two trigonometric constants can be calculated prior to use, so we won't be concerned with there being radicals.

The second oscillator, which changes frequency while in operation, will use the first-order binomial expanded version of the coupled standard quadrature oscillator (33–43). We are using this oscillator since it allows for easy frequency and amplitude control; remember the amplitude control is required here, somewhat for accumulated numerical errors, but mainly since this matrix (33–43) does not strictly obey the Barkhausen criteria. Fortunately, the AGC can more than compensate for the oscillator's gain as long as k is small, which means low frequencies with this oscillator, and hence the use of the frequency translation in the overall modulator.

Since the input to the modem is a sequence of bits (1200 bps), we need to do several things before we let it control the oscillator's frequency. First we need to perform an antipodal mapping of the data. That is, map one bits to a value of +1, and map zero bits to −1. Next we need to resample these 1200 values per second to 8000 per second since this is the modem's sampling rate. Basically a multirate filter is used to interpolate by 20 and decimate by 3. The lowpass filter used in this process will be selected to offer ISI (intersymbol interference) rejection. A polyphase raised cosine filter will suffice. And finally, we will scale the input to the oscillator, so it will emit the correct frequencies. Since we are doing an FM process, the scaling just sets the modulation index. The scaling can be combined into the sample rate conversion process. If a polyphase filter method is used, the coefficients for each of the filters just get scaled to set the modulation index.

The fixed-frequency matrix is set to operate at 1700 Hz (i.e., the average of the two desired frequencies). Numerically, the fixed frequency matrix is just:

$$\begin{bmatrix} \cos\left(\dfrac{2\pi1700}{8000}\right) & \sin\left(\dfrac{2\pi1700}{8000}\right) \\ -\sin\left(\dfrac{2\pi1700}{8000}\right) & \cos\left(\dfrac{2\pi1700}{8000}\right) \end{bmatrix} \approx \begin{bmatrix} 0.233445 & 0.972370 \\ -0.972370 & 0.233445 \end{bmatrix} \qquad (33\text{–}47)$$

Now for the variable-frequency part: Since we desire to deviate between +400 and −400 Hz, we find the scaling for the frequency input to the oscillator by $k = \sin(2\pi400/8000) = 0.309017$. So our ISI-filtered antipodal values need to range between ±0.309017. The frequency input, which is updated 8000 times per second, becomes the value k used in (33–43). The two unique values in matrix (33–43) are calculated 8000 times per second and then used in the variable frequency oscillator iteration. And of course the initial state values for the oscillator combination are simply:

$$\begin{bmatrix} \dfrac{\sqrt{2}}{2} \\ 0 \end{bmatrix} \approx \begin{bmatrix} 0.707107 \\ 0 \end{bmatrix} \qquad (33\text{–}48)$$

These are found by scaling the result of (33–27) by sqrt(P_0), and P_0 is chosen so (33–42) may be used for amplitude control.

For each iteration then, we just

1. Perform antipodal mapping, resample, ISI rejection filter, and scale the 1200 bps data to create "k" for this iteration.

2. "Matrix multiply" the state variables by the 1700 Hz fixed-frequency matrix.

3. "Matrix multiply" the state variables by the variable-frequency matrix—the value of k was determined above.

4. Measure the power: $P = (x_1)^2 + (x_2)^2$.

5. Calculate the gain: $G = (3/2) - P$.

6. Scalar multiply the state variables by the gain, G.

33.12   CONCLUSIONS

We have explored the basic theory of recursive digital oscillators with a bent towards the practical, and from there we have looked at some common oscillators. Then we added some mechanisms for controlling their amplitude and adjusting their frequency. Finally, we showed a brief example of how these oscillators and their control mechanisms may be used to make FSK modulators. I hope I have piqued the reader's interests, and I encourage you to go and develop your own oscillators using the rules and techniques presented here.

33.13 REFERENCES

[1] J. DIEFENDERFER, *Principles of Electronic Instrumentation*. Saunders College Publishing, Philidelphia, PA, 1979, pp. 185.

[2] M. FRERKING, *Digital Signal Processing in Communication Systems*. Kluwer Academic Publishers, Norwell, MA, 1993. pp. 214–217.

[3] S. FRIEDBERG and A. INSEL, *Introduction to Linear Algebra with Applications*. Prentice Hall, Englewood Cliffs, NJ, 1986, pp. 253–276.

[4] R. HIGGINS, *Digital Signal Processing in VLSI*. Prentice Hall, Englewood Cliffs, NJ, 1990, pp. 529–532.

[5] S. LEON, *Linear Algebra with Applications*. 2nd Ed. MacMillan, New York, 1986, pp. 230–259.

[6] A. OPPENHEIM and R. SCHAFER, *Discrete-Time Signal Processing*. Prentice-Hall, Englewood Cliffs, NJ, 1989, pp. 342–344.

[7] J. SMITH and P. COOK "The second order digital waveguide oscillator," *International Computer Music Conference*, San Jose, CA, Oct 1992, pp. 150–153.

[8] SMITH III, et al. "System and method for real time sinusoidal signal generation using waveguide resonance oscillators," U.S. Patent #5701393, December 23, 1997.

[9] C. TURNER, "A Discrete Time Oscillator for a DSP Based Radio," *SouthCon/96 Conference Record*, Orlando, FL, IEEE 1996, pp. 60–65.

EDITOR COMMENTS

To elaborate on the very useful Figure 33–7 coupled standard quadrature oscillator in a fixed-frequency mode of operation, with $k = \sin(\theta)$ and knowing that $\cos^2(\theta) + \sin^2(\theta) = 1$ we can write

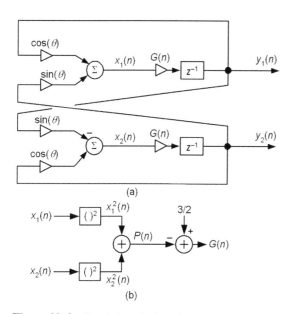

Figure 33–8   Coupled standard quadrature oscillator with AGC.

$$\sqrt{1-k^2} = \cos(\theta).$$

Thus we can redraw the oscillator as shown in Figure 33–8(a). In that figure we have included the implementation of the automatic gain control (AGC) discussion in Section 33.9. Because $b = \cos(\theta)$ and $c = -\cos(\theta)$, (33–39) becomes

$$P(n) = x_1^2(n) - \frac{b}{c}x_2^2(n) = x_1^2(n) + x_2^2(n)$$

and the amplitude correction factor in (33–42) becomes

$$G(n) = \frac{3}{2} - P(n) = \frac{3}{2} - [x_1^2(n) + x_2^2(n)].$$

The computation of the $G(n)$ amplitude correction factor is shown in Figure 33–8(b).

Chapter 34

Direct Digital Synthesis: A Tool for Periodic Wave Generation

Lionel Cordesses

Technocentre, Renault

Discrete-time oscillators are the subject of intensive research. From Colpitts oscillators (Chapter 7 of [1]) to phase locked-loops [2], methods have been proposed to improve stability, frequency resolution, and spectral purity. Among the all-digital approaches such as the one presented in [3], *direct digital frequency synthesis* (DDS) appeared in 1971 [4]. Three years later, this technique was embedded in a commercial unit measuring group delay of telephone lines [5]. DDS is now available as integrated circuits and it outputs waveforms up to hundreds of megahertz.

While DDS is slowly gaining acceptance in new system designs, methods used to improve the quality of the generated waveform are seldom used, even nowadays. The purpose of this chapter is to give an overview of the basics of DDS, along with simple formulas to compute bounds of the signal characteristics. Moreover, several methods—some patented—are presented to overcome some of the limits of the basic DDS with a focus on improving output signal quality.

34.1 DIRECT DIGITAL SYNTHESIS: AN OVERVIEW

The DSP operation we want to perform is to generate a periodic, discrete-time waveform of known frequency F_o. The waveform may be a sinewave, as in [3]. It can also be a sawtooth wave, a triangle wave, a square wave, or any periodic waveform. We will assume that the sampling frequency F_s is known and constant. Before proceeding with the theory of operation, we summarize why direct digital synthesis is a valuable technique:

Streamlining Digital Signal Processing: A Tricks of the Trade Guidebook, Second Edition. Edited by Richard G. Lyons.
© 2012 the Institute of Electrical and Electronics Engineers. Published 2012 by John Wiley & Sons, Inc.

- The tuning resolution can be made arbitrarily small to satisfy almost any design specification.
- The phase and the frequency of the waveform can be controlled in one sample period, making phase modulation feasible.
- The DDS implementation relies upon integer arithmetic, allowing implementation on virtually any microcontroller.
- The DDS implementation is always stable, even with finite-length control words. There is no need for an automatic gain control.
- The phase continuity is preserved whenever the frequency is changed (a valuable tool for tunable waveform generators).

34.2 THEORY OF OPERATION AND IMPLEMENTATION

The implementation of DDS is divided into two distinct parts as shown in Figure 34–1; a discrete-time *phase generator* (the accumulator) outputting a phase value *ACC*, and a *phase-to-waveform converter* outputting the desired DDS signal.

From a Sampling Frequency to a Phase

The implementation of the DDS relies upon integer arithmetic. The size of the accumulator (or word length) is N bits. Assuming that the period of the output signal is 2π radians, the maximum phase is represented by the integer number 2^N. Let us denote Δ_{ACC} the phase increment related to the desired output F_o frequency. It is coded as an integer number with $N - 1$ bits.

During one sample period T_s, the phase increases by Δ_{ACC}. It thus takes T_o to reach the maximum phase 2^N:

$$T_o = \frac{1}{F_o} = \frac{2^N T_s}{\Delta_{ACC}} \qquad (34\text{–}1)$$

We can rewrite (34–1) in terms of frequency F_o, as a function of Δ_{ACC}:

$$F_o = \frac{F_s}{2^N} \Delta_{ACC}. \qquad (34\text{–}2)$$

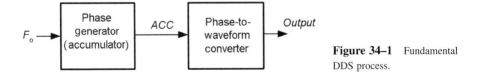

Figure 34–1   Fundamental DDS process.

The phase increment Δ_{ACC}, rounded to the nearest integer ($\lfloor x \rfloor$ is the integer part of x), is given by:

$$\Delta_{ACC} = \left\lfloor F_o \frac{2^N}{F_s} + 0.5 \right\rfloor. \tag{34-3}$$

Equation (34–2) is the basic equation of any DDS system. One can infer from (34–2) the tuning step $\Delta_{Fo,min}$, which is the smallest step in frequency that the DDS can achieve (remember that Δ_{ACC} is an integer).

$$\Delta_{Fo,min} = F_o(\Delta_{ACC} + 1) - F_o(\Delta_{ACC})$$
$$= \frac{F_s}{2^N}(\Delta_{ACC} + 1 - \Delta_{ACC}) = \frac{F_s}{2^N}. \tag{34-4}$$

Equation (34–4) allows the designer to choose the number of bits (N) of the accumulator ACC. This number N is often referred to as the frequency tuning word length [6]. It is reckoned thanks to:

$$N = \left\lfloor \log_2 \left(\frac{F_s}{\Delta_{Fo,min}} \right) + 0.5 \right\rfloor. \tag{34-5}$$

The minimum frequency, $F_{o,min}$, the DDS can generate is given by (34–2) with $\Delta_{ACC} = 1$, the smallest phase increment that still increases the phase ($\Delta_{ACC} = 0$ does not increase the phase). $F_{o,min}$ is

$$F_{o,min} = \frac{F_s}{2^N}. \tag{34-6}$$

The maximum frequency $F_{o,max}$ the DDS can generate is given by the uniform sampling theorem (Nyquist, Shannon, see, for instance, Chapter 9 of [7]):

$$F_{o,max} = \frac{F_s}{2}. \tag{34-7}$$

From a practical point of view, a lower $F_{o,max}$ is often preferred, $F_{o,max} = F_s/4$, for example. The lower $F_{o,max}$ is, the easier the analog reconstruction using a lowpass filter.

From a Phase to a Waveform

The phase is coded with N bits in the accumulator. Thus, the waveform can be defined with up to 2^N phase values. In case 2^N is too large for a realistic implementation, the phase to amplitude converter uses fewer bits than N. Let us note P as the number of bits used as the phase information (with $P \leq N$). The output waveform values can be stored in a lookup table (LUT) with 2^P entries: the output value is

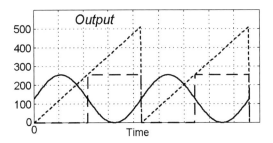

Figure 34–2 Signals generated by software DDS: sinewave, squarewave, and sawtooth signals.

computed as *Output = LUT(ACC)*, which is implemented in the phase-to-waveform converter in Figure 34–1; other output waveform generation techniques, based upon approximations, are presented later.

DDS can generate a sinewave with an offset *b* and a peak amplitude *a*. The content of the LUT, containing the DDS output values, is computed for the index *i* ranging from 0 to $(2^P - 1)$ using:

$$LUT(i) = \left\lfloor b - a\sin\left(\frac{2\pi i}{2^P}\right) + 0.5 \right\rfloor. \tag{34–8}$$

Using the LUT computed for $P = 9$, $a = 127.5$, and $b = 127.5$, the output waveform for $F_s = 44,100$ Hz and $F_o = 233$ Hz is plotted as the solid curve in Figure 34–2.

One might want to generate two quadrature signals: one just has to read both *LUT(i)* and $LUT(i + 2^P/4)$, which, in turn, correspond to the sine and to the cosine functions.

A squarewave can be had with no computational overhead because that waveform is already available as the most significant bit of the phase accumulator *ACC*, as shown by the dashed curve in Figure 34–2. The most significant bit toggles every π radians, since the accumulator represents 2π radians. We must point out that this square wave is corrupted by phase jitter [8] of one sampling period T_s. This phase jitter is caused by the sampling scheme used to synthesize the waveform. To quote [38]:

> [T]he output of the direct digital synthesizer can occur only at a clock edge. If the output frequency is not a direct submultiple of the clock, a phase error between the ideal output and the actual output slowly increases (or decreases) until it reaches *one clock period*, at which time the error returns to zero and starts to increase (or decrease) again.

A sawtooth signal is also available with no computational overhead. The linearly increasing phase accumulator *ACC* value is stored modulo 2^N, thus leading to a sawtooth signal as shown by the dotted curve in Figure 34–2. The LUT is not used in this case, or it is the identity function: *Output = ACC*. With the use of logic gates, a triangular output waveform can be generated from the sawtooth.

34.3 QUANTIZATION EFFECTS

Quantization occurs on both the *ACC* phase information and on the *Output* amplitude information. The DDS is now redrawn including this effect. The number of bits used by each variable is written below the variables on Figure 34–3.

Phase Quantization

Phase quantization occurs when the phase information *ACC* is truncated from *N* bits to *P* bits as shown in Figure 34–3. The reason behind this quantization is to keep the memory requirements of the phase-to-waveform converter quite low: When implemented as a LUT, the size of the memory is $2^P \times M$ bits. A realistic value for *N* is 32, but this would lead to a $2^{32} \times M$ memory, which is not realistic. Thus we quantize the phase information Φ to *P* bits, as it decreases the number of entries of the LUT.

   Unfortunately, the phase quantization introduces noise on the phase signal Φ. It leads to *phase noise* (see Chapter 7 of [1], and Chapter 3 of [9]) and it produces unwanted spurious spectral components in the DDS output signals, often referred to as *spurs*. The difference between the carrier level (which is the desired signal) and the maximum level of spurs, is called *spurious free dynamic range* (SFDR). A simplified formula given in [10] to estimate the maximum level of the spurs S_{max} when the carrier level is 0 dB, is:

$$S_{max} = -\text{SFDR} = -6.02P + 3.92 \text{ dB}. \tag{34–9}$$

For a detailed derivation of the exact formulas (including the frequency and the SFDR of spurs), the reader is referred to [11] and [12].

Amplitude Quantization

The output of the Phase to Waveform Converter is quantized to *M* bits, *M* being the word length of the *Output* amplitude word. This quantization results in a signal-to-noise ratio (in this case, it is a noise-to-signal ratio) [10] usually approximated by:

$$SNR = -6.02M - 1.76 \text{ dB}. \tag{34–10}$$

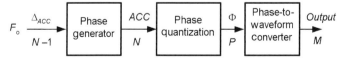

Figure 34–3   DDS including quantization.

This bound also limits the performance of the DDS, as the output spectrum will exhibit a $-6.02M - 1.76$ dB noise floor. Thanks to (34–9) and (34–10), one can infer (see [10]):

$$P = M + 1. \tag{34–11}$$

Thus, $S_{max} = -6.02(M + 1) + 3.992$ dB $= -6.02M - 2.028$ dB and $SNR = -6.02M - 1.76$ dB leading to $S_{max} < SNR$. This inequality means that the unwanted signals are caused by the amplitude quantization, and not by the phase truncation. Knowing (34–11), we can now focus on improving the SFDR of a DDS.

34.4 IMPROVING SFDR BY SINEWAVE COMPRESSION

There are many techniques to improve the SFDR of a DDS. The easiest one would be to increase the phase wordlength. Due to (34–9) and (34–10), we can increase P (and thus M according to Equation (34–11))in order to meet the technical specifications. The only drawback of this approach is the total amount of LUT memory, $2^P \times M$ bits. For small P (such as $P = 9$ bits, and $M = 8$ bits), implementing the LUT with a memory leads to simple, low-cost, hardware; see [13] and [8] for a realization based upon this method.

For higher values of P, the memory requirements become impractical at high frequency, or for embedded system implementations. To circumvent this impediment, the solution is to compress the sine waveform, thus reducing memory consumption. Two methods are reported in the next sections. One is based upon symmetry and the other on sinewave approximations.

A Quarter of a Sinewave

Instead of storing the whole sinewave $f(\Phi) = \sin(\Phi)$ for $0 \leq \Phi \leq 2\pi$, one can store the same function for $0 \leq \Phi \leq \pi/2$ and use symmetry to get the complete 2π waveform range. This approach only uses 2^{P-2} entries in the LUT, leading to a LUT-size compression ratio of $4:1$. The full sinewave can be reconstructed at the expense of some hardware (see [5], [14], and [9]). From here out, we will only deal with a quarter of a sinewave. Next we discuss four methods of approximating a sinewave.

Sinewave Approximations

The first sinewave approximation method goes as follows: instead of storing $f(\Phi) = \sin(\Phi)$ using M bits, one can store $g(\Phi) = \sin(\Phi) - 2\Phi/\pi$, hence the name *sine-phase difference algorithm* found in [14]. It has been shown in [14] that this new function g only needs $M - 2$ bits to get the same amplitude quantization for the

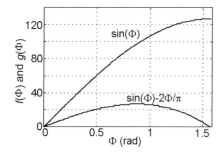

Figure 34–4 Sine-phase difference LUT example ($P = 9$, $M = 8$).

sinewave (see Figure 34–4 for an example). The only drawback is the need for an adder at the output of the LUT.

The second sinewave approximation method is called the *Sunderland technique*. This method, named after its author [15], makes use of trigonometric identities. It has been used for $P = 12$ and it uses the following identity:

$$\sin(A + B + C) = \sin(A + B)\cos(C) + \cos(A)\cos(B)\sin(C) - \sin(A)\sin(B)\sin(C). \tag{34–12}$$

The 12 bits of the phase are:

- A, the four most significant bits (with $0 \le A \le \pi/2$)
- B, the following 4 bits (with $0 \le B \le (\pi/2)/(2^4)$)
- C, the four least significant bits (with $0 \le C \le (\pi/2)/(2^8)$)

Equation (34–12) is then approximated by:

$$\sin(A + B + C) \approx \sin(A + B) + \cos(A)\sin(C). \tag{34–13}$$

Using two LUTs (one for $\sin(A + B)$ and one for $\cos(A)\sin(C)$) leads to a significant amount of compression. The $\sin(A + B)$ LUT uses $2^8 \times 11$ bits ($P = 12$ thus $M = P - 1 = 11$). The second LUT is filled with small numbers, thus requiring less than M bits (actually four bits, see [15]). Finally, the compression ratio of this architecture is 51:1 (see [16] for a comparison of various compression methods).

Several improvements to this architecture have been presented (see [14]) and the compression ratio of the modified Sunderland technique leads to a 59:1 compression ratio [16]. The same method has been used in [17] with a 128:1 compression ratio, and in [18] with a 165:1 compression ratio.

The third sinewave approximation method involves first-order Taylor series expansions. Let us introduce δ_Φ with $\delta_\Phi \ll \Phi$. The Taylor series expansion of the sine function is:

$$\sin(\Phi + \delta_\Phi) \approx \sin(\Phi) + \delta_\Phi \cos(\Phi). \tag{34–14}$$

Instead of storing the sine function, the key idea presented in [19] and described in [20] proposes to use two coarse LUTs storing $\sin(\Phi)$ and $\cos(\Phi)$. Moreover, the sine-phase difference algorithm can be used to store efficiently sin(Φ), further decreasing the size of the LUT. The compression ratio obtained with this method is $64:1$ [19] and even reaches $67:1$ [16].

Another method has been introduced in [21] where the sine function is approximated by linear interpolation. A few samples (16 in [22]) of the sine function are stored in a LUT, and the values are computed using linear interpolation. The compression ratio, computed thanks to data given in [22], is: $10 \times 2^{11}/960 \approx 21:1$. Another implementation of the linear solution does not rely on a LUT. Using notations as in [23], the sine function is written as:

$$\sin(\Phi + \delta_\Phi) \approx y_0 + m_0(\Phi - \delta_\Phi). \tag{34-15}$$

The solutions presented in [24] and used in [23] carefully impose a power of two number of segments, thus using the most significant bits of δ_Φ as an address. Moreover, to further decrease complexity, the lengths of all the segments are equal. The implementation of (34-15) only uses one multiplication and one addition, without any complex address decoder. There is no LUT in this design. According to their authors, such an approach reaches the performance of other methods, showing a 60 dBc SFDR for $P = 12$ bits phase [23]. The method is now patented [25].

The fourth sinewave approximation method involves higher-order Taylor series expansions. In [26], the Taylor series expansion of the sine function is:

$$\sin(\Phi + \delta_\Phi) \approx \sin(\Phi) + \delta_\Phi \cos(\Phi) - (\delta_\Phi)^2 \sin(\Phi)/2. \tag{34-16}$$

The compression ratio, given in [27], is $110:1$ for $P = 12$ bits and a SFDR of 85 dBc. Higher-order interpolation is also used for sinewave compression: parabolic interpolation is presented in [28]. Only interpolation coefficients V_1, V_2, and V_3 are stored in the LUT, and the value of the sine function is reckoned thanks to:

$$\sin(\Phi + \delta_\Phi) \approx V_1(\Phi) + \delta_\Phi V_2(\Phi) + (\delta_\Phi)^2 V_3(\Phi). \tag{34-17}$$

The compression ratio obtained using (34-17) is given in [28] for a 64 dBc SFDR at $M = 11$ bits is $157:1$. As in the previous first-order Taylor series method, there is a counterpart of the LUT-less method. Based upon a quadruple angle equality:

$$\cos(4\Phi) = 1 - 8\sin^2(\Phi)[1 - \sin^2(\Phi)]. \tag{34-18}$$

With $0 \le \Phi \le \pi/8$, (34-18) is approximated in [29] by:

$$\cos(4\Phi) \approx 1 - 8\Phi^2(1 - \Phi^2). \tag{34-19}$$

Equation (34-19) is implemented with multipliers and adders only.

There are still other approaches to approximating a sinewave. A phase-to-sinewave converter can be implemented thanks to an angle-rotation algorithm, such as the CORDIC (*COordinate Rotation DIgital Computer*) algorithm [30]. A DDS based on this method is described in [31], and the effect of finite precision on the CORDIC converter is analyzed in [32]. Another method relies upon a nonlinear digital-to-analog converter that implements the sinewave generation [33].

There are trade-offs with each of the above sinewave approximation techniques, and no single technique is best for all DDS applications. In what follows, we discuss additional tricks used to optimize DDS performance by maximizing SFDR.

34.5 IMPROVING SFDR THROUGH SPUR REDUCTION TECHNIQUES

The easiest methods to reduce the level of DDS spurs, discussed in Part 34.4, is to increase the accuracy of the phase-to-waveform converter. The limit of this approach has been mentioned; it is mainly technological (lookup table size).

We now review three simple and effective methods to reduce the spur level of the sinewave DDS, along with the corresponding spectra computed from simulated DDS outputs.

The Odd Number Approach

The worst-case spur level is given by (34–9). Making Δ_{ACC} an odd number improves the SFDR by 3.9 dB [14]. The repetition period of the accumulator T_{ACC}, often referred to as grand repetition period, is given by:

$$T_{ACC} = \frac{2^N}{GCD(2^N, \Delta_{ACC})}. \tag{34–20}$$

with $GCD(x,y)$ standing for greatest common divisor of x and y [34]. When $GCD(2^N, \Delta_{ACC}) = 1$, as it will be whenever Δ_{ACC} is an odd number, then $T_{ACC} = 2^N$, which *spreads* the spurs over the entire spectrum (otherwise they are aliased to a frequency within the spectrum, as described in [14] and [12]). As an example, we computed the output spectra with a fast Fourier transform (FFT), the length of which is equal to T_{ACC}, for the two cases of $\Delta_{ACC} = 13{,}248$ and for $\Delta_{ACC} = 13{,}249$, with $N = 16$, $P = 9$, and $M = 8$. The odd number for Δ_{ACC} leads to an increase of $54.2 - 50.3 = 3.9$ dB in SFDR.

The Phase Dithering Approach

In another method to spread the spurs throughout the available bandwidth, one can add a dither signal [35] to the *ACC* phase values as shown in Figure 34–5. The dither signal can be a pseudo-random noise sequence (generated, for example, with binary shift registers and exclusive-or gates, and having a repetition period much greater than the output signal period) whose word width is B bits providing noise values in the range of 0 and 2^B. Choosing $B = N - P$, the spurs do follow a 12-dB-per-phase-bit law [36], instead of the 6-dB-per-phase bit of (34–9), thus allowing a smaller LUT for the same SFDR.

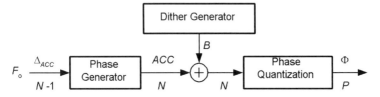

Figure 34–5   DDS including phase dither.

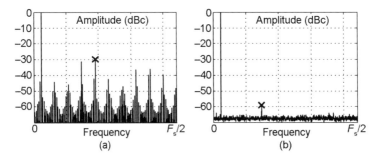

Figure 34–6   DDS output spectra: (a) without dither; (b) with dithering.

An output spectrum is given without any dither signal in Figure 34–6(a) with $F_s = 44,100$ Hz, $\Delta_{ACC} = 1657$, $N = 16$, $P = 5$, and $M = 16$, and with the dither signal ($B = N - P = 16 - 5 = 11$ bits) applied in Figure 34–6(b). The spectra have been computed for 10 output signals that have been averaged following the method described in [36]. The high-resolution, 16-bits LUT has been chosen so as to focus only on the 5 bits of phase quantization, as in [36]. The drawback of this dithering method is the increase of the noise floor, but that's a small price to pay for such a large increase in SFDR.

Other dithering methods are available, such as amplitude dithering, and both phase and amplitude dithering (see [36] and [37]). The phase dithering approach has also been applied to squarewave signal DDS [38].

The Noise Shaping Approach

The key idea of the noise shaping approach, to improve our DDS SFDR, is to filter out the quantization noise introduced by the phase quantization step in Figure 34–3. This quantization can be viewed as a special case of noise addition [39], as depicted in Figure 34–7(a). The quantization noise signal n can be recovered from the following equations:

$$\Phi = n + ACC, \text{ and } e_Q = ACC - \Phi. \tag{34–21}$$

Thus $e_Q = -n$. In the *noise shaping* approach, this quantization noise signal $-n$ is fed back, through a filter G, to the *ACC* signal as shown in Figure 34–7(b). The transfer

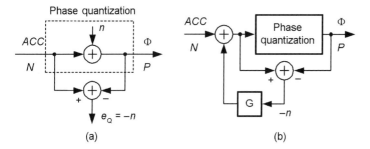

Figure 34–7 Quantization and noise shaping: (a) quantization model; (b) noise shaping implementation.

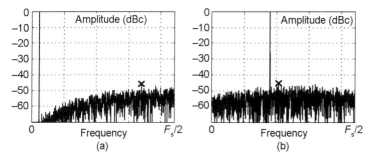

Figure 34–8 DDS spectra with first-order noise shaping and $G = z^{-1}$: (a) $\Delta_{ACC} = 1657$; (b) $\Delta_{ACC} = 13,249$.

function of interest is the one from n to Φ, as the noise added to the phase signal Φ will eventually lead to phase noise. The phase signal is then:

$$\Phi = n(1 - G) + ACC. \tag{34–22}$$

From (34–22), one can infer that the phase signal Φ is affected by the filtered noise signal n, $(1 - G)$ being the transfer function of the filter. The choice of G will lead to different results, as we shall see.

A *first-order noise shaping* approach is to use the simple transfer function G proposed in [10] being the finite impulse response (FIR) filter $G = z^{-1}$. Here z is the symbol of the z-transform used for discrete-time systems. The function G can be implemented as a single delay register, and $1 - G$ has a zero at $z = 0$ (zero Hz): It acts as a discrete time differentiator. The system filters out the low-frequency components of the noise signal n, but high-frequency signals, greater than 8 kHz, are amplified. This simple approach prevents the filter from rejecting high-frequency components of the noise signal n. Thus justifying the statement that one should use this filter for low frequency F_o signals [40].

An example of the output spectrum is given in Figure 34–8(a) (the parameters are the same as in Figure 34–6(a)). The SFDR is greater than 60 dB near the carrier, but it decreases with the frequency, and eventually reaches 46.8 dBc. The overall behavior is compliant with the above $G = z^{-1}$ analysis. At a higher frequency ($\Delta_{ACC} = 13{,}249$), the noise close to the carrier is less filtered, as one can see on Figure 34–8(b).

To implement *higher-order noise shaping*, a more complex filter can be used instead of $G = z^{-1}$. A second order FIR filter has been proposed in [41]:

$$1 - G = 1 + b_1 z^{-1} + b_2 z^{-2}. \tag{34-23}$$

Careful choice of b_1 and b_2 can lead to a double zero at zero: $1 - G = 1 - 2z^{-1} + z^{-2} = (1 - z^{-1})^2$, which improves the rejection at zero Hz. When the noise shaping is applied to the amplitude signal instead of the phase signal, other values (often integer values, so as to ease implementation, see [41]) are preferred, and this filter can even be tuned online.

Multiple-zeros filters are also of interest, for example, when one wants to reject a known frequency such as $2 \times F_o$ [40]. A tunable notch filter is added at the expense of a more complex feedback structure. The transfer function becomes:

$$\begin{aligned} 1 - G &= (1 - z^{-1})(1 + bz^{-1} + z^{-2}) \\ &= 1 - (1-b)z^{-1} + (1-b)z^{-2} - z^{-3} \end{aligned} \tag{34-24}$$

with $b = -2\cos[2\pi(2F_o/F_s)]$. At low frequencies, as shown in Figure 34–9(a), the spectrum looks like the one with the first-order noise shaping. But at a higher frequency (the same as on Figure 34–8(b)), the improvement close to the carrier is clear as shown in Figure 34–9(b).

Note that the same filter (with a zero at a specific frequency) can be implemented by feeding the error signal back to the accumulator. This structure, proposed and patented in [40], is presented in Figure 34–10. F is the transfer function of the accumulator (an integrator: $ACC(k) = ACC(k - 1) + \Delta_{ACC}(k - 1)$), given by:

$$F = \frac{z^{-1}}{1 - z^{-1}} \tag{34-25}$$

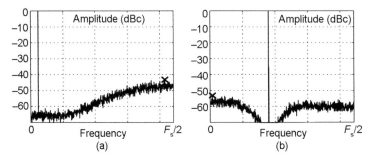

Figure 34–9   DDS spectra with higher-order noise shaping: (a) same parameters as Figure 34–6(a); (b) same parameters as Figure 34–8(b).

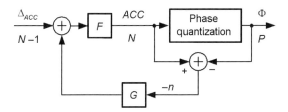

Figure 34–10   Noise shaping: analog devices' approach.

with the output phase Φ is given by $\Phi = (1 - FG)n + F\Delta_{ACC}$. One can chose G so as to ensure $(1 - FG) = (1 - z^{-1})(1 + bz^{-1} + z^{-2})$ as in (24).

34.6 CONCLUSIONS

Direct digital synthesis is a useful tool for generating periodic waveforms. We presented the basic idea of this synthesis technique, and then focused on the quality of the sinewave a DDS can create, introducing the SFDR quality parameter. Next we presented effective methods to increase the SFDR through sinewave approximations, hardware schemes such as dithering and noise shaping, and an extensive list of references. When the desired output is a digital signal, the signal's characteristics can be accurately predicted using the formulas given in this chapter. When the desired output is an analog signal, the reader should keep in mind that the performance of the DDS is eventually limited by the performance of the digital to analog converter and the follow-on analog filter [42].

We hope that this discussion will incite engineers to use DDS, either integrated circuits DDS or software-implemented DDS. From the author's experience, this technique has proved valuable when frequency resolution is the challenge, particularly when using low-cost microcontrollers.

34.7 REFERENCES

[1] U. ROHDE, J. WHITAKER, and T.T.N. BUCHER, *Communications Receivers*, 2nd ed. ISBN 0-07-053608-2. McGraw Hill, 1997.

[2] U. ROHDE, *Digital PLL Frequency Synthesizers: Theory and Design*. ISBN 0-13-214239-2. Prentice-Hall, 1983.

[3] C. TURNER, "Recursive Discrete-Time Sinusoidal Oscillators," *IEEE Signal Processing Magazine*, vol. *20*, no. 3, May 2003, pp. 103–111.

[4] J. TIERNEY, C. RADER, and B. GOLD, "A Digital Frequency Synthesizer," *IEEE Transactions on Audio and Electroacoustics*, vol. *19*, no. 1, March 1971, pp. 48–57.

[5] D. GUEST, "Simplified Data-Transmission Channel Measurements," *Hewlett-Packard Journal*, November 1974, pp. 15–24.

[6] *A Technical Tutorial on Digital Signal Synthesis*. Analog Devices, Inc., 1999.

[7] H. BLINCHIKOFF and A. ZVEREV, *Filtering in the Time and Frequency Domains*. Noble Publishing, 2001.

[8] E. MCCUNE, "Create Signals Having Optimum Resolution, Response, and Noise," *EDN Magazine*, March 14 1991, pp. 95–108.

[9] J. CRAWFORD, *Frequency Synthesizer Design Handbook*. ISBN 0-89006-440-7. Artech House, 1994.

[10] P. O'LEARY and F. MALOBERTI, "A Direct-Digital Synthesizer with Improved Spectral Performance," *IEEE Transactions on Communications*, vol. *39*, no. 7, July 1991, pp. 1046–1048.

[11] V. KROUPA, V. CIZEK, J. STURSA, and H. SVANDOVA, "Spurious Signals in Direct Digital Frequency Synthesizers Due to the Phase Truncation," *IEEE Transactions on Ultrasonics, Ferroelectrics and Frequency Control*, vol. *47*, no. 5, September 2000, 1166–1172.

[12] F. CURTICAPEAN and J. NIITTYLAHTI, "Exact Analysis of Spurious Signals in Direct Digital Frequency Synthesisers Due to Phase Truncation," *Electronics Letters*, vol. *39*, no. 6, March 2003, pp. 499–501.

[13] B-G. GOLDBERG, "Digital Frequency Synthesizer." U.S. Patent US4,752,9029, Scitech Electronics, Inc, San Diego, CA, July 1985.

[14] H. NICHOLAS, H. SAMUELI, and B. KIM, "The Optimization of Direct Digital Frequency Synthesizer Performance in the Presence of Finite Word Length Effects," *Proceedings of the 42nd Annual Frequency Control Symposium*, June 1988, pp. 357–363.

[15] D. SUNDERLAND, R. STRAUCH, S. WHARFIELD, H. PETERSON, and C. COLE, "CMOS/SOS Frequency Synthesizer LSI Circuit for Spread Spectrum Communications," *IEEE Journal of Solid-State Circuits*, vol. *19*, no. 4, August 1984, 497–506.

[16] K. ESSENWANGER and V. REINHARDT, "Sine Output DDSs. A Survey of the State of the Art," *Proceedings of the 1998 IEEE International Frequency Control Symposium*, May 1998, pp. 370–378.

[17] H. NICHOLAS and H. SAMUELI, "A 150-MHz Direct Digital Frequency Synthesizer in 1.25-μm CMOS with −90-dBc Spurious Performance," *IEEE Journal of Solid-State Circuits*, vol. *26*, no. 2, December 1991, pp. 1959–1969.

[18] G. KENT and N. SHENG, "A High Purity, High Speed Direct Digital Synthesizer," *Proceedings of the 1995 IEEE International 49th. Frequency Control Symposium*, June 1995, pp. 207–211.

[19] DDS Tutorial: Technical Report V3, Scitech Electronics, Inc., San Diego, CA, 1991.

[20] B-G. GOLDBERG, "Source of Quantized Samples for Synthesizing Sinewaves," U.S. Patent US5,321,642, Scitech Electronics, Inc., San Diego, CA, June 1994.

[21] R. FREEMAN, "Digital Sine Conversion Circuit for Use in Direct Digital Synthesizers," U.S. Patent US4,809,205, Rockwell International Corporation, El Segundo, CA, November 1989.

[22] A. BELLAOUAR, M. OBRECHT, A. FAHIM, and M. ELMASRY, "A Low-Power Direct Digital Frequency Synthesizer Architecture for Wireless Communications," *Proceedings of the IEEE 1999 Custom Integrated Circuits*, May 1999, pp. 593–596.

[23] J. LANGLOIS and D. AL-KHALILI, "A Low Power Direct Digital Frequency Synthesizer with 60 dBc Spectral Purity," *Proceedings of the 12th ACM Great Lakes Symposium on VLSI*, ACM Press, 2002, pp. 166–171.

[24] J. LANGLOIS and D. AL-KHALILI, "Piecewise Continuous Linear Interpolation of the Sine Function for Direct Digital Frequency Synthesis," *IEEE Radio Frequency Integrated Circuits (RFIC) Symposium*, June 2003, pp. 579–582.

[25] D. AL-KHALILI and J. LANGLOIS, "Phase to Sine Amplitude Conversion System and Method," European Patent EP1286258, Canada Min. Nat. Defence, February 2003.

[26] L. WEAVER Jr. and R. KERR, "High Resolution Phase to Sine Amplitude Conversion," U.S. Patent US4,905,177, Qualcomm, Inc., San Diego, CA, January 1988.

[27] J. VANKKA, "Methods of Mapping from Phase to Sine Amplitude in Direct Digital Synthesis," *IEEE Transactions on Ultrasonics, Ferroelectrics and Frequency Control*, vol. *44*, no. 2, March 1997, pp. 526–534.

[28] A. ELTAWIL and B. DANESHRAD, "Piece-wise Parabolic Interpolation for Direct Digital Frequency Synthesis," *Proceedings of the IEEE Custom Integrated Circuits Conference*, May 2002, pp. 401–404.

[29] C. WANG, H. SHE, and R. HU, "A ROM-less Direct Digital Frequency Synthesizer By Using Trigonometric Quadruple Angle Formula," *9th International Conference on Electronics, Circuits and Systems*, vol. *1*, September 2002, pp. 65–68.

[30] G. GIELIS, R. van de PLASSCHE, and J. van VALBURG, "A 540-MHz 10-b Polar-to-Cartesian Converter," *IEEE Journal of Solid-State Circuits*, vol. *26*, no. 11, November 1991, pp. 1645–1650.

[31] A. MADISETTI, A. KWENTUS, and A. WILLSON, "A 100-MHz, 16-b, Direct Digital Frequency Synthesizer with a 100-dBc Spurious-Free Dynamic Range," *IEEE Journal of Solid-State Circuits*, vol. *34*, no. 8, August 1999. pp. 1034–1043.

[32] C. KANG and E. SWARTZLANDER Jr., "An Analysis of the CORDIC Algorithm for Direct Digital Frequency Synthesis," *The IEEE International Conference on Application-Specific Systems, Architectures and Processors*, July 2002, pp. 111–119.

[33] A. MCEWAN and S. COLLINS, "Analogue Interpolation Based Direct Digital Frequency Synthesis," *Proceedings of the 2003 International Symposium on Circuits and Systems ISCAS '03.*, vol. 1, May 2003, pp. 621–624.

[34] J. GARVEY and D. BABITCH, "An Exact Spectral Analysis of a Number Controlled Oscillator Based Synthesizer," *Proceedings of the 44th Annual Symposium on Frequency Control*, May 1990, pp. 511–521.

[35] L. SCHUCHMAN, "Dither Signals and Their Effect on Quantization Noise," *IEEE Transactions on Communications*, vol. *12*, no. 4, December 1964, pp. 162–165.

[36] M. FLANAGAN and G. ZIMMERMAN, "Spur-Reduced Digital Sinusoid Synthesis," *IEEE Transactions on Communications*, vol. *43*, no. 7, July 1995, 2254–2262.

[37] J. VANKKA, "Spur Reduction Techniques in Sine Output Direct Digital Synthesis," *Proceedings of the 1996 IEEE Frequency Control Symposium*, June 1996, pp. 951–959.

[38] C. WHEATLEY III and D. PHILLIPS, "Spurious Suppression in Direct Digital Synthesizers," *Proceedings of the 35th Annual Frequency Control Symposium*, May 1981, pp. 428–435.

[39] H. SPANG III and P. SCHULTHEISS, "Reduction of Quantizing Noise by Use of Feedback," *IEEE Transactions on Communications*, vol. *10*, no. 4, December 1962, pp. 373–380.

[40] D. RIBNER and S. KIDAMBI, "Direct-Digital Synthesizers," European Patent EP1037379, Analog Devices Inc., Norwood, MA, September 2000.

[41] J. VANKKA, "A Direct Digital Synthesizer with a Tunable Error Feedback Structure," *IEEE Transactions on Communications*, vol. *45*, no. 4, April 1997, pp. 416–420.

[42] T. HIGGINS, "Analog Output System Design for a Multifunction Synthesizer," *Hewlett-Packard Journal*, February 1989, pp. 66–69.

Chapter 35

Implementing a ΣΔ DAC in Fixed-Point Arithmetic

Shlomo Engelberg
Jerusalem College of Technology

$\mathbf{T}$his chapter describes a simple sigma delta (ΣΔ) digital-to-analog converter (DAC) that is suitable for use on the simplest of microprocessors to generate constant-level, or slowly varying, control voltages by taking advantage of a ΣΔ network's intrinsic limit cycle behavior. We describe the properties of the converter, show that it is inherently stable, and explain how the user can control the DC level of the DAC's analog output. Finally, we discuss the implementation of such a DAC on an industry-standard microprocessor.

35.1 THE ΣΔ DAC PROCESS

The block diagram of a simple ΣΔ DAC is shown in Figure 35–1. The input to the system is a number presented in digital form, and the "conversion" is done within the microprocessor. The single-bit binary logic-level output of the chip, $d(t)$, is applied to an analog lowpass filter.

The output $y(n)$ of the comparator in Figure 35–1 satisfies the equation $y(n) = f(z(n))$ where:

$$f(z) = \begin{cases} 3A/2 & z \geq A \\ A/2 & z < A \end{cases},$$

and A is a positive number chosen to be, in our application, half the peak amplitude of the $x(n)$ input.

As $z(n)$ is the output of the summation, it is clear that $z(n) = z(n-1) + e(n)$. From Figure 35–1 we see that $e(n) = x(n) - y(n-1) = x(n) - f(z(n-1))$, with $e(n)$ the error and $x(n)$ the input. Putting it all together, we find that:

Streamlining Digital Signal Processing: A Tricks of the Trade Guidebook, Second Edition. Edited by Richard G. Lyons.
© 2012 the Institute of Electrical and Electronics Engineers. Published 2012 by John Wiley & Sons, Inc.

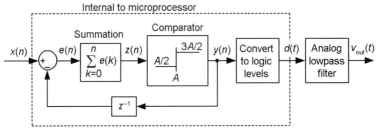

Figure 35–1   A simple ΣΔ DAC.

$$z(n) = z(n-1) + x(n) - f(z(n-1)). \tag{35–1}$$

If $x(n) \equiv c$ is a constant that satisfies $A/2 \le c \le 3A/2$ and if $z(0)$ satisfies the condition $0 \le z(0) < 2A$ then (using mathematical induction) it is not hard to show that, for all $n \ge 0$, the elements $z(n)$ satisfy:

$$0 \le z(n) < 2A. \tag{35–2}$$

Let us consider the practically important case in which A is an even integer and c is constrained to be an integer. This is often the case when implementing a ΣΔ DAC using a simple microprocessor. Under these conditions, $z(n)$ is constrained to be an integer between 0 and $2A - 1$. Additionally, from (35–1) we see that $z(n)$ can be thought of as the state of a finite-state machine. As the next state of the machine is a function of the previous state, we find that if the machine ever repeats a state, then from that point on the state is periodic. As the machine has $2A$ possible states, it is clear that the maximal period that the ΣΔ DAC is capable of producing is $2A$ elements long. Using more sophisticated arguments, it is possible to show that the maximal period that the ΣΔ DAC is capable of producing is actually only A elements long. We find that for a constant $x(n)$ input our feedback loop produces periodic oscillations—limit cycles. (For more information on limit cycles in ΣΔ-type circuits, see [1]–[3] and the references therein.)

It is easy to see that the average value of one cycle of the discrete time system's $y(n)$ output must be equal to the constant value at the input to the system. Suppose that this were not the case. Then at the end of each period the output of the summation block, the system's state, $z(n)$, would change. But this is inconsistent with the system's state being periodic. Thus, the average value must be the same as that of the constant input. The lowpass filter that produces the final output of the system will produce an output that tends to the value of the constant being input.

35.2 ΣΔ DAC PERFORMANCE

To understand the frequency-domain behavior of this simple ΣΔ DAC, we can perform a z-domain analysis by representing the summation block by $S(z)$ and representing the comparator as an additive noise source, $Q(z)$. These representations allow us to analyze the DAC as shown in Figure 35–2(a) and write

$$Y(z) = Q(z) + S(z)[X(z) - z^{-1}Y(z)]. \tag{35-3}$$

Solving (35-3) for $Y(z)$ yields

$$Y(z) = \frac{X(z)S(z)}{1 + S(z)z^{-1}} + \frac{Q(z)}{1 + S(z)z^{-1}}. \tag{35-4}$$

Because the summation block's z-domain expression is $S(z) = 1/(1 - z^{-1})$, (35-4) can be simplified to

$$Y(z) = X(z) + (1 - z^{-1})Q(z). \tag{35-5}$$

The $(1 - z^{-1})$ factor in (35-5) is a simple first-difference differentiator, depicted in Figure 35-2(b). The magnitude of the differentiator's frequency response is

$$|H_d(\omega)| = 2\left|\sin\left(\frac{\omega}{2}\right)\right|. \tag{35-6}$$

This function has a zero at $\omega = 0$. The magnitude of the differentiator's frequency response, $|H_d(\omega)|$, is plotted in Figure 35-2(c)—and it is the frequency response of a highpass filter.

Equations (35-5) and (35-6) tell us that the digital portion of the DAC passes the input, $x(n)$, without change while the quantization noise is converted to

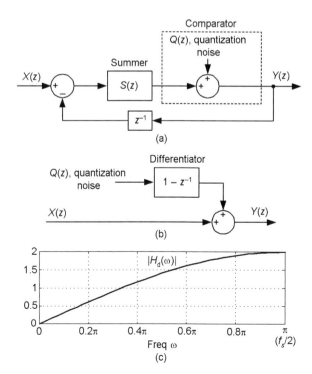

(a)

(b)

(c)

Figure 35-2 Z-domain analysis of a ΣΔ DAC: (a) block diagram; (b) differentiator; (c) magnitude of the differentiator's frequency response.

high-frequency noise by the feedback loop. The analog lowpass filter will have little effect on a low-frequency input signal while almost completely removing the high-pass noise. (The effect that the feedback loop of a $\Sigma\Delta$ converter has on the quantization noise is called *noise shaping*.) We find that a $\Sigma\Delta$ DAC converts low-frequency digital signals into analog signals with minimal error.

35.3 IMPLEMENTATION

We now consider implementing a $\Sigma\Delta$ DAC using a microprocessor from the (extremely popular) 8051 family. The goal is to implement the DAC with a minimum of instructions. The 8051 family supports unsigned addition and subtraction of 8-bit numbers. Selecting $A = 128$, we find that we can implement a 7-bit DAC. The only tricky part is making sure that no intermediate calculations become negative or exceed 255.

When calculating $z(n)$ we consider the two cases $z(n-1) \geq 128$ and $z(n-1) < 128$ separately. In the first case, we calculate $z(n)$ by first computing $f(z(n-1)) - c = 192 - c$. This number must be between 0 and 128. Then we calculate $z(n-1) - (f(z(n-1)) - c)$. Because of the assumption about $z(n-1)$ this number will be greater than or equal to zero and less than or equal to 255. Thus there is no need to borrow or carry during this calculation.

If $z(n-1) < 128$, then we first calculate $c - f(z(n-1))$. This number must be greater than or equal to zero—there can be no need to borrow here. Next we calculate $z(n) = z(n-1) + (c - f(z(n-1)))$ and we have again avoided any need for carry or borrow. Additionally checking whether $z(n-1)$ is greater than or equal to 128 is easy to do on a microprocessor that is a member of the 8051 family. Some of the registers in an 8051 are bit addressable. By moving $z(n-1)$ into such a register and checking whether or not its most significant bit is set, one has checked the condition.

The next-to-final stage is to output $d(t)$. The easy way to do this is to use one of the general purpose input/output (I/O) pins that all members of the 8051 family have. Rather than outputting the binary values 64 or 192, one outputs a logical low voltage or a logical high voltage, respectively. This introduces a scaling factor—but so do all DACs. The problem here is that the values taken by the general purpose I/O pins are not generally 0 and V_{cc} volts—they are voltages near these values. Though the values output are reasonably constant at any given pin, they need not be the same on all pins of a given port—to say nothing of the pins on different ports. (The choice of an output pin is usually not considered very important. Here changing pins may change the value actually seen at the output of the lowpass filter.) A $\Sigma\Delta$ DAC implemented this way should have 7-bit resolution but will not have 7-bit accuracy. Practically speaking one may find that one does not need the absolute accuracy. Alternatively, one could add hardware before the analog lowpass filter that would accommodate level shifting.

The final stage in the $\Sigma\Delta$ DAC is an analog lowpass filter. To avoid the need for a complicated, high-quality, lowpass filter one must make sure that the sampling speed of the digital portion is high enough. One sets the sampling speed—the length of the delay in the loop—by setting the rate at which an (internal) timer interrupts the microcontroller. If the sampling speed is high enough and if the input $x(n)$ signal is sufficiently low-frequency then one does not need an expensive lowpass filter. Suppose, for example, that one would like to convert the constant digital signal $x(n) = c$ to an analog voltage. Suppose that an 8-bit microprocessor is being used and a 7-bit DAC is desired. Then the maximal period of the output, $y(n)$, of the system is 128 clock cycles. Suppose that your clock is interrupting every 10 μs—that you perform a new calculation every 10 μs. (This is practical with many of the modern 8052-based processors.) Then the frequency of the noise riding on the constant starts at about 800 Hz. If one uses a simple RC lowpass analog filter with a 25 Hz cutoff frequency, one will obtain good results.

Also note that one of the advantages of the sigma-delta DAC relative to a pulse width modulation (PWM) DAC is that for many constant-valued inputs the period of the sequences $z(n)$ and $y(n)$ will be much less than 128 samples long and the frequency of the noise that "rides" on the desired output will be higher than the frequency of the noise produced by the PWM DAC. In such cases, an inexpensive lowpass filter will work well.

In order to demonstrate how a $\Sigma\Delta$ DAC behaves, we consider several examples. In Figure 35–3 the signals $z(n)$, $y(n)$, and $d(t)$ are shown when $A = 128$ and $x(n) = 188$. There we find that $z(n)$ remains between 0 and 255 and the average value of $y(n)$ is equal to 188, as the theory predicts.

When $x(n) = A = 128$, the output of the digital portion, $y(n)$, oscillates between the values $A/2$ (64) and $3A/2$ (192), and has an average value of 128—as it should. In this situation a low-cost analog lowpass filter will work very well indeed.

Figure 35–4 shows $z(n)$, $y(n)$, and $d(t)$ when $A = 128$ and $x(n) = 68$. Again $z(n)$ remains between 0 and 255 and the average value of $y(n)$ is equal to the value of the $x(n)$ input.

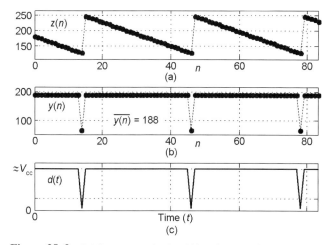

Figure 35–3   DAC sequences for $A = 128$ and $x(n) = 188$.

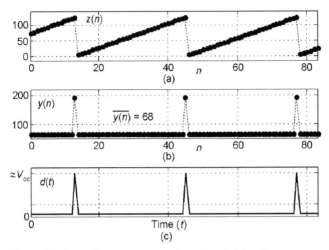

Figure 35–4   DAC sequences when $A = 128$ and $x(n) = 68$.

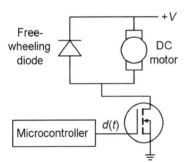

Figure 35–5   Unidirectional speed control of a DC motor.

Additionally, in each figure the period of the output is significantly less than the maximum period that the theory permits—and this means that our lowpass filtering will be more effective than we predicted.

If one is interested in controlling the speed of a DC motor control, one can take care of the problem of the not-very-well-defined logic levels and the analog lowpass filtering in a particularly simple fashion. DC motors are themselves lowpass filters—so no lowpass filtering is required. Additionally, the DC motor requires an input that is capable of driving the motor—and not the signal-level outputs of a microcontroller. One way to produce such a final output signal is to use the logic-level outputs of the microcontroller to control some form of power amplifier—and if done properly, the power amplifier can take care of the level shifting too.

If one is interested in using the motor to turn in one direction only, then connecting a MOSFET with a freewheeling diode to the output of the microcontroller, as shown in Figure 35–5, will do the job. As the MOSFET's resistance when turned on is very close to zero, the voltage driving the DC motor will be almost precisely $+V$. If one is interested in bidirectional control, then an H-bridge DC motor motion control chip (such as the National Semiconductor LM18201, or the Texas Instruments L293) is called for.

35.4 CONCLUSIONS

We have presented the properties of a simple $\Sigma\Delta$ DAC. We have shown that, when one uses fixed-point arithmetic, a constant input leads to a limit cycle whose average value is precisely the value of the constant. We have also shown that it is possible to actually implement the digital portion of DACs of this type in a simple fashion using microprocessors from the simple and popular 8051 family. Finally, we considered some implementation issues as well.

35.5 REFERENCES

[1] V. FRIEDMAN, "The Structure of the Limit Cycles in Sigma Delta Modulation," *IEEE Transactions on Communications*, vol. *36*, no. 8, August 1988, pp. 972–979.

[2] R. GRAY, "Oversampled Sigma-Delta Modulation," *IEEE Transactions on Communications*, vol. *35*, no. 5, May 1987, pp. 481–489.

[3] D. REEFMAN, J. REISS, E. JANSSEN, and M. SANDLER, "Description of Limit Cycles in Sigma-Delta Modulators," *IEEE Transactions on Circuits and Systems I—Regular Papers*, vol. *52*, no. 6, June 2005, pp. 1211–1223.

Chapter 36

Efficient 8-PSK/16-PSK Generation Using Distributed Arithmetic

Josep Sala
Technical University of Catalonia

This chapter describes a computationally efficient technique using distributed arithmetic (DA) to implement the digital pulse shaping filters required in generating phase shift keyed (PSK) signals for digital communications systems. DA constitutes an efficient multiplierless procedure for digital filtering in terms of reduced quantization noise. Nevertheless, optimization is usually required to reduce the exponential size of the associated lookup tables (LUT). Optimization techniques are either general (i.e., partitioning, where the filter is implemented as the addition of smaller subfilters), or specific (this chapter's material), when so allowed by the structure of the input signal or the filter itself. LUT size is critical in 8-PSK and 16-PSK finite impulse response (FIR) pulse shaping. Hence, here we show that 8-PSK and 16-PSK can be split into the superposition of simpler constellations, which leads to reduced complexity in DA implementations.

36.1 BACKGROUND

Pulse shaping (symbol filtering) constitutes the most complex part of digital modulators. Although some approaches use trivial coefficients for complexity reduction [1], DA is more flexible for implementing filters of any shape. Therefore, before describing the specifics of 8-PSK and 16-PSK optimization, and for the sake of completeness, a brief summary of DA and the structure of baseband digital modulators is provided. A detailed description of DA and related optimizations may be found in [2], [3]. In radix-2 DA, the output $y[n]$ from a FIR filter of impulse response h_n,

Streamlining Digital Signal Processing: A Tricks of the Trade Guidebook, Second Edition. Edited by Richard G. Lyons.
© 2012 the Institute of Electrical and Electronics Engineers. Published 2012 by John Wiley & Sons, Inc.

when the input sequence $x[n]$ is quantized to B bits, may be evaluated from the following decomposition,

$$y[n] = \sum_{k=0}^{L-1} h_k \cdot x[n-k] = \sum_{k=0}^{L-1} h_k \sum_{m=0}^{B-1} x_{n-k,m} w_m 2^{-m}$$

$$= \sum_{m=0}^{B-1} 2^{-m} w_m \cdot \overbrace{\left(\sum_{k=0}^{L-1} h_k x_{n-k,m} \right)}^{T_m} \qquad (36\text{--}1)$$

where $x_{n-k,m}$ constitutes the bit encoding of the input sample $x[n-k]$ and w_m the two's complement sign encoding of each binary weight ($w_0 = -1$ and $w_m = +1$ for all $m \geq 1$). All possible combinations $T_m = h_0 x_{n,m} + h_1 x_{n-1,m} + \ldots + h_{L-1} x_{n-L+1,m}$ are quantized and stored in a LUT of size 2^L, which is addressed by the L-bit-long word $\mathbf{x}_{n,m} = (x_{n,m}, \ldots, x_{n-L+1,m})$ (optimizations [3] based on offset binary coding of the input data or filter partitioning may be used to reduce LUT size). Thus, quantizing T_m rather than h_n allows to improve the signal-to-quantization-noise power ratio. After B lookups and $B-1$ shift and accumulations (summation over m), the output $y[n]$ is generated.

A baseband linear modulator performs pulse shaping with h_n as,

$$b[n] = \sum_{k=-\infty}^{+\infty} s[k] \cdot h_{n-k \cdot N_{SS}} \qquad (36\text{--}2)$$

with N_{SS} the number of samples per symbol at the modulator output $b[n]$, and $s[k]$ the input symbol sequence. This can be recast into a bank of N_{SS} subfilters, each corresponding to the convolution with each of the different N_{SS} decimated versions of the shaping pulse $h_{n1}[n_2] = h_{n1+N_{SS} \cdot n2}$, with $0 \leq n_1 \leq N_{SS} - 1$ and $-\infty \leq n_2 \leq +\infty$. Hence,

$$b[n_1 + n_2 N_{SS}] = \sum_{k=-\infty}^{+\infty} s[k] \cdot h_{n_1}[n_2 - k] \qquad (36\text{--}3)$$

This allows us to reduce the memory space, as only N_{SS} LUTs associated with the different $h_{n1}[n_2]$ are needed to compute the corresponding $b_{n1}[n_2] = b[n_1 + n_2 N_{SS}]$. Each decimated filter $h_{n1}[n_2]$ has either $L = \lceil N_C/N_{SS} \rceil$ or $L = \lfloor N_C/N_{SS} \rfloor$ coefficients, with a global LUT size $2^L N_{SS}$.

36.2   GENERATION OF 8-PSK AND 16-PSK MODULATIONS

For BPSK, QPSK, ASK, or QAM modulations, DA is straightforward as the input symbols $s[k]$ can be exactly quantized using very few bits (and hence, very few DA shifts and accumulations are necessary). Nevertheless, 8-PSK and 16-PSK do not yield easily to implementation in DA as their symbols do not coincide exactly with a finite (regular) quantization grid. An excessive number of bits is required to yield quantized symbols $s[k]$ with negligible amplitude and phase residual errors. Taking 8-PSK, an alternative to symbol quantization is to generate a LUT of size 4^L to store all possible outputs (8^L for 16-PSK), as the 8-PSK symbols have four different projections on the I-axis (and the Q-axis) as shown in Figure 36–1 using the small arrows p1 to p4. Unfortunately, for low L, the memory size is already too large, as medium-to-low roll-off factors of the shaping pulse may require a substantial number of coefficients.

Although filter partitioning may be applied, we propose a previous optimization named constellation superposition, which helps reduce the memory space to a large extent. We prove that the 8-PSK constellation can be decomposed into the exact addition of two QPSK constellations of different scale. The resulting constellation has 16 symbols, out of which 8 coincide exactly with an 8-PSK constellation. Accordingly, the 16-PSK constellation can be split into the exact addition of two 8-PSK constellations, or alternatively, into the addition of four QPSK constellations of different scale. As symbol shaping is a linear operation, decomposing 8-PSK or 16-PSK into the addition of simpler constellations is equivalent to decomposing the modulator output into the addition of simpler modulators. These decompositions are depicted in Figures 36–1 and 36–2. A geometrical proof is used to derive these results.

36.3 OPTIMIZED GENERATION OF 8-PSK MODULATIONS

Figure 36–1 shows the decomposition of 8-PSK into the addition of two QPSK constellations: the vertices A, B, C, D constitute the larger QPSK constellation, where the upper left vertex A is defined as the midpoint of segment A1-A2 (and so on for the other vertices B, C, and D). A second smaller QPSK constellation, centered at A, is used to generate A1 and A2 (which coincide with the wanted 8-PSK constellation), and two other unused symbols a1 and a2. The same procedure applies to the other vertices B, C, and D. Therefore, we may write the 8-PSK constellation as $8PSK = QPSK_1 + \xi \cdot QPSK_2$, with $0 < \xi < 1$ being the scale factor between the larger $QPSK_1$ and the smaller $QPSK_2$ constellations. If we let ρ denote the radius of the 8-PSK constellation, then we may calculate ξ considering triangle A2-O-A. Note that angle A2-O-A (between segments A2-O and O-A) is half the angle between two neighboring 8-PSK symbols, which is $\pi/8$. Hence, segments A2-A and O-A are the sides of a right-angled triangle, of hypotenuse ρ, and also the radius of constellations $QPSK_2$ and $QPSK_1$, respectively. Hence, the ratio of segment lengths $|A2 - A|/|O - A|$ equals ξ, which yields $\xi = \tan(\pi/8)$.

In Figure 36–1 the four-bit labels used to identify the two QPSK pseudo-symbols are related to the three-bit code of the 8-PSK constellation according to the procedure described in the forthcoming section on pseudosymbol generation for 8-PSK/16-PSK constellations.

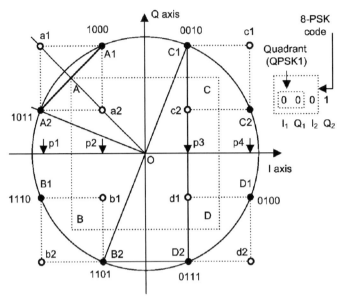

Figure 36–1   Construction of the 8-PSK constellation in terms of two constituent QPSK constellations.

Thus, two LUTs of equal size are required, storing the $QPSK_1$ and $QPSK_2$ outputs, respectively. The scale factor ξ is absorbed into the contents of the LUT corresponding to modulator $QPSK_2$. Very simple logic generates the pseudo-QPSK symbols to the two constituent QPSK modulators. Hence, the complexity of 8-PSK modulation is twice that of QPSK, which equals the ratio of their corresponding number of symbols: $8/4 = 2$.

An alternative approach [4] for the generation of 8-PSK (and 16-PSK) considers the rotation and combination of QPSK constellations as: $8PSK = C_1 + \exp[j\pi/4]C_2$. Nevertheless, this requires that the constituent constellations C_1 and C_2 be QPSK, extended with the zero symbol (five symbols per constellation). This is necessary as C_1 need be zero (modulator 1) when the 8-PSK symbol is generated by $\exp[j\pi/4]C_2$ (modulator 2). If DA is used to perform pulse shaping, the 0-symbol extension increases LUT size.

A useful result we will apply to the 16-PSK case is angle B2-C1-D2. Note that this angle is the same as that between segment O-C1 and the Q-axis, which yields $\pi/8$.

36.4   OPTIMIZED GENERATION OF 16-PSK MODULATIONS

Figure 36–2 shows the decomposition of 16-PSK into the addition of two 8-QPSK constellations. Grouping the 16 symbols into pairs {A1, A2} up to {H1, H2} and

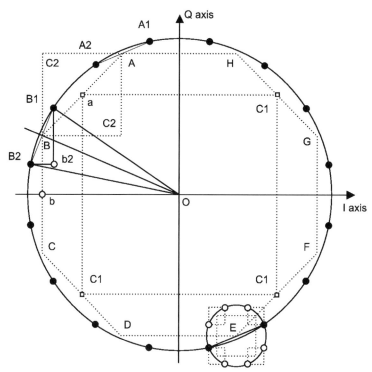

Figure 36–2   Construction of the 16-PSK constellation in terms of four constituent QPSK constellations or two constituent 8-PSK constellations.

using A up to H to denote the midpoint of segments A1-A2 up to H1-H2, respectively, the set of points {A, B, C, D, E, F, G, H} defines a regular octagon whose vertices constitute an 8-PSK constellation. A smaller 8-PSK constellation may be centered at each vertex of the octagon so that two of their eight symbols coincide with the symbols of the 16-PSK constellation (see example at vertex E). To this purpose, and considering center B, it is sufficient to show that angle B2-B1-b2 is precisely $\pi/8$, which, as proved previously, is the case for 8-PSK (see angle B2-C1-D2 in Figure 36–1). The reasoning follows in three steps:

1. Segments B2-B1 and B-O, crossing at B, are perpendicular.
2. Angle b-O-B (equal to angle B2-O-B1) is $2\pi/16 = \pi/8$, therefore its complementary angle b-B-O is $\pi/2 - \pi/8$.
3. Angle B2-B-b, which is the same angle as B2-B1-b2, is complementary to b-B-O (because of the perpendicularity of segments B2-B1 and B-O), which equals $\pi/2 - (\pi/2 - \pi/8) = \pi/2$.

Hence, an 8-PSK constellation centered at B generates symbols B1 and B2 on the 16-PSK constellation. This is easily extended to the other seven octagon vertices

due to the symmetries of the 16-PSK constellation (see example at vertex E). Therefore, we may write the 16-PSK constellation as: 16PSK = [8PSK]$_1$ + μ[8PSK]$_2$.

To compute μ, note that segment B2-B is the radius ρ_2 of the smaller 8-PSK
constellation, while the opposite side O-B of the same right-angled triangle B2-B-O
is the radius ρ of the larger 8-PSK constellation. As angle B2-O-B is precisely
π/16 (half the angle between neighboring 16-PSK symbols), its tangent is
μ = ρ_2/ρ = tan(π/16). Therefore, a complete expansion in terms of QPSK constellations yields,

$$16\,\text{PSK} = \left[\text{QPSK}_1 + \tan\left(\frac{\pi}{8}\right)\cdot\text{QPSK}_2 \right] + \tan\left(\frac{\pi}{16}\right)\cdot\left[\text{QPSK}_3 + \tan\left(\frac{\pi}{8}\right)\cdot\text{QPSK}_4 \right]$$

$$(36\text{-}4)$$

In summary, two 8-PSK modulators of equal complexity are required, generating
the [8PSK]$_1$ and [8PSK]$_2$ outputs, respectively, plus additional combinatorial logic
to perform scaling by tan(π/16). Alternatively, four QPSK modulators may be used,
with all scaling factors included in their respective LUTs. Very simple logic generates the pseudo-QPSK or pseudo-8-PSK symbols to the constituent modulators.
Hence, the complexity of 16-PSK modulation is roughly four-fold that of QPSK,
which equals the ratio of their corresponding number of symbols: 16/4 = 4.

The embedding of the 16-PSK constellation into the irregular 256-QAM constellation generated by the four constituent QPSK modulators is shown in Figure
36–3.

Figure 36–4 shows one possible implementation of a 16-PSK modulator where
all four constituent QPSK modulators are implemented in parallel for maximum
throughput. A small LUT (16 positions) or combinatorial function is required to

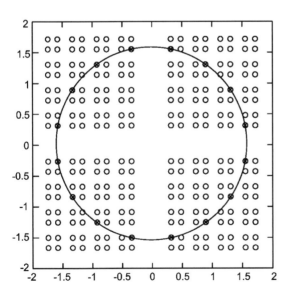

Figure 36–3 Embedding of the
16-PSK constellation in the irregular
256-QAM constellation.

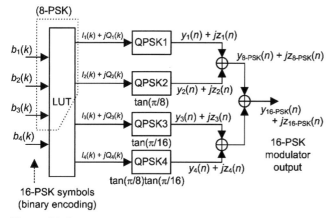

Figure 36–4 One version of 16-PSK modulator implementation.

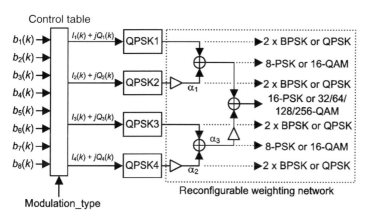

Figure 36–5 Multipurpose modulator architecture.

translate each 16-PSK symbol to its expression into a set of four QPSK pseudosymbols. The scaling factors are incorporated in each QPSK LUT.

Figure 36–5 shows how a slight modification in the 16-PSK implementation in Figure 36–4 yields a flexible architecture allowing us to generate a number of different modulations. The setting of the control table and of the multiplication factors α_i in the weighting network may generate a wide variety of modulations. For example, setting $\alpha_1 = \alpha_2 = \alpha_3 = 1/2$ allows to generate the 256-QAM modulation. Note that all QAM modulations up to 256-QAM can be generated with the addition of up to four QPSK constellations, where the multiplication factors are conveniently chosen to be either 0 or 1/2 (and the control table is conveniently configured to generate the corresponding pseudosymbols). All QPSK modulators are identical, which makes hardware re-use possible.

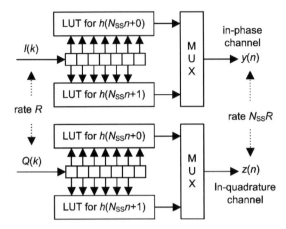

Figure 36–6   Structure of one of the constituent QPSK modulators in Figures 36–4 and 36–5.

Figure 36–6 shows the internal structure of one of the constituent QPSK modulators in Figures 36–4 and 36–5 for the case $N_{SS} = 2$ samples per symbol and a 13-coefficient shaping pulse $h(n)$. A bit-level shift register is used to address the LUTs storing the outputs corresponding to the two different decimation phases of the shaping filter. The outputs are multiplexed to yield the modulator output corresponding to each decimation phase.

36.5   PSEUDOSYMBOL GENERATION

We describe a symbol labeling procedure suitable for generating the pseudo-QPSK symbols to the constituent QPSK modulators: this requires 4- and 8-bit *labels* for 8-PSK and 16-PSK symbols, respectively, which are not to be mistaken with the original 3- and 4-bit *codes* of the corresponding constellations. Let the two bits identifying a QPSK symbol be denoted b_1, b_2. Let the corresponding QPSK symbol be denoted $I + jQ = S(b_1) + jS(b_2)$, where the sign function S of a digital bit b is defined as $\{S(0) = 1, S(1) = -1\}$. In this way, the opposite symbol in the constellation is the negation of the corresponding bits. Let the 3 and 4 bits corresponding to one 8-PSK or 16-PSK symbol be denoted b_1, b_2, b_3 and b_1, b_2, b_3, b_4, respectively. Then, setting $I_2 = S(b_3)$, the 8-PSK symbol can be constructed as,

$$I^{8\text{-PSK}} + jQ^{8\text{-PSK}} = (I_1 + jQ_1) + \tan\left(\frac{\pi}{8}\right) \cdot \left(I_2 + j\overbrace{(-I_1Q_1I_2)}^{Q_2}\right) \qquad (36\text{-}5)$$

with $Q_2 = -I_1Q_1I_2$. This can be verified in Figure 36–1. The first pseudosymbol (QPSK$_1$): $I_1 + jQ_1$ selects the quadrant (symbols A, B, C, or D). The second pseudosymbol (QPSK$_2$): $I_2 + jQ_2$ selects which of the two allowed pseudosymbols for that quadrant has to be generated. Note that any $I_2 + jQ_2$ of the smaller QPSK$_2$ constellation is perpendicular to $I_1 + jQ_1$ in all cases (their relative angle is $\pm\pi/2$ radians)

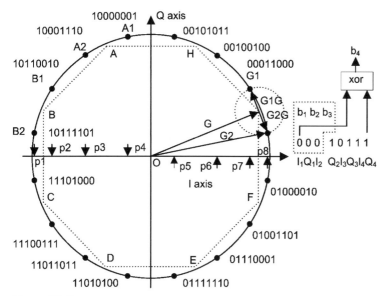

Figure 36–7 Construction of 8-bit labels for identifying the four constituent QPSK pseudosymbols in the 16-PSK constellation.

(see, for example, symbol A of QPSK$_1$ and symbols A1 and A2 in Figure 36–1). Hence, the term I_2 can be used to control the sign of $I_2 + jQ_2$, which allows us to determine $Q_2 = -I_1Q_1I_2$ in Figure 36–1. Therefore, the three independent terms I_1, Q_1, I_2 can be associated with bits b_1, b_2, b_3, respectively, and we may set $I_1 = S(b_1)$, $Q_1 = S(b_2)$, $I_2 = S(b_3)$.

With these thoughts in mind, Figure 36–7 shows the construction of the 8-bit labels for the identification of the four constituent QPSK pseudosymbols in the 16-PSK constellation. Their relationship between the original 4-bit code of a 16-PSK symbol is shown. The small arrows p1 to p8 show the eight possible values of the real (and imaginary) component of the 16-PSK symbols.

We note that to generate any 16-PSK symbol as the addition of two 8-PSK symbols, the two constituent 8-PSK symbols are perpendicular, for example, in Figure 36–7 where symbol G1 is generated as the addition of G and G1G. Therefore, we observe in the graph of Figure 36–1, that for any 8-PSK symbol, either the perpendicular symbol at +90° or the one at −90°, has a binary *label*, which is a left-to-right flipped version of the first 8-PSK symbol's binary *label*. For example, symbols C2 and A1 in Figure 36–1 have left-to-right flipped *labels*. A1 is found at +90° from C2, and its opposite D2 at −90° from C2 has then a *label* that is the bit-inversion of A1's *label*. It can be easily checked that this property holds for all 8-PSK symbols. Because of perpendicularity between the two constituent 8-PSK symbols of any 16-PSK symbol, the fourth 16-PSK bit b_4 can be used to control the sign of the second constituent (and perpendicular) 8-PSK symbol with $S_4 = S(b_4)$, such that final 16-PSK symbol labelling can be expressed as,

$$I^{16\text{-PSK}} + jQ^{16\text{-PSK}} = \left(I_1^{8\text{-PSK}} + jQ_1^{8\text{-PSK}}\right) + \tan\left(\frac{\pi}{16}\right) \cdot \left(I_2^{8\text{-PSK}} + jQ_2^{8\text{-PSK}}\right)$$

$$I_1^{8\text{-PSK}} + jQ_1^{8\text{-PSK}} = (I_1 + jQ_1) + \tan\left(\frac{\pi}{8}\right) \cdot (I_2 - jI_1Q_1I_2)$$

$$I_2^{8\text{-PSK}} + jQ_2^{8\text{-PSK}} = S_4 \cdot \left[(-I_1Q_1I_2 + jI_2) + \tan\left(\frac{\pi}{8}\right) \cdot (Q_1 + jI_1)\right] \tag{36–6}$$

$$= (I_3 + jQ_3) + \tan\left(\frac{\pi}{8}\right) \cdot (I_4 + jQ_4)$$

It can be checked by visual inspection of (36–6) that, except for the sign control S_4, the four real/imaginary terms in $I_{1,8\text{-PSK}} + jQ_{1,8\text{-PSK}}$ are just those of $I_{2,8\text{-PSK}} + jQ_{2,8\text{-PSK}}$ flipped left-to-right. This procedure can be examined in Figure 36–7, where the addition of the first constituent 8-PSK symbol (vector G) with the second constituent (perpendicular) 8-PSK symbol (vector G2G), generates the 16-PSK symbol G2. Note that for symbol G1 the first group of four bits of its *label*, 00011000, is a left-to-right flipped version of the second group of four bits, which identify, respectively, the first and second constituent 8-PSK symbols.

The *label* for symbol G2 is 00010111, whose second half is the inversion of G1's second half, as controlled by b_4. Note that b_4 can be obtained from the product $I_1Q_4 = (I_1)^2S_4 = S_4 = S(b_4)$, which corresponds to taking the exclusive-OR of the first and last bit of the 8-bit *label*. Along with $I_1 = S(b_1)$, $Q_1 = S(b_2)$, $I_2 = S(b_3)$, the four bits identifying a 16-PSK symbol can be recovered.

It is easy to verify by generating any complex symbol according to (36–5) or (36–6) that the radius of the 8-PSK and 16-PSK constellations becomes, respectively, $r_{8\text{-PSK}} = \sqrt{2} \cdot [\cos(\pi/8)]^{-1}$ and $r_{16\text{-PSK}} = \sqrt{2} \cdot [\cos(\pi/8)\cos(\pi/16)]^{-1}$.

36.6 COMPLEXITY EVALUATION

In this section, we will compare the complexity of two alternative DA implementations with the proposed QPSK constellation superposition method. To generate the N_{SS} complex samples in one output symbol, we require the N_{SS} subfilters (decimated versions of the original pulse shape) denoted $h_{n_1}[n_2] = h[n_1 + n_2N_{SS}]$ in (36–3). The final complexity measure results from the addition of the complexities of the N_{SS} subfilters, which depends on their respective lengths. For a pulse of N_C coefficients, we set $N_C = N_1N_{SS} + N_2$, with $N_1 = \lfloor N_C/N_{SS} \rfloor$, so that we get $N_2 = N_C - N_{SS}N_1$ subfilters of $N_1 + 1$ coefficients plus $N_{SS} - N_2$ subfilters of N_1 coefficients. Let $P_M(L)$ denote the number of memory positions in a DA method M for implementing a real subfilter of length L operating on a real channel. Then, the total number of required memory positions $P_{\text{all},M}$ required for processing either the in-phase or the in-quadrature channel becomes,

$$P_{\text{all},M} = N_2 \cdot P_M(N_1 + 1) + (N_{SS} - N_2) \cdot P_M(N_1). \tag{36–7}$$

EXAMPLE

For a 17-coefficient filter $\boldsymbol{h} = [h_0 h_1 h_2 \ldots h_{15} h_{16}]$ and $N_{SS} = 4$ samples per symbol, the four different decimations of the filter are: $\boldsymbol{h}_0 = [h_0 h_4 h_8 h_{12} h_{16}]$ of length $N_1 + 1 = \lfloor 17/4 \rfloor + 1 = 5$, plus $N_{SS} - N_2 = 4 - 1 = 3$ other subfilters $\boldsymbol{h}_1 = [h_1 h_5 h_9 h_{13}]$, $\boldsymbol{h}_2 = [h_2 h_6 h_{10} h_{14}]$, and $\boldsymbol{h}_3 = [h_3 h_7 h_{11} h_{15}]$ of length $N_1 = 4$. Hence, $P_{\text{all},M} = P_M(5) + 3P_M(4)$.

Now, we will compare $P_M(L)$ for three different methods M1, M2, and M3, also using additional optimization procedures (memory partitioning).

1. **Architecture M1.** A single LUT is used for the in-phase and in-quadrature channel to store all possible outputs of the modulator. If we look at either channel, the 8-PSK and 16-PSK constellations can take four (from p1 to p4 in Figure 36–1) and eight (from p1 to p8 in Figure 36–7) possible values, so that, respectively, the corresponding LUT size is 4^L and 8^L for storing all outputs of an L-coefficient filter. This is clearly unattainable for even moderate filter lengths. Therefore, filter partitioning techniques have to be applied where the filter is split into the addition of two or more smaller sub-filters. For example, filter $\boldsymbol{h}_0 = [h_0 h_4 h_8 h_{12} h_{16}]$ ($L = 5$) can be split into the addition of subfilters $\boldsymbol{h}_{01} = [h_0 h_4 h_8 00]$ ($L_1 = 3$) and $\boldsymbol{h}_{02} = [000 h_{12} h_{16}]$ ($L_2 = 2$). Then, instead of one large LUT, we add the outputs of two smaller LUTs. For a (L_1, L_2) partitioning of an L-coefficient filter with $L = L_1 + L_2$ (partitioning by two), the final number of memory positions after partitioning is reduced in the 8-PSK case from 4^L to $4^{L1} + 4^{L2}$ (in 16-PSK from 8^L to $8^{L1} + 8^{L2}$). In our case, from $4^5 = 1024$ to $4^3 + 4^2 = 80$ (in 16-PSK from 32768 to 576). The most suitable $(L_1, \ldots, L_r)$ partitioning with $L = L_1 + \ldots + L_r$ will depend in each case on constraints of the implementation platform. In fact, if for an L-coefficient filter we apply partitioning by $r = L$, we get L different LUTs of size 4 (8-PSK) or 8 (16-PSK) corresponding to the different multiplications of the in-phase/in-quadrature component of the complex symbol with each of the respective L coefficients. The number of memory positions required by method M1 for a filter of length L and partitioning by r is expressed as,

$$P_{M1}(L, r) = L_2 \cdot C^{L_1+1} + (r - L_2) \cdot C^{L_1}$$
$$L = L_1 \cdot r + L_2$$
$$L_1 = \lfloor L/r \rfloor \tag{36–8}$$
$$L_2 = L - r \cdot L_1$$

where $C = 4$ (8-PSK) or $C = 8$ (16-PSK). This partitioning corresponds to generating L_2 subfilters of length $L_1 + 1$ (C^{L_1+1} memory positions) plus $r - L_2$ subfilters of length L_1 (C^{L_1} memory positions). An aspect associated with memory partitioning is the number of additions required for adding up all subfilter LUT outputs, which is precisely $N_+(r) = r - 1$. Although partitioning reduces memory size, it introduces latency associated with the number of additions, and if one is not careful, an increased level of quantization noise. The minimum memory size is achieved for the maximum number of additions (L memories of size C), in partitioning by $r = L$. In DA implementations, partitioning may be used to trade-off additions with number of memory positions (which depends on additional constraints of the implementation platform).

2. **Architecture M2 (QPSK constellation superposition).** In this architecture we use the constellation partitioning techniques previously exposed. In the 8-PSK case (two constituent QPSK modulators), and for the five-coefficient subfilter $\boldsymbol{h}_0 = [h_0 h_4 h_8 h_{12} h_{16}]$, the number of required memory positions is $2 \times 2^5 = 64$ ($4 \times 2^5 = 128$ for 16-PSK, four constituent QPSK modulators). If the same partitioning as in the previous example had been applied, the final number of memory positions would be $2 \times (36 + 2^2) = 24$ (8-PSK) and $4 \times (36 + 2^2) = 48$ (16-PSK). The number of memory positions required by method M2 for a filter of length L and partitioning by r is expressed as

$$P_{M2}(L, r) = (C/2) \cdot \left(L_2 \cdot 2^{L_1 + 1} + (r - L_2) \cdot 2^{L_1}\right)$$
$$L = L_1 \cdot r + L_2$$
$$L_1 = \lfloor L/r \rfloor \tag{36-9}$$
$$L_2 = L - r \cdot L_1$$

Factor $C/2 = 2$(8-PSK) or 4(16-PSK) is accounting for the presence of the two- and four-constituent QPSK modulators in the specified modulations. The number of additions associated with this method (8-PSK) is $N_{+,M2}(r)_{\text{8-PSK}} = 2(r - 1) + 1 = 2r - 1$: twice the number of additions associated with partitioning by r (two QPSK constituent modulators) plus the final addition of the two QPSK modulators. In the 16-PSK case, we get $N_{+,M2}(r)_{\text{16-PSK}} = 4(r - 1) + 3 = 4r - 1$. This expression can be generalized in terms of C as $N_{+,M2}(r) = (C/2)r - 1$.

The following two factors should also be taken into account when evaluating complexity (nevertheless, as they can be applied to any of the three implementations described, they do not contribute to the comparison of their respective complexities):

1. For any partitioning strategy in either M1 or M2, the LUTs used by the in-phase or in-quadrature channels are equal. Hence, they may be shared depending on the clock rate.

2. The contents of any LUT have symmetry, as they store modulations with BPSK symbols (± 1's): The binary address $A = [a_0 a_1 \ldots a_{L-1}]$ and its negation $\text{neg}(A) = [\text{neg}(a_0) \, \text{neg}(a_1) \ldots \text{neg}(a_{L-1})]$ contain words opposite in sign, such that $\text{LUT}(A) = -\text{LUT}(\text{neg}(A))$. With very simple logic, this helps reduce the total memory size to 50%. This symmetry is described in [3].

We may now compare architectures M1 and M2 in terms of:

1. Memory size versus the partitioning order r. Evaluations for the 8- and 16-PSK cases are provided by (36–8) and (36–9), respectively.

2. Memory size for the same number of additions. Necessarily, this will imply different partitioning orders r for M1 and M2. This happens when

$$N_{+,\mathcal{M}2}(r') = (C/2) \cdot r' - 1 = N_{+,\mathcal{M}1}(r) = r - 1 \tag{36–10}$$

For comparison of type 2, we get from the previous equation that $r = (C/2)r'$. Under this constraint, the respective memory sizes of M1 in (36–8) and M2 in (36–9) become,

$$
\begin{aligned}
P_{\mathcal{M}2}(L, r') &= (C/2) \cdot \big[(L - r' \cdot \lfloor L/r' \rfloor) \cdot 2^{\lfloor L/r' \rfloor + 1} + \\
&\quad + (r' - L + r' \cdot \lfloor L/r' \rfloor) \cdot 2^{\lfloor L/r' \rfloor} \big] \\
&= (C/2) \cdot (L - r' \cdot \lfloor L/r' \rfloor + r') \cdot 2^{\lfloor L/r' \rfloor}
\end{aligned}
$$

$$
\begin{aligned}
P_{\mathcal{M}1}\!\left(L, \frac{C}{2}r'\right) &= \left(L - \frac{C}{2}r' \cdot \left\lfloor L \middle/ \frac{C}{2}r' \right\rfloor \right) \cdot C^{\left\lfloor L \middle/ \frac{C}{2}r' \right\rfloor + 1} + \\
&\quad + \left(\frac{C}{2}r' - L + \frac{C}{2}r' \left\lfloor L \middle/ \frac{C}{2}r' \right\rfloor \right) C^{\left\lfloor L \middle/ \frac{C}{2}r' \right\rfloor} \\
&= \left((C-1)\left(L - \frac{C}{2}r' \left\lfloor L \middle/ \frac{C}{2}r' \right\rfloor \right) + \frac{C}{2}r' \right) C^{\left\lfloor L \middle/ \frac{C}{2}r' \right\rfloor}
\end{aligned} \tag{36–11}
$$

Particularizing for 8-PSK and 16-PSK, we get,

$$
\begin{aligned}
P_{\mathcal{M}2}(L, r')_{8\text{-PSK}} &= 2 \cdot (L - r' \cdot \lfloor L/r' \rfloor + r') \cdot 2^{\lfloor L/r' \rfloor} \\
P_{\mathcal{M}1}(L, 2r')_{8\text{-PSK}} &= (3(L - 2r' \cdot \lfloor L/2r' \rfloor) + 2r') \cdot 2^{2\lfloor L/2r' \rfloor} \\
P_{\mathcal{M}2}(L, r')_{16\text{-PSK}} &= 4 \cdot (L - r' \cdot \lfloor L/r' \rfloor + r') \cdot 2^{\lfloor L/r' \rfloor} \\
P_{\mathcal{M}1}(L, 4r')_{16\text{-PSK}} &= (7(L - 4r' \cdot \lfloor L/4r' \rfloor) + 4r') \cdot 2^{3\lfloor L/4r' \rfloor}
\end{aligned} \tag{36–12}
$$

We note that architecture M2 allows to save memory space by reusing similar LUTs (provided the clock is fast enough): If the multiplication factors $\tan(\pi/8)$, $\tan(\pi/16)$, and $\tan(\pi/8)\tan(\pi/16)$ are brought outside the LUTs (as in Figure 36–5), the four constituent QPSK modulators have all equal LUTs, which can be reused, allowing a reduction of 50% (8-PSK) and 75% (16-PSK) if each LUT is reused $C/2 = $ two (8-PSK) or $C/2 = $ four (16-PSK) times. Tables 36–1 and 36–2 present these results in terms of filter length L and level of partitioning r', considering the proposed LUT reuse factor $C/2$.

Table 36–1 provides the 8-PSK percentage of physical memory size of M2 (based on the LUT reuse factor $C/2$) with respect to M1 for the same number of additions N_+: $(C/2)^{-1}P_{M2}(L, r')/P_{M1}(L, (C/2)r')$. Note that if LUT reuse is not implemented (multiply the percentages by two), there exists still a slight improvement in memory size. Short lengths are used for modulation pulses with a high roll-off factor.

Table 36–2 provides the 16-PSK percentage of physical memory size of M2 (based on the LUT reuse factor $C/2$) with respect to M1 for the same number of additions N_+: $(C/2)^{-1}P_{M2}(L, r')/P_{M1}(L, (C/2)r')$. The memory size of architecture M2 is shown in brackets. Those combinations of filter length and partitioning order

Table 36–1   8-PSK Percentage of Physical Memory Size of M2

	$r' = 1$	$r' = 2$	$r' = 3$	$r' = 4$
$L = 6$	0.5000	0.4000	0.5000	
$L = 7$	0.4000	0.4615	0.4444	
$L = 8$	0.5000	0.5000	0.4167	0.5000
$L = 9$	0.4000	0.4286	0.4000	0.4545
$L = 10$	0.5000	0.4000	0.4444	0.4286
$L = 11$	0.4000	0.4615	0.4762	0.4118
$L = 12$	0.5000	0.5000	0.5000	0.4000
$L = 13$	0.4000	0.4286	0.4444	0.4348
$L = 14$	0.5000	0.4000	0.4167	0.4615
$L = 15$	0.4000	0.4615	0.4000	0.4828
$L = 16$	0.5000	0.5000	0.4444	0.5000
$L = 17$	0.4000	0.4286	0.4762	0.4545
$L = 18$	0.5000	0.4000	0.5000	0.4286

Table 36–2   16-PSK Percentage of Physical Memory Size of M2

	$r' = 1$	$r' = 2$	$r' = 3$	$r' = 4$
$L = 6$	0.4433 [64]			
$L = 7$	0.6400 [128]			
$L = 8$	1.0000 [256]	0.5000 [32]		
$L = 9$	0.7273 [512]	0.4000 [48]		
$L = 10$	0.8889 [1024]	0.3636 [64]		
$L = 11$	1.2800 [2048]	0.4138 [96]		
$L = 12$	2.0000 [4096]	0.4444 [128]	0.5000 [48]	
$L = 13$	1.4545 [8192]	0.5581 [192]	0.4211 [64]	
$L = 14$	1.7778 [16384]	0.6400 [256]	0.3846 [80]	
$L = 15$	2.5600 [32768]	0.8421 [384]	0.3636 [96]	
$L = 16$	4.0000 [65536]	1.0000 [512]	0.4000 [128]	0.5000 [64]
$L = 17$	2.9091 [131072]	0.8000 [768]	0.4255 [160]	0.4348 [80]
$L = 18$	3.5556 [262144]	0.7273 [1024]	0.4455 [192]	0.4000 [96]

where M2 does not reduce memory size are underlined (note that this only happens for long filter lengths, which are only required for atypical, very small roll-off factors of the modulation pulse). Nevertheless, subsequent partitioning (larger r') does already offer an improvement in memory size for the same number of additions.

We observe that when comparing M1 and M2 with respect to the same number of additions, architecture M2 (8-PSK, Table 36–1) shows a slight improvement, which is substantial in the case of LUT reuse. We also observe that architecture M2

(16-PSK, Table 36–2) provides reductions in memory size with respect to M1 when compared at the same number of additions.

So far we have seen that M2 leads to savings in complexity when memory is reused. We may compare it now with the classical distributed arithmetic scheme described in the introduction, which is also based on memory reuse.

> **3. Architecture M3.** This is a typical DA processor, where input symbols are quantized to B bits. The wordlength B assigned to the quantized input symbols must be sufficiently long to reduce the quantization error of 8-PSK or 16-PSK to a level similar to that in M1 or M2. If we use Q for the memory reuse factor, the DA processor performs $Q = B$ iterations (memory lookup and shift-accumulation operations) over the same memory. For the same partitioning strategy of M2, the physical memory is the same (it stores BPSK modulated outputs, with $P_{M3}(L) = 2^L$ when no partitioning is used). Nevertheless, for M2 the memory reuse factor was found to be $Q = C/2 = 2$(8-PSK) and $Q = C/2 = 4$(16-PSK). It is clear that M3 requires $Q = B \gg 4$ for a good quantization signal-to-noise ratio (SNR). Hence, although the physical memory in M3 is the same as in M2, more iterations are required.

For a typical wordlength of $B = 8$ bits, the classical DA-8-PSK scheme requires four times as many iterations as the optimized 8-PSK scheme M2 (two iterations), whereas the DA-16-PSK scheme requires twice as many iterations as the optimized 16-PSK scheme M2 (four iterations). Hence for 8-PSK and 16-PSK we can modulate at 25% and 50% of the cost achievable with M3.

The relationship that may be established between M2 and M3 is the fact that M2 can be viewed as a quantization of input symbols on an irregular quantization grid (see the grid shown in Figure 36–3 as an irregular 256-QAM constellation), in contrast to method M3, which uses a regular quantization grid, and thus requires more iterations (a finer grid) to achieve a similar quantization SNR. Hence, the equivalent in method M2 of the shift-accumulate operations on the memory lookups in method M3 is found in the output weighting network of the four LUTs in Figure 36–5.

36.7 CONCLUSIONS

We have shown a construction method for 8- and 16-PSK symbols based on a superposition of two and four QPSK constellations, respectively, that leads to reductions in the physical memory size of DA implementations of the corresponding modulators. The comparison has been established versus the partitioning order of (36–8) and (36–9) and for the same number of additions (Tables 36–1 and 36–2).

The proposed scheme is also very flexible in the sense that it can operate as a multipurpose modulator, which, with minimum hardware reconfigurability, can generate a number of different linear modulations. The structure is shown in Figure 36–5, where the choice of the adequate three external multiplications provides for generation of: BPSK, QPSK, 8-PSK, 16-PSK, 16-QAM, 32-QAM, 64-QAM,

128-QAM, and 256-QAM. An input interface module allows us to translate the input symbol encoding to the respective QPSK pseudosymbol encoding.

Architecture M1 has been used as a benchmark for comparison as, disregarding complexity (LUT size), M1 is associated with best performance in terms of quantization SNR (without partitioning, it stores the direct quantization of the modulator outputs). In this comparison, architecture M2's SNR is degraded with respect to M1 only a fraction of one quantization bit (1 bit = 6 dB) as shown in the results presented in Figure 36–8.

Figure 36–8(a) shows the SNR loss (in dB) of architecture M2 with respect to M1: $L(M1, M2) = SNR_{M1,B_o} - SNR_{M2,B_o}$ dB, for a final rounding to $B_o = 9$ bits (circles) and $B_o = 10$ bits (crosses). Memory reductions allowed by architecture M2 are traded-off with a small degradation in quantization SNR with respect to M1, where a loss of 6 dB in quantization SNR corresponds to one-bit resolution. Thus, incurred degradation is just a fraction of a quantization bit. Results obtained with quantization analyses are usually sensitive to low-level optimizations: different choices for partitioning or the configuration of the arithmetic. In this simulation,

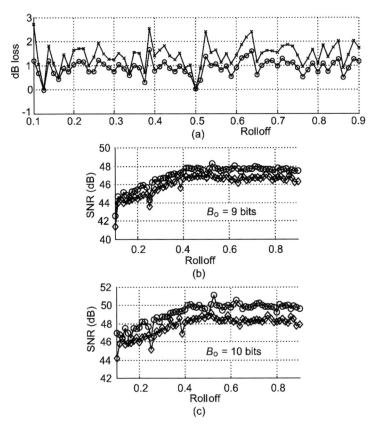

Figure 36–8 SNR comparison of M2 and M1 architectures.

filter partitioning assigns contiguous samples of the original pulse shape to each subfilter.

Figures 36–8(b) and (c) show the quantization SNR versus rolloff (the shaping pulse is a square-root raised-cosine filter) for architectures M2 (diamonds) and M1 (circles), when partitioning is adjusted for both to have the same number of additions and $r' = 1$. The wordlength of the LUT values is 8 bits and the summation network of LUT outputs performs a final rounding to $B_o = 9$ bits in Figure 36–8(b), and $B_o = 10$ bits in Figure 36–8(c). The pulse shape is sampled at four samples per symbol and has $7 \times 4 + 1 = 29$ coefficients, which corresponds to entry ($L = 7$, $r' = 1$) in Table 36–2. The additive complexity is four additions.

This work was financed by the Spanish/Catalan Science and Technology Commissions and FEDER funds from the European Commission: TEC2004-04526, TIC2003-05482, and 2005SGR-00639.

36.8 REFERENCES

[1] D. KLYMYSHYN and D. HALUZAN, "FPGA Implementation of Multiplierless M-QAM Modulator," *Electronic Letters*, vol. *38*, issue 10, 9 May 2002, pp. 461–462.

[2] A. PELED and B. LIU, "A New Hardware Realization of Digital Filters," *IEEE Trans. on ASSP*, vol. *ASSP-22*, no. 6, December 1974, pp. 456–461.

[3] S. WHITE, "Applications of Distributed Arithmetic to Digital Signal Processing: A Tutorial Review," *IEEE ASSP Magazine*, July 1989, pp. 4–19.

[4] M. RUPP and J. BALAKRISHNAN, "Efficient Chip Design for Pulse Shaping," *2nd Workshop on Signal Processing Advances in Wireless Communications (SPAWC)*, 9–12 May 1999, pp. 304–307.

Chapter 37

Ultra-Low-Phase Noise DSP Oscillator

Fred Harris
San Diego State University

Many DSP algorithms in analysis and communication systems require a complex sinusoid to accomplish various signal rotation tasks. Examples include the discrete and fast Fourier transforms, digital up/down conversions, and operations in communication carrier recovery loops [1]. Often the desired sine and cosine samples of the complex sinusoid are generated by a computationally efficient oscillator, called a *direct digital synthesizer* (DDS) and implemented with the CORDIC algorithm [2]. In a CORDIC DDS the phase noise (phase angle error) is inversely proportional to the number of CORDIC iterations performed. As such, to improve the accuracy (i.e., reduce the phase noise) of the sine and cosine samples, additional CORDIC iterations need to be performed.

This chapter describes a novel complex oscillator, which is based on an interesting variation of the traditional CORDIC DDS and produces sine and cosine output samples of any specified angle. Our oscillator provides drastically improved phase noise performance (relative to a traditional CORDIC DDS) without the need for additional CORDIC iteration processing. In addition, our oscillator supports real-time output sample-by-sample digital frequency control.

In describing our enhanced complex oscillator, we first present the arithmetic processing needed to generate sine and cosine samples. Next we show how that processing is implemented using the CORDIC algorithm. Following that we detail the trick used to reduce oscillator phase noise errors. Finally we show how to guarantee a stable oscillator output amplitude and present an example of our oscillator's ultra-low-phase noise performance.

Streamlining Digital Signal Processing: A Tricks of the Trade Guidebook, Second Edition. Edited by Richard G. Lyons.
© 2012 the Institute of Electrical and Electronics Engineers. Published 2012 by John Wiley & Sons, Inc.

37.1 OSCILLATOR OUTPUT SEQUENCE GENERATION

Consider a complex oscillator with output samples $w(n)$ given by

$$w(n) = \exp(j\theta) \cdot w(n-1), \quad \text{where } n = 0, 1, 2, \ldots \text{ and } w(-1) = 1 + j0. \quad (37\text{–}1)$$

An oscillator that implements (37–1) operates at a digital frequency of $\theta = 2\pi f_C/f_S$ radians/sample. Frequency θ represents the oscillator's change in angle per unit time index, and frequencies f_C and f_S are the oscillator's output cyclic frequency and sample rate, respectively, in hertz. Using $w(n) = x(n) + jy(n)$ we can write

$$x(n) = x(n-1)\cos(\theta) - y(n-1)\sin(\theta): \quad x(-1) = 1$$
$$y(n) = x(n-1)\sin(\theta) + y(n-1)\cos(\theta): \quad y(-1) = 0. \quad (37\text{–}2)$$

Sequences $x(n)$ and $y(n)$ are the desired cosine and sine oscillator outputs, respectively. The remainder of this chapter describes how to accurately generate these two sequences.

The traditional structure of a complex oscillator that implements (37–2) is illustrated in the block diagram of Figure 37–1(a). Next, from this structure we can obtain an alternate form, which is illustrated in Figure 37–1(b), by factoring the cosine terms from (37–2) as

$$x(n) = [x(n-1) - y(n-1)\tan(\theta)]\cos(\theta)$$
$$y(n) = [y(n-1) + x(n-1)\tan(\theta)]\cos(\theta) \quad (37\text{–}3)$$

Then, by rearranging the computations in Figure 37–1(b) we can draw our desired Figure 37–1(c) oscillator wherein the $\tan(\theta)$ multiplications can be efficiently implemented by the CORDIC algorithm, without the need for multipliers.

37.2 CORDIC ALGORITHM-BASED PROCESSING

CORDIC tan(Θ) Computation

To understand how processing is implemented using the CORDIC algorithm, consider the structure illustrated in Figure 37–1(c). If the digital frequency θ is a known fixed value, we can compute $\tan(\theta)$ and $\cos(\theta)$ offline. On the other hand, if the digital frequency θ is to be programmable or continuously variable, we have the problem of computing the $\tan(\theta)$ and $\cos(\theta)$ terms for an arbitrary angle θ. We respond to this task by replacing the $\tan(\theta)$ multiplications in (37–3) with a sequence of elementary rotations known as the CORDIC rotate. We convert a multiplication by $\tan(\theta)$ to trivial shift-and-add operations by selecting rotation angles θ_k that satisfy

$$\tan(\theta_k) = 2^{-k}, \quad k = 0, 1, 2, \ldots, K-1. \quad (37\text{–}4)$$

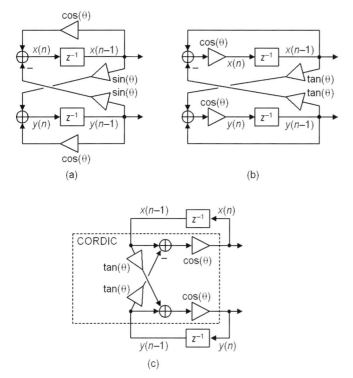

Figure 37–1   Block diagrams of oscillator networks: (a) traditional form; (b) alternate form; (c) desired recursive CORDIC form.

In practice the θ_k angles, which are explicitly listed in Table 37–1, are stored in a lookup table memory. Any angle θ in the first quadrant can be approximated by a binary (± 1) weighted sum of the angles, such as

$$\theta = \sum_{k=0}^{9} \alpha_k \theta_k + \theta_{\text{Rem}}; \alpha_k = \pm 1 \qquad (37\text{–}5)$$

where θ_{Rem} is a residual error. (Values of θ larger than 90° are accommodated by multiplying with j.) For example, a 10-term approximation results in an angle error less than atan(2^{-10}), which is approximately equal to 0.056 degrees. Later we'll see how this residual error θ_{Rem} is folded into the rotation as a final clean-up correction to greatly improve the performance of our DDS.

To successively approximate the desired tan(θ) in K rotations (iterations) we use a state machine that cycles through the sequence of binary shifts and adds to approximate the desired product. An initial $-\theta$ value stored in the angle accumulator is driven by the CORDIC system towards zero by adding or subtracting the successive angles accessed from the arctangent lookup table in Figure 37–2.

Table 37–1   CORDIC θ_k Rotation Angles for Different k

k	2^{-k}	$\theta_k = \text{atan}(2^{-k})$, [degrees]	$\cos(\theta_k) = \dfrac{1}{\sqrt{1+2^{-2k}}}$
0	1.0	45.0000000	0.70710678
1	0.5	26.5650510	0.89442719
2	0.25	14.0362430	0.97014250
3	0.125	7.1250160	0.99227787
4	0.0625	3.5763340	0.99805257
5	0.03125	1.7899110	0.99951207
6	0.015625	0.8951737	0.99987795
7	0.0078125	0.4476142	0.99996948
8	0.00390625	0.2238105	0.99999237
9	0.001953125	0.1119057	0.99999809

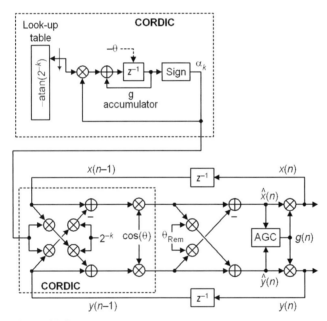

Figure 37–2   Recursive CORDIC DDS with automatic level control.

CORDIC cos(Θ) Computation

As the CORDIC algorithm proceeds to compute $\tan(\theta)$, the $\cos(\theta)$ multiply operation in Figure 37–1(c) remains to be performed. To limit the number of arithmetic operations, instead of executing a $\cos(\theta_k)$ multiplication during each CORDIC rotation, we perform the $\cos(\theta)$ multiply only once at the end of the rotation sequence. For

θ in the first quadrant, recalling that $\cos(\theta_k) = 1/\text{sqrt}[1 + \text{atan}^2(\theta_k)]$, the $\cos(\theta)$ factor for $K = 10$ CORDIC rotations is given by

$$\cos(\theta) = \prod_{k=0}^{9} \cos(\theta_k) = \prod_{k=0}^{9} \frac{1}{\sqrt{1+2^{-2k}}} = 0.596495. \tag{37-6}$$

So, at the end of the CORDIC rotations, we multiply each CORDIC output by $\cos(\theta) = 0.596495$. (Note that the right-most column in Table 37–1 contains the individual $\cos(\theta_k)$ factors whose cumulative product is 0.596495.) As an added benefit, the $\cos(\theta)$ multiply exactly compensates for the undesired output amplitude increase inherent in the CORDIC algorithm.

37.3 LOW-PHASE NOISE CORDIC OSCILLATOR DESIGN

With these notes in mind, we now obtain the structure of our final recursive CORDIC oscillator as shown in Figure 37–2. In this oscillator, for each complex output sample the negative of the desired frequency angle θ is inserted in the angle accumulator. The CORDIC rotation engine performs (for instance) $K = 10$ iterations of binary shift-and-add operations of the ordered pairs $[x(n - 1), y(n - 1)]$ while trying to zero the content of the angle accumulator by adding or subtracting the angles $\theta_k = \text{atan}(2^{-k})$ stored in the arctangent lookup table. Based on the sign of the current-iteration angle accumulator content, the sign function in Figure 37–2, whose output is ±1, determines whether an angle addition or angle subtraction takes place. The $\cos(\theta)$ multiply is applied once, at the end (after the 10th rotation) rather than once per rotation. After the 10th CORDIC rotation there is assuredly a nonzero residual angle θ_{Rem} in the angle accumulator. So our DSP trick is that we perform a clean-up rotation by advancing the CORDIC's complex output angle by the correction angle θ_{Rem}. This phase angle correction, whose derivation is described in [3], is what suppresses first-order phase noise in the DDS.

37.4 OUTPUT STABILIZATION

Upon analysis of the network in Figure 37–1(a) we find that its z-domain single-pole location resides on a Cartesian grid with the real part of the pole equal to $\cos(\theta)$ and the imaginary part of the pole equal to $\sin(\theta)$. Because the sine and cosine (or the tangent and cosine) in the network are transcendental numbers, and the arithmetic implementing the network has finite precision, the system pole can not lie precisely on the unit circle. Thus the pole is either inside or outside the unit circle and the response of the network to an initial condition is an exponentially decaying or growing sinusoid. To use our DDS as a quadrature signal generator we must incorporate an automatic gain control (AGC) loop to stabilize its output amplitude as we want to keep the system pole on the unit circle. The AGC that we employ is shown in Figure 37–2, where the data-dependent $g(n)$ gain correction, needed to ensure the oscillator's output magnitude remain at unity, is given by

$$g(n) = \frac{3 - [\hat{x}^2(n) + \hat{y}^2(n)]}{2} \tag{37-7}$$

where $\hat{x}$ and $\hat{y}$ are the outputs of the θ_{Rem} rotation operation [3]–[5]. The $g(n)$ gain term in (37–7), applied to the output of the θ_{Rem} phase angle correction process, operates in the direction to correct the amplitude error caused by the final phase correction, and the amplitude increase or decrease due to the pole position error relative to the unit circle.

37.5 EXAMPLE

Consider an example where the CORDIC algorithm ran 10 iterations, and the arithmetic used 20-bit multipliers. The spectrum of the oscillator's complex sinusoid stabilized with the AGC mechanism in (37–7) is shown in Figure 37–3(a), which also illustrates the ultra-low-phase noise of our oscillator. The AGC gain factor $g(n)$ in (37–7) less its nominal unity value is shown in Figure 37–3(b).

   An interesting aspect of the recursive CORDIC is that for a fixed-frequency sinusoid, the angle accumulator is initialized with the same angle value for each successive time sample. Thus the sequence of add-subtract iterations in the CORDIC

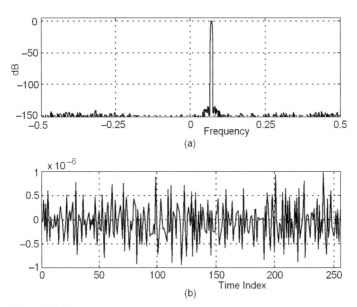

Figure 37–3   Complex CORDIC oscillator: (a) spectrum; (b) small angle and finite-arithmetic gain correction time series.

is identical for each computed trigonometric sample. The memory of the recursive CORDIC resides in the network states rather than in the traditional phase accumulator that forms and presents a sequence of phase angles modulo 2π to the CORDIC's angle accumulator. Thus the phase error sequence is a constant for the recursive CORDIC; it is always the same angle error residing in the angle accumulator. Consequently there is no line structure in the spectrum of the recursive CORDIC, and the phase error correction is not applied to suppress phase error artifacts but rather to complete the phase rotation left incomplete due to the residual phase term in the angle accumulator. This is a very different DDS!

37.6 IMPLEMENTATION

As a practical note, there are truncating quantizers between the AGC multipliers and the feedback delay element registers. As such, the truncation error circulates in the registers and contributes an undesired DC component to the complex sinusoid output. This DC component can (and should) be suppressed by using a sigma delta-based DC cancellation loop between the AGC multipliers and the feedback delay elements [6].

37.7 CONCLUSIONS

We modified the traditional recursive DDS complex oscillator structure to a tangent/cosine configuration. The $\tan(\theta)$ computations were implemented by CORDIC rotations avoiding the need for multiply operations. To minimize output phase angle error, we applied a post-CORDIC clean-up angle rotation. Finally, we stabilized the DDS output amplitude by an AGC loop. The phase-noise performance of the DDS is quite remarkable and we invite you, the reader, to take a careful look at its structure. A MATLAB-code implementation of the DDS is made available at http://booksupport.wiley.com.

37.8 REFERENCES

[1] C. DICK, F. HARRIS, and M. RICE, "Synchronization in Software Defined Radios—Carrier and Timing Recovery Using FPGAs," *IEEE Symposium on Field-Programmable Custom Computing Machines*, Napa Valley CA, April 17–19, 2000, pp. 195–204.
[2] J. VALLS, et al., "The Use of CORDIC in Software Defined Radios: A Tutorial," *IEEE Communications Magazine*, vol. *44*, no. 9, September 2006, pp. 46–50.
[3] F. HARRIS, C. DICK, and R. JEKEL, "An Ultra Low Phase Noise DDS," Software Defined Radio Conference (SDR-2006), Orlando, FL, 13–15 November 2006.
[4] R. LYONS, *Understanding Digital Signal Processing*, 3rd ed. Upper Saddle River, NJ, Prentice Hall, 2010, pp. 786–788.
[5] C. TURNER, "Recursive Discrete-Time Sinusoidal Oscillators," *IEEE Signal Processing Magazine*, vol. *20*, no. 3, May 2003, pp. 103–111.
[6] C. DICK and F. HARRIS, "FPGA Signal Processing Using Sigma-Delta Modulation," *IEEE Signal Proc. Magazine*, vol. *17*, no. 1, January 2000, pp. 20–35.

Chapter 38

An Efficient Analytic Signal Generator

Clay S. Turner
Pace-O-Matic, Inc.

Sometimes in DSP applications one has a real-valued signal and needs to transform it into an analytic signal. Interest in analytic signals stems from their having one-sided Fourier transforms and the ease with which one may calculate their instantaneous amplitudes and frequencies. Traditionally, generating analytic signals is done by augmenting the real signal with its Hilbert transform as shown in Figure 38–1(a). Specifically if our real signal is $i(t)$ and its Hilbert transform is $q(t)$, then our analytic signal is simply $i(t) + jq(t)$, where $j^2 = -1$ [1].

One shortcoming of the Figure 38–1(a) analytic signal generation scheme is that the Hilbert transformer cannot be designed to have exactly unity gain, as we desire, over its full passband. Therefore the two paths exhibit unmatched frequency magnitude responses. (For completeness, we mention that reference [2] proposed an analytic signal generation scheme where both its $i(t)$ and $q(t)$ channels have identical frequency magnitude responses; however, that scheme does not exhibit a linear-phase frequency response.) Since the absolute phase associated with a true analytic signal is rarely needed, we can generate a pseudoanalytic signal by using a pair of unity gain filters having a phase response difference of 90°. This generality suggests that there are many such possible pairs of filters, and there are! Reference [3] presents an analytic signal generation method where a halfband filter is created via a Remez algorithm and the result is modulated by a complex sinusoid. A limitation of their method applies to the length of filters because the design algorithm encounters numerical problems for large filter lengths.

This chapter presents a simple way to construct a pair of quasi-linear-phase bandpass filters that have identical magnitude responses and differ in phase by 90° that can be used for analytic signal generation as shown by the quadrature filter in Figure 38–1(b). Furthermore, these filters have useful symmetry properties that

Streamlining Digital Signal Processing: A Tricks of the Trade Guidebook, Second Edition. Edited by Richard G. Lyons.
© 2012 the Institute of Electrical and Electronics Engineers. Published 2012 by John Wiley & Sons, Inc.

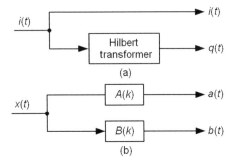

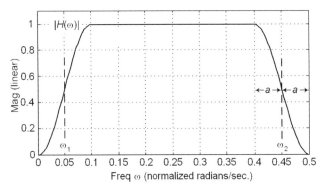

Figure 38–1 Analytic signal generation methods: (a) traditional; (b) proposed.

Figure 38–2 Desired analytic signal generation filters' magnitude response.

significantly reduce their computational complexity and coefficient storage requirements, and there is no inherent length limitation to the size of the filters.

38.1 ANALYTIC SIGNAL GENERATION APPROACH

Since the filter pair's phase difference is key to the design, we will elect to use FIR filters with linear phase. And as is well known, to have linear phase, FIR filters must have impulse responses that are either odd or even symmetric about their midpoint [4]. This fact will prove very useful momentarily.

38.2 DEFINING THE FREQUENCY RESPONSE

We start by defining our two filters' identical frequency magnitude response for positive frequencies. Then we will exploit the aforementioned symmetry rules of linear phase filters to find their impulse responses. We construct our bandpass response using two pieces of a sinusoid joined together with a horizontal line as shown in Figure 38–2.

In this filter design we will specify three parameters: The two half-amplitude points ω_1 and ω_2, and the transition half-bandwidth a. For this example they are, respectively, 0.05, 0.45, and 0.05 radians per second. The transition region width is $2a = 0.1$ radians per second.

Algebraically, the positive frequency, $\omega \geq 0$, response of $H(\omega)$ is given by:

$$H(\omega) = \begin{cases} 0, & \omega < \omega_1 - a \\ \sin^2\left\{\dfrac{\pi}{4a}[\omega-(\omega_1-a)]\right\}, & \omega_1 - a \leq \omega \leq \omega_1 + a \\ 1, & \omega_1 + a \leq \omega \leq \omega_2 - a \\ \cos^2\left\{\dfrac{\pi}{4a}[\omega-(\omega_2-a)]\right\}, & \omega_2 - a \leq \omega \leq \omega_2 + a \\ 0, & \omega > \omega_2 + a. \end{cases} \quad (38\text{--}1)$$

38.3   IMPULSE RESPONSE DERIVATIONS

To find our filters' impulse responses, we will now use the fact that they are related to the $H(\omega)$ frequency response via the Fourier transform, and in finding the Fourier transform we will exploit the time symmetry demanded by the linear phase constraint. Our in-phase filter will have even time symmetry, thus its frequency response will also be even symmetric. So we perform an even extension $H(\omega)$ and then find its continuous inverse Fourier transform $I(t)$. By doing this, the in-phase impulse response is given by:

$$I(t) = 2\int_0^\infty H(\omega)\cos(\omega t)d\omega = 2\pi^2 \cos(at)\frac{\sin(\omega_1 t)-\sin(\omega_2 t)}{t\left(4a^2 t^2 - \pi^2\right)}. \quad (38\text{--}2)$$

This method is similar to that used to find the impulse responses of *raised cosine* and *root raised cosine* filters [5]. Likewise, we find our quadrature filter's $Q(t)$ impulse response by forming an odd extension of $H(\omega)$ and finding its inverse Fourier transform. Thus,

$$Q(t) = 2\int_0^\infty H(\omega)\sin(\omega t)d\omega = 2\pi^2 \cos(at)\frac{\cos(\omega_2 t)-\cos(\omega_1 t)}{t\left(4a^2 t^2 - \pi^2\right)}. \quad (38\text{--}3)$$

Figure 38–3 shows the $I(t)$ and $Q(t)$ filters' impulse responses. Due to space limitations here, the derivations of (38–2) and (38–3) are made available at http://booksupport.wiley.com.

38.4   ROTATING FILTER IMPULSE RESPONSES

These filters essentially meet our design criteria, but when we sample the $I(t)$ and $Q(t)$ impulse responses, certain asymmetries in the resulting frequency response will

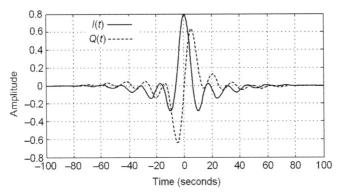

Figure 38–3   In-phase and quadrature phase filter impulse responses.

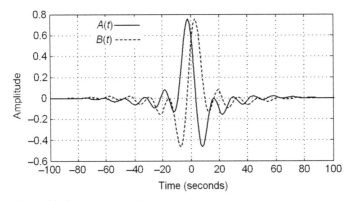

Figure 38–4   Rotated (±45°) filters' impulse responses.

occur, potentially reducing the filters' magnitude and phase performance. So what we do is transform our 0° and 90° filters into a −45° and +45° pair via a 45° rotation. This is done by

$$\begin{bmatrix} A(t) \\ B(t) \end{bmatrix} = \begin{bmatrix} \cos(45°) & -\sin(45°) \\ \sin(45°) & \cos(45°) \end{bmatrix} \begin{bmatrix} I(t) \\ Q(t) \end{bmatrix} \qquad (38\text{–}4)$$

where $A(t)$ and $B(t)$ are the rotated in-phase and quadrature phase impulse responses. This simplifies to:

$$A(t) = \frac{I(t) - Q(t)}{\sqrt{2}} \qquad (38\text{–}5)$$

$$B(t) = \frac{I(t) + Q(t)}{\sqrt{2}}. \qquad (38\text{–}6)$$

The resulting $A(t)$ and $B(t)$ impulse responses are shown in Figure 38–4.

If we apply the transformations, (38–5) and (38–6), to (38–2) and (38–3), respectively, we obtain:

$$A(t) = \frac{2\pi^2 \cos(at)}{t(4t^2 a^2 - \pi^2)}\left[\sin\left(\omega_1 t + \frac{\pi}{4} \right) - \sin\left(\omega_2 t + \frac{\pi}{4} \right) \right] \qquad (38\text{–}7)$$

$$B(t) = \frac{2\pi^2 \cos(at)}{t(4t^2 a^2 - \pi^2)}\left[\sin\left(\omega_1 t - \frac{\pi}{4} \right) - \sin\left(\omega_2 t - \frac{\pi}{4} \right) \right]. \qquad (38\text{–}8)$$

This transformation allows us to eliminate half the coefficients. Both filters will now share the same set of coefficients because $B(t) = A(-t)$. Filter B's coefficients are just a time-reversed version of filter A's coefficients, and this "mirrored in time" relation will cut the coefficient memory requirement in half as well as make the two frequency magnitude responses identical.

38.5 COMPUTING FIR FILTER COEFFICIENTS

When using (38–7) to compute the filter coefficients, one may encounter a divide by zero error, which we resolve by applying L'Hospital's rule. Doing this, we obtain the following complete formulation for $A(t)$:

$$A(t) = \begin{cases} \sqrt{2}(\omega_2 - \omega_1), & t = 0 \\ a\left\{ \sin\left[\frac{\pi}{4}\left(\frac{a + 2\omega_2}{a} \right) \right] - \sin\left[\frac{\pi}{4}\left(\frac{a + 2\omega_1}{a} \right) \right] \right\}, & t = \frac{\pi}{2a} \\ a\left\{ \sin\left[\frac{\pi}{4}\left(\frac{a - 2\omega_1}{a} \right) \right] - \sin\left[\frac{\pi}{4}\left(\frac{a - 2\omega_2}{a} \right) \right] \right\}, & t = \frac{-\pi}{2a} \\ \frac{2\pi^2 \cos(at)}{t\left(4a^2 t^2 - \pi^2 \right)}\left[\sin\left(\omega_1 t + \frac{\pi}{4} \right) - \sin\left(\omega_2 t + \frac{\pi}{4} \right) \right], & \text{otherwise.} \end{cases} \qquad (38\text{–}9)$$

Now we need to temporally sample $A(t)$ to generate our filters' coefficients. Using N to denote the length of the FIR filters' impulse responses, and $k = 0, 1, 2, \ldots, N - 1$ as the index of the impulse response samples (note that N can be either even or odd), we find our in-phase filter coefficients by evaluating the continuous expression in (38–9) at the time instants defined by:

$$A[k] = A\left[2\pi\left(k - \frac{N-1}{2} \right) \right]. \qquad (38\text{–}10)$$

Again, the quadrature phase coefficients, $B[k]$, are a reversed-order version of the in-phase $A[k]$ coefficients in (38–10).

As an example, when $N = 50$ and $\omega_1 = 0.05$, $\omega_2 = 0.45$, and $a = 0.05$, we obtain the frequency magnitude response shown in Figure 38–5(a) for our analytic signal generator.

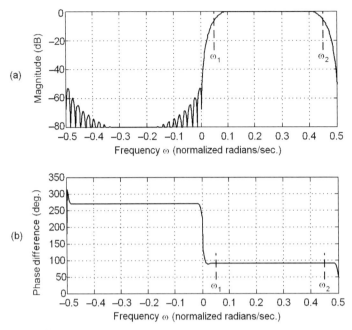

Figure 38–5   Analytic signal generator; (a) frequency magnitude response; (b) A[k] and B[k] filters' phase difference.

As Figure 38–5(a) shows, our analytic signal generator has a flat frequency response over its positive-frequency passband, and the A[k] and B[k] filters have a phase response differential that differs by 90° as seen in Figure 38–5(b). The filters' peak-peak phase-difference error (deviation from 90°) is a mere 0.04° over the passband range of $0.1 \leq \omega \leq 0.4$.

Since we know that FIR filters with strictly linear phase must have either even or odd symmetry and our transformed filters do not possess this symmetry, then they can't be truly linear phase. However, their deviations from linear phase are small and only occur for frequencies around DC and at half of the sampling rate—places where the magnitude response is nearly zero.

Our filter transformations also bring about a neat property. When the desired frequency response is chosen to be symmetric about one-fourth of the sampling rate, and our filter lengths are chosen to be even, then half of the filter's coefficients will be zero! So this, in combination with the mirror property, means our pair of filters will have half the computational effort of a pair of arbitrary filters of the same length.

38.6   CONCLUSION

We proposed a simple method for generating a pair of high-performance phase-orthogonal FIR filters used to generate analytic signals. The filters have a 90° phase differential, quasi-linear phase responses, and flat magnitude responses in their passbands. Due to their time reversal symmetries, only one set of filter coefficients need be stored in memory. If the frequency response of even-length filters is chosen to be centered at one-fourth the sampling rate, then half of the coefficients will be zero. And finally, because the filters coefficients are defined by a closed-form formula, there is no filter length limitation imposed by a numerical process that fails to converge.

MathCAD and MATLAB code, modeling this dual quadrature filters technique, is made available at http://booksupport.wiley.com.

38.7 REFERENCES

[1] D. GABOR, "Theory of Communications," *Trans. Inst. Electr. Eng.*, vol. *93*, no. 26, November 1946, pp. 429–457.

[2] C. RADER, "A Simple Method for Sampling In-Phase and Quadrature Components," *IEEE Trans. Aerospace and Electronic Syst.*, vol. *20*, no. 6, November 1984, pp. 821–824.

[3] A. REILLY, G. FRAZER, and B. BOASHASH, "Analytic Signal Generation—Tips and Traps," *IEEE Trans. on Signal Processing*, vol. *42*, no. 11, November 1994, pp. 3241–3245.

[4] J. MCCLELLAN and T. PARKS, "A Unified Approach to the Design of Optimal FIR Linear-Phase Digital Filters," *IEEE Transactions on Circuit Theory*, vol. *CT-20*, no. 6, November 1973, pp. 697–701.

[5] C. TURNER, "Raised Cosine and Root Raised Cosine Formulae," May 2007, [Online: http://www.claysturner.com/dsp/Raised%20Cosine%20and%20Root%20Raised%20Cosine%20Formulae.pdf.]

Part Five

Assorted High-Performance DSP Techniques

Chapter 39

Frequency Response Compensation with DSP

Laszlo Hars
Seagate Research

In modern telecommunication systems there are situations when test instruments must work over hundreds of narrowband-frequency channels. However it is difficult and expensive to build hardware test equipment with flat frequency response over their full operational frequency range. In this situation simple FIR filters can be applied for gain compensation to improve an instrument's frequency response flatness. This chapter describes a very fast method for run-time design of these filters, while minimizing storage requirements, based on test instrument calibration data [1]–[7].

39.1 FILTER TABLE

When a particular center frequency for a channel-under-test is selected, a simple FIR filter can be used to improve test instrument gain flatness. However, computing and storing the necessary filter coefficients is often impractical. The frequency responses of the filters depend on the test center frequency, and this would require storing thousands of sets of the filter coefficients along with a table telling us which filter is needed for each center frequency. The filter coefficients have to be determined at calibration time. If the center frequency can be chosen with high precision this precalculated filter approach is unattractive because a set of filter coefficients is needed for each center frequency, requiring huge storage space and a very long calibration time.

Streamlining Digital Signal Processing: A Tricks of the Trade Guidebook, Second Edition. Edited by Richard G. Lyons.
© 2012 the Institute of Electrical and Electronics Engineers. Published 2012 by John Wiley & Sons, Inc.

39.2 RUN-TIME FILTER DESIGN

A better gain compensation filtering approach is to measure the amplitude characteristics of the test instrument on a sufficiently dense frequency grid when the instrument gets calibrated. Only this table has to be stored. Using interpolation (linear, cubic or spline—depending the smoothness of the frequency response curve), the required gain compensation filter response can be determined with a handful of arithmetic operations at run time. The compensation filter can be very short if the sampling rate and center frequency are chosen appropriately. Given the required compensation filter gain, we developed a closed-form expression for the filter coefficients allowing us to compute them with only a handful of operations. All together the calculations are so fast that even arbitrary frequency hopping can be implemented with slow, low-power DSPs.

39.3 CALIBRATION TABLES

In most applications, the analog input signal has to be attenuated or amplified by circuits that are themselves not perfect, either. In theory, we would need a frequency response table for each possible attenuation setting. In practice, however, the frequency response varies only at smaller attenuation values; higher values provide good decoupling between otherwise interfering circuit parts. As such, in practice roughly 10 compensation tables are often enough for even high-precision measurements. The effects of different attenuation devices are cumulative at proper design (at least at higher attenuation values), which further reduces the number of necessary tables. Temperature compensation can be incorporated, too. The measured ambient or internal temperature represents another dimension for the family of tables.

39.4 FIR VERSUS IIR FILTERS

IIR filters have more complicated formulas for their gain response than FIR filters, therefore real-time calculation of the IIR coefficients takes longer. We normally also need constant group delay in the passband, which is more difficult to achieve with IIR filters. The filters must be stable; that is, the roots of the denominator of the transfer function of IIR filters must all lie inside the complex unit circle, and this is another difficulty to be dealt with. However, the same long IIR filters can somewhat better approximate a given gain curve. If this curve has very sharp peaks, notches, or edges, IIR filters are needed. Also, the FIR filters can have large group delays. If the group delay must be small or even negative, IIR filters have to be used.

The little better approximation of the desired amplitude curve by an IIR filter can be balanced by applying a longer FIR filter, whose coefficient calculations are simpler. The filtering takes about the same time because of the faster FIR filter code. Therefore FIR filters are a good choice. They must be short, having less than 10 coefficients, otherwise the calculation of the coefficients gets complicated.

39.5 LINEAR PHASE FIR FILTERS

The frequency response function of an N-tap FIR filter with coefficient sequence c_0, $c_1, \ldots, c_{N-1}$ is given by

$$H(\omega) = \sum_{k=0}^{N-1} c_k \cdot e^{-jk\omega} \tag{39–1}$$

If the filter has linear phase response (constant group delay), the coefficient sequence must be symmetric or antisymmetric. We want a filter with relative flat amplitude response, that is, close to unity gain everywhere, also at DC. If the coefficient sequence is antisymmetric ($c_k = -c_{N-1-k}$), the DC gain is 0; therefore we need a symmetric coefficient sequence. The filter delay is $(N-1)/2$. It is easier to accommodate an integer number of filter sample delays, which is what we have if N is odd. In this case the response function becomes

$$H(\omega) = e^{-j(N-1)\omega/2} \times \left[c_{(N-1)/2} + 2 \sum_{k=(N+1)/2}^{N-1} c_k \cdot \cos\left[\left(k - \frac{N-1}{2}\right)\omega\right] \right]. \tag{39–2}$$

The factor $e^{-j(N-1)\omega/2}$ represents the delay having a constant magnitude of 1, and the real factor in the brackets gives the (signed) amplitude response. The length of the filter can be chosen to be $N = 7$, and it has four free coefficients. We need the desired gain to be a given value at three different frequencies, and one degree of freedom remains to enforce a smooth amplitude response curve. Let the coefficient sequence be $[c, b, a, d, a, b, c]$. From (39–2), the amplitude response of the corresponding FIR filter is

$$A(\omega) = d + 2a \cdot \cos(\omega) + 2b \cdot \cos(2\omega) + 2c \cdot \cos(3\omega) \tag{39–3}$$

39.6 SAMPLING AND SIGNAL FREQUENCY

Usually, if some signal conversion is performed before the amplitude response correction we can simplify the design and save processing time. We need mixing and decimation to reduce the sampling rate f_{Samp} to a little above double the signal bandwidth. This is the minimum, which preserves all the information of the original signal (Nyquist theorem). It usually involves a bandpass filter step, too, removing those disturbing signals that would alias to the useful frequency band. This frequency band is best located around the center frequency f_0, where $f_0 = f_{Nyq}/2 = f_{Samp}/4$ (or at $3f_{Nyq}/2$), such that no signal component aliases back into the useful frequency band at another location. (On the normalized scale where the Nyquist frequency is 1, $\omega = \pi \cdot f$ and $f_0 = 1/2$.) Using a normalized frequency axis, the FIR filter's amplitude response can be expressed

$$A(f) = d + 2a \cdot \cos(\pi f) + 2b \cdot \cos(2\pi f) + 2c \cdot \cos(3\pi f). \tag{39–4}$$

39.7 FILTER DESIGN

When applying the compensation filter we do not want to change the amplitude at f_0, the center of the frequency band of the signal. (If necessary, we adjust the overall gain outside of the filter.) For the frequency response compensation we specify the gains g_1 and g_2 of the filter at two other frequencies, say $f_1 = 1/4$ and $f_2 = 3/4$ at both sides of f_0. (They can be chosen closer or further away from f_0, according to the need to have more accurate compensation close to the center frequency or less accurate compensation over the whole band.) These represent three constraints, unity gain at f_0 and gains g_1 and g_2 at f_1 and f_2, for the filter response that is a function of four free coefficients. Using (39–4) we can express the three amplitude constraints as:

$$1 = d + 2a \cdot \cos(\pi/2) + 2b \cdot \cos(\pi) + 2c \cdot \cos(3\pi/2). \tag{39–5}$$

$$g_1 = d + 2a \cdot \cos(\pi/4) + 2b \cdot \cos(\pi/2) + 2c \cdot \cos(3\pi/4). \tag{39–6}$$

$$g_2 = d + 2a \cdot \cos(3\pi/4) + 2b \cdot \cos(3\pi/2) + 2c \cdot \cos(9\pi/4). \tag{39–7}$$

A very important additional requirement is that the filter must not have large ripple, (i.e., its amplitude response must be smooth). This can be guaranteed if we mandate a fourth constraint to require the slope of the response curve at f_0 be the same as that of the secant line connecting the points $[f_1,g_1]$ and $[f_2,g_2]$. The slope of the curve, the derivative, tells how steep the response curve is in the neighborhood of a given point. It is the same as the slope of the tangent line of the filter response curve. The slopes of the secant lines $[f_1,g_1]$ and $[f_2,g_2]$ around a point $[f_0,1]$ approximate the slope of the tangent $A'(f_0)$ of the (hopefully smooth) function A in (39–4). If we require an exact equality, it intuitively ensures some kind of smoothness. Of course, you can specify more complicated conditions, but in our case the following equality proved sufficient. That is:

$$\frac{g_1 - g_2}{f_1 - f_2} = A'(f_0) = \frac{d[A(f)]}{df} \tag{39–8}$$
$$= -2\pi a \cdot \sin(\pi f_0) - 4\pi b \cdot \sin(2\pi f_0) - 6\pi c \cdot \sin(3\pi f_0).$$

Because $f_1 - f_2 = -1/2$, and $f_0 = 1/2$, we have

$$g_1 - g_2 = \pi a \cdot \sin(\pi/2) + 2\pi b \cdot \sin(\pi) + 2\pi c \cdot \sin(3\pi/2). \tag{39–9}$$

This last requirement, constraint (39–9), now gives us a linear system of four equations for the four unknown coefficients of the filter.

Evaluating (39–5) through (39–7) and (39–9) for the real values of the trigonometric functions gives:

$$1 = d - 2b. \tag{39–10}$$

$$g_1 = d + \sqrt{2a} - \sqrt{2c}. \tag{39–11}$$

$$g_2 = d - g_2 = d - \sqrt{2a} + \sqrt{2c}. \qquad (39\text{--}12)$$

$$g_1 - g_2 = \pi a - 3\pi c. \qquad (39\text{--}13)$$

Solving those equations, we get a simple solution for our filter coefficients:

$$a = (g_1 - g_2)\left(\frac{3\sqrt{2}}{8} - \frac{1}{2\pi}\right). \qquad (39\text{--}14)$$

$$b = \frac{g_1 + g_2}{4} - \frac{1}{2}. \qquad (39\text{--}15)$$

$$c = (g_1 - g_2)\left(\frac{\sqrt{2}}{8} - \frac{1}{2\pi}\right). \qquad (39\text{--}16)$$

$$d = \frac{g_1 + g_2}{2}. \qquad (39\text{--}17)$$

So, based on predetermined test instrument calibration data stored as an array of g_1 and g_2 values versus center frequency, the 7-tap FIR filter's (39–14) through (39–17) coefficients are computed and used in real-time as new center frequencies are assigned during test instrument operation.

39.8 MATLAB SIMULATION

The following function, in MATLAB code, calculates the desired filter coefficients based on the desired gains at the frequencies $f_0 = 1/2$, $f_1 = 1/4$, and $f_2 = 3/4$.

```
function v = relatflt(dB1,dB2)
%relatflt(dB1,dB2) len=7 FIR filter of response:
% 0 dB gain at f0 = 1/2
% dB1 = gain compensation in dB at f1 = 1/4
% dB2 = gain compensation in dB at f2 = 3/4
g1 = 10^(dB1/20); % convert dB gain to linear
g2 = 10^(dB2/20); % convert dB gain to linear
a = (g1-g2)*0.37117514279802; % 3sqrt(2)/8 - 1/2/pi
b = (g1+g2)*0.25 - 0.5;
c = (g1-g2)*0.01762175220474; % sqrt(2)/8 - 1/2/pi
d = (g1+g2)*0.5;
v = [c b a d a b c];
```

Using the above code to compute and plot filter responses of linear compensation curves with dB gains of [0.4,–0.4], [0.2,–0.2], [0,0], [–0.2,0.2], [–0.4,0.4], we have those shown in Figure 39–1. The desired gain compensation values are indicated by the dots.

Example curves with decreasing gains of [0.5,–0.4], [0.3,–0.4], [0.1,–0.4], [–0.1,–0.4], [–0.3,–0.4], in dB, are provided in Figure 39–2.

Finally, example curves with increasing gains of [0.3,0.4], [0.1,0.4], [–0.1,0.4], [–0.3,0.4], [–0.5,0.4], in dB, are shown in Figure 39–3.

They agree completely with the gain compensation we wanted.

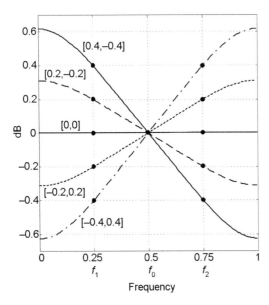

Figure 39–1   Example gain compensation curves for the $N = 7$ FIR filter.

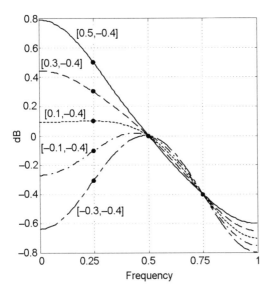

Figure 39–2   Example gain compensation curves for decreasing gain versus frequency.

39.9   IMPLEMENTATION IN C

Calculation of the FIR filter coefficients is straightforward in the C language as shown below. The floating point variables **d1** and **d2** specify the desired filter gains in dB at $f_1 = 1/4$ and $f_2 = 3/4$. The filter coefficients are stored after calculation in the array **Filt[Len]**, with **Len = 7**.

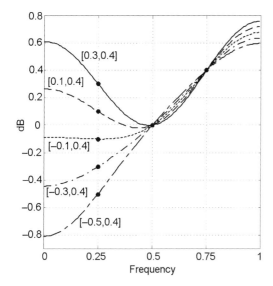

Figure 39–3 Example gain compensation curves for increasing gain versus frequency.

```
#define Len 7
#define C ((Len-1)/2)
/* from dB to ratio: */
float g1 = pow( 10.0, d1 / 20.0);
float g2 = pow( 10.0, d2 / 20.0);
float Filt[Len];
/* 3*sqrt(2)/8 - 1/2/pi = 0.37117514279802 */
/* sqrt(2)/8 - 1/2/pi = 0.01762175220474 */
Filt[C-1] = Filt[C+1] = (g1-g2)*0.37117514279802;
Filt[C-2] = Filt[C+2] = (g1+g2)*0.25 - 0.5;
Filt[C-3] = Filt[C+3] = (g1-g2)*0.01762175220474;
Filt[ C ] = (g1+g2)*0.5;
```

39.10 EXTENSIONS

The designed filters work well even beyond a ±6 dB compensation range. These large flatness errors, however, should not occur. They indicate serious test instrument hardware faults. If the shape of the compensation curve needs to be more complex, we can easily add to the constraints a second pair of frequencies with specified gains. We should request the slope of the response curve at the innermost frequency points be equal to the slope of the secant going through the surrounding specified curve points. We need now a 15-tap filter of the form: a_7, a_6, a_5, a_4, a_3, a_2, a_1, a_0, a_1, a_2, a_3, a_4, a_5, a_6, a_7.

Let $f_k = k\pi/6$ be the five frequencies where the gain will be specified, thus $k = 1$, $2, \ldots, 5$, and the corresponding gain values are g_k normalized at the center with $g_3 = 1$. We have five equations describing the correction gains, similar in form to (39–5) through (39–7):

$$g_k = a_0 + 2\sum_{i=1}^{7} a_i \cos(i\pi/6), \quad k = 1, \ldots, 5, \tag{39-18}$$

and another three, requiring the correct slopes, similar in form to (39–9):

$$g_k - g_{k+2} = (2\pi/3)\sum_{i=1}^{7} a_i \sin(i\pi/6), \quad k = 1, 2, 3. \tag{39-19}$$

The solution of the corresponding linear system of equations goes similarly as before:

$$a_0 = \frac{-3\sqrt{3}(g_1 - 2g_3 + g_5) + (5g_1 + 3g_2 - 4g_3 + 3g_4 + 5g_5)\pi}{12\pi}. \tag{39-20}$$

$$a_1 = \frac{[3(g_2 - g_4) + \sqrt{3}(g_1 - g_5)](-2 + \pi)}{4\pi}. \tag{39-21}$$

$$a_2 = \frac{(g_1 - 2g_3 + g_5)(-3\sqrt{3} + 4\pi)}{12\pi}. \tag{39-22}$$

$$a_3 = \frac{[3(g_2 - g_4) + \sqrt{3}(g_1 - g_5)](-3 + \pi)}{6\pi}. \tag{39-23}$$

$$a_4 = \frac{(g_1 - 2g_3 + g_5)(-3\sqrt{3} + 2\pi)}{12\pi}. \tag{39-24}$$

$$a_5 = \frac{72g_4 + 27\sqrt{3}g_5 + 36g_2(-2 + \pi) - 36g_4 - 5\sqrt{3}g_5 + \sqrt{3}g_1(-27 - 5\pi)}{72\pi}. \tag{39-25}$$

$$a_6 = \frac{-3\sqrt{3}(g_1 - 2g_3 + g_5) + (g_1 + 3g_2 - 8g_3 + 3g_4 + g_5)\pi}{24\pi}. \tag{39-26}$$

$$a_7 = \frac{-3[3g_1 + 4\sqrt{3}(g_2 - g_4) - 3 - g_5] + [g_1 + 6\sqrt{3}(g_2 - g_4) - g_5]\pi}{24\sqrt{3}\pi}. \tag{39-27}$$

It is more complex, requiring a little more processor power for the run-time design, but the work is still manageable. The following MATLAB code calculates the desired 15-tap filter coefficients with the like terms in (39–20) through (39–27) evaluated and $g_3 = 1$.

```
function v = flt15(dB1,dB2,dB4,dB5)
%flt15(dB1,dB2,dB4,dB5) len=15 FIR filter response
% dB1 = gain in dB at f1 = 1/6
% dB2 = gain in dB at f2 = 2/6
% 0 dB gain at f3 = 3/6
% dB4 = gain in dB at f4 = 4/6
% dB5 = gain in dB at f5 = 5/6
g1 = 10^(dB1/20);  g4 = 10^(dB4/20);
g2 = 10^(dB2/20);  g5 = 10^(dB5/20);
```

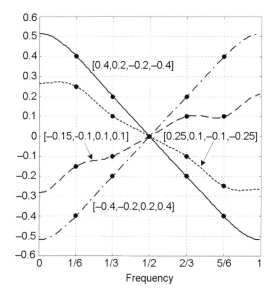

Figure 39–4 Example gain compensation curves for a 15-tap FIR filter.

```
a0 = 0.27883*(g1+g5) + 0.25000*(g2+g4) -0.05767;
a1 = 0.15735*(g1-g5) + 0.27254*(g2-g4);
a2 = 0.19550*(g1+g5) -0.39100;
a3 = 0.01301*(g1-g5) + 0.02254*(g2-g4);
a4 = 0.02883*(g1+g5) -  0.05767;
a5 =-0.08647*(g1-g5) + 0.18169*(g2-g4);
a6 =-0.02725*(g1+g5) + 0.12500*(g2+g4) -0.19550;
a7 =-0.04486*(g1-g5) + 0.09085*(g2-g4);
v =[a7 a6 a5 a4 a3 a2 a1 a0 a1 a2 a3 a4 a5 a6 a7];
```

Figure 39–4 shows a few example compensation response curves of 15-tap FIR filters designed using the above code. Again, the desired gain compensation values are indicated by the dots.

39.11 CALIBRATION TABLES

Usually the input signal gets attenuated and/or amplified before the analog-to-digital conversion to assure maximum digital resolution. These attenuator-amplifier circuits affect each other to a different degree, dependent on the selected attenuation. In theory, we need a frequency response calibration table for each possible attenuation value. If an instrument needs to be calibrated at several different attenuation levels over the whole frequency range, it will consume a lot of time and cost. However, proper design reduces the number of necessary calibration runs. In practice, without significant effort in the hardware design the frequency characteristics change only at smaller attenuation values, because higher attenuations provide good decoupling between otherwise interfering circuit parts. The affects of decoupled cascaded attenuation devices are cumulative (additive when measured in dB), which further reduces the number of necessary tables.

Temperature compensation can be incorporated, too. The measured ambient or internal temperature represents another dimension for the family of tables. If the temperature dependency is smooth (linear or close to that), and does not change the shape of the frequency response curve, one extra table is enough for high-precision compensation. The correction can be done by a temperature-dependent multiplication factor (additive in dB).

39.12   REFERENCES

[1] A. POULARIKAS, *Formulas and Tables for Signal Processing*, CRC Press—IEEE Press, New York, 1999.
[2] N. KALOUPTSIDIS, *Signal Processing Systems: Theory and Design*, John Wiley & Sons, New York, 1997.
[3] J. PROAKIS and D. MANOLAKIS. *Digital Signal Processing: Principles, Algorithms and Applications*, 3rd ed. Prentice Hall, Englewood Cliffs, NJ, 1996.
[4] N. FLIEGE, *Multirate Digital Signal Processing: Multirate Systems, Filter Banks, Wavelets*, John Wiley & Sons, New York, 1995.
[5] W. PRESS, S. TEUKOLSKY, W. VETTERLING, and B. FLANNERY. *Numerical Recipes in C*, Cambridge University Press, Cambridge, MA, 1992.
[6] L. RABINER and B. GOLD. *Theory and Application of the Digital Signal Processing*, Prentice Hall, Englewood Cliffs, NJ, 1975.
[7] A. OPPENHEIM and R. SCHAFER. *Digital Signal Processing*, Prentice Hall, Englewood Cliffs, NJ, 1975.

Chapter 40

Generating Rectangular Coordinates in Polar Coordinate Order

Charles Rader
Retired, formerly with MIT Lincoln Laboratory

When we deal with two-dimensional data, we almost always locate the data using either of two coordinate systems, rectangular coordinates or polar coordinates. Converting from one representation to the other can be a major nuisance. For example, suppose we have a function $f(r, \theta) = g(x, y) = g(r\cos\theta, r\sin\theta)$ and we can only measure the data using a polar coordinate measuring device even though we want to do our processing using rectangular coordinates. Here we present an elegant algorithm for computing rectangular indices in order of increasing polar angle.

40.1 CENTERED GRID ALGORITHM

For example, suppose I am standing on the origin of a piece of ruled graph paper with intersections at $(x, y) = (m, n)$; $0 \le m \le L$; $0 \le n \le M$ and I am holding a laser pointer. It is initially pointed along the x axis and illuminating the set of graph intersection points $(x, y) = (1, 0), (2, 0), \ldots, (L, 0)$. As I move my laser pointer slowly counterclockwise, the next intersection I illuminate will be $(L, 1)$, as shown in Figure 40–1, and shortly after that it will illuminate $(L - 1, 1)$, and so on.

Consider the illustrated case of $L = M = 5$. The table and the diagram in Figure 40–2 gives the order in which points will be illuminated (the solid dots—the open dots are shadowed by the solid dots on the same radial line).

This example illustrates that the order in which data becomes available with a polar coordinate measuring device is most unnatural from the point of view of a

Streamlining Digital Signal Processing: A Tricks of the Trade Guidebook, Second Edition. Edited by Richard G. Lyons.
© 2012 the Institute of Electrical and Electronics Engineers. Published 2012 by John Wiley & Sons, Inc.

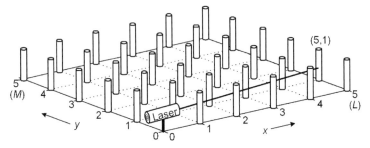

Figure 40–1 Rectangular coordinates and laser pointer illuminating point (5, 1).

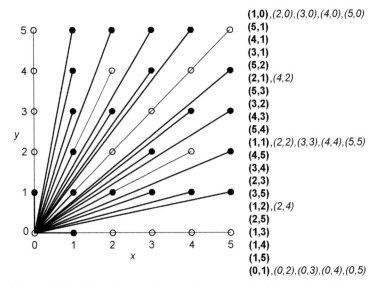

Figure 40–2 Illustrating the polar coordinates order of grid points (m, n); $0 \leq m$, $n \leq 5$.

rectangular coordinate system. If L is fairly large, this order becomes more and more obscure.

Normally, that won't matter. As soon as we make a measurement, we can store it in a random access memory such that $g(m, n)$ is in memory location $Kn + m$ (where K is any integer larger than L and will most likely be chosen as a power of two). Once the data are in the large memory, the rectangular coordinate system representation is achieved, and the order in which the data were accessed does not matter.

But suppose it does matter! I can think of two reasons we might care about the order. First, after we have measured the data at, say, (3, 5) we next need to point the laser beam in the direction of (1, 2) (and, on the same radial line, (2, 4)). That is, we need to point in the direction with $\theta = \tan^{-1}(2/1)$. In general, we need to know the pointing angles $\theta = \tan^{-1}(n/m)$ in their natural order. (Once we have reached that angle, the radii r of the intersections will be all the multiples of sqrt($m^2 + n^2$) with m not larger than L.)

Second, maybe we are simply given the data in a polar coordinate order, measured say every 5°. It will be necessary to use some sort of interpolation procedure to obtain the values of the function on rectangular coordinates. But interpolation procedures compute the value of a function based on nearby values. In the simplest case, we might want the value assigned to (m, n) to be the nearest measurement in r, θ. Or, we might want to use a weighted sum of the two closest points, and so forth. If we were to perform these interpolations for (m, n) taken in a natural rectangular order, we would need to read each appropriate polar coordinate datum into local memory several times. For example, data at angle 25° provides the nearest neighbors for (2, 1), (4, 2), (6, 3), (7, 3), (8, 4). But if we schedule the interpolation for points (m, n) in their natural polar coordinate order, we only need to keep in local memory the measurements along two successive angles, find the nearest neighbor (or several nearest neighbors) for all (m, n) pairs between those two angles, then drop the data for the smaller angle and read in data for the next larger angle, and so on.

For these reasons, it would be nice if we had an algorithm that generates the points (m, n) of a rectangular coordinate system in the natural order in which they would be obtained in a polar coordinate system.

Admittedly, if this need arises, it is easily met by constructing a table, just like in Figure 40–2. We present here an attractive alternative.

Let's note that in the table, whenever the point (m, n) occurs, other points (km, kn) in the rectangle lie on the same line so long as neither $kn > M$ nor $km > L$. Since every intersection in the square appears once, it follows that we can limit our attention to producing intersections (m, n) for which m, n have no common divisor. When the laser points at (m, n), the point nearest to the origin in that beam direction, it also points at other intersections (km, kn) at the same angle, and we only need to identify the first intersections, the one nearest to the origin.

Here is the algorithm for generating the mutually prime (m, n) pairs in order of increasing θ, using MATLAB notation:

```
a=0;b=1;c=1;d=L; %initialization
m=[b d];
n=[a c];
while ~((c==M)&(d==1)) % test for end
  Z = floor((L+b)/d);
  e = Z*c-a;
  f = Z*d-b;
  if e<=M
      % report pair (e,f)
      m = [m f];
      n = [n e];
  end
  b=d;d=f;a=c;c=e;
end
plot(m,n,'-.o')
axis([0 L 0 M])
```

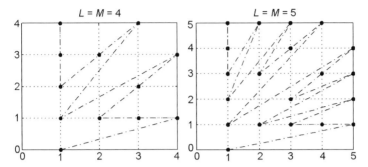

Figure 40–3   Algorithm (m, n) pairs.

Figure 40–3 shows the result of the code for $L = M = 4$ and $L = M = 5$.

Of course, the place in the program where the pair m, n is reported is the place where another routine can do whatever needs to be done with that pair.

As presented above, the algorithm generates (m, n) points in $0° \leq \theta \leq 90°$. But we have some flexibility. The routine can be initialized with any two successive angles $\theta_1 = \tan^{-1} n_1/m_1$, $\theta_2 = \tan^{-1} n_2/m_2$ and it can be ended at another angle $\theta_k = \tan^{-1} n_k/m_k$ we expect it to reach.

Therefore it is also possible to control the algorithm as a subroutine call. The subroutine inputs are L, and the last two pairs it reported, (a, b) and (c, d). The subroutine computes $Z = (L + b)/d$ and returns $e = Zc - a$ and $f = Zd - b$. However, the points with $e > M$ must not be left out, so M is not used within the subroutine.

As the laser beam angle increases, more and more points are skipped by the test $e \leq M$. Assuming $M \geq L$, the algorithm will never skip for angles less than 45°. So if we are only interested in $0 \leq \theta \leq \pi/4$, M becomes irrelevant and we can use

```
a=0;  b=1;  c=1;  d=L;  %initialization
% report pair a,b
% report pair c,d
while ~((c==1)&(d=1))  % test for end
  Z = floor((L+b)/d);
  e = Z*c-a;
  f= Z*d-b;
  % report pair e,f
  b=d;  d=f;  a=c;  c=e;
end
```

It is possible to use the 45° generating routine to fill the entire square by exploiting a symmetry. First we generate (m, n) points as above, from $(0, 1)$ to $(1, 1)$. Then we replace the initialization statement by

```
a=1;b=1;c=L-1;d=L;  %initialization
```

and replace the stop test by

```
while ~((c==0)&(d==1))  % end test
```

If we now continue the program it generates the original set of (m, n) points in reverse order. *Yes, the algorithm runs backwards!* But now we can swap (m, n)

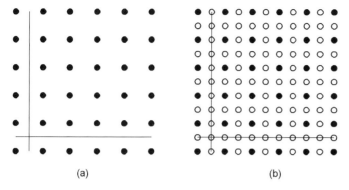

Figure 40–4   Origin centered among four grid points.

→ (n, m) and the resulting points fill the 45° to 90° slice of the desired rectangle. To go past 90° we run the original form and swap $(m, n) \rightarrow (-n, m)$.

We mention a curious property of the points this algorithm returns. If (m_{i-1}, n_{i-1}) and (m_i, n_i) are any two successive points, the triangle that they form with the origin $(0, 0)$ will always have area 1/2. Hence, the orbit obeys one of Kepler's laws of planetary motion, sweeping out equal area in equal time.

For a hardware implementation, it is convenient that the routine uses only non-negative integers, none larger than L.

40.2 NONCENTERED GRID

As we have presented the algorithm so far, the origin of the rectangular coordinate system is at $(0, 0)$. Another possibility is that the origin is midway between four grid points, as shown in Figure 40–4(a).

The grid points are located at $(m + 1/2, n + 1/2)$. But we can still visit them in their natural polar coordinate order.

First, imagine that we have overlaid the grid with a finer grid, as shown in Figure 40–4(b), and doubled the scale. The grid points of interest are now those at $(m', n') = (2m + 1, 2n + 1)$ (e.g., both m' and n' odd), but our algorithm will generate the all the points (m, n). If m, n are both odd, report out $(m', n') = (m, n)$, which stands for itself and for all its odd multiples lying on the same radial line. But if either m or n is even, the point (m, n) is ignored and we go on to the point the algorithm generates next. (The algorithm will never report out m and n both even because the pairs it finds are always mutually prime.)

40.3 TRIANGULAR GRID

Another possibility is that the grid, including the point at the origin, is not square but triangular. With a triangular grid, every other row of grid points is displaced by half a unit. Once again, we can interpolate other grid points so that they fill out a rectangular array, as shown in Figure 40–5.

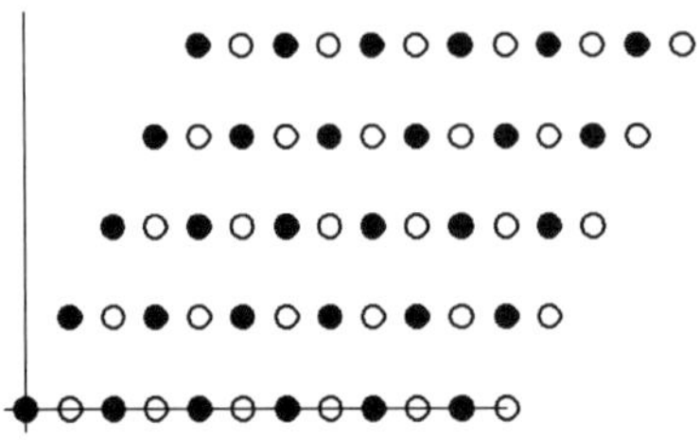

Figure 40–5   A triangular grid with interpolated grid points.

The points of interest (m', n') now have either both coordinates odd or both coordinates even. So when we run the algorithm and get a candidate (m, n) there are three cases:

> m **odd**, n **odd** Report $(m', n') = (m, n)$. The points on the same radial line are (km, kn) for all integers k.
>
> Otherwise: Report $(m', n') = (2m, 2n)$. The points on the same radial line are $(2km, 2kn)$ for all integers k.

40.4 CONCLUSIONS

We introduced the situation where we need the indices (m, n) within an L by M rectangle, of a rectangular coordinate system, to be sorted in the order of increasing angle $\theta = \tan^{-1}(n/m)$. If both L and M are small, we would probably accomplish this with a lookup table, but a surprisingly simple and elegant algorithm can be used instead and can be used for much larger L and M. It uses only integer arithmetic and a few simple tests. The algorithm can quickly identify the grid points between any two angles, or the polar coordinate measurements we need to compute rectangular coordinate measurements by interpolation.

Chapter 41

The Swiss Army Knife of Digital Networks

Richard Lyons   Amy Bell
Besser Associates   Institute for Defense Analysis

This chapter describes a general discrete-signal network that appears, in various forms, inside many signal processing applications. Practicing DSP engineers are well advised to become acquainted with this network. Figure 41–1 shows how the network's structure has the distinct look of a digital filter—a comb filter followed by a second-order recursive network. However, we do not call this general network a filter because its capabilities extend far beyond simple filtering. Through a series of examples we illustrate the network's ability to be reconfigured to perform a surprisingly large number of useful functions based on the values of its seven control parameters (coefficients).

The general network in Figure 41–1 has a transfer function of

$$H(z) = (1 - c_1 z^{-N}) \frac{b_0 + b_1 z^{-1} + b_2 z^{-2}}{1 / a_0 - a_1 z^{-1} - a_2 z^{-2}}. \tag{41–1}$$

From here out we'll use DSP filter lingo and call the second-order recursive subnetwork in Figure 41–1 a *biquad* because its transfer function is the ratio of two quadratic polynomials. The tables in this chapter list various signal processing functions performed by the network based on the a_n, b_n, and c_1 coefficients. Variable N is the order (the number of unit-delay elements) of the comb filter. Included in the tables are depictions of the network's impulse response, z-plane pole/zero locations, as well as frequency-domain magnitude and phase responses. The frequency axis in those tables is normalized such that a value of 0.5 represents a frequency of $f_s/2$ where f_s is the sample rate in hertz.

Streamlining Digital Signal Processing: A Tricks of the Trade Guidebook, Second Edition. Edited by Richard G. Lyons.
© 2012 the Institute of Electrical and Electronics Engineers. Published 2012 by John Wiley & Sons, Inc.

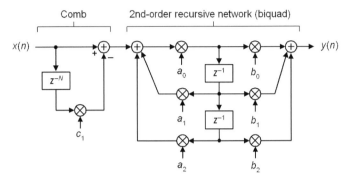

Figure 41–1   General discrete-signal processing network.

41.1 GENERAL FUNCTIONS

Moving Averager (*N*-Point)

Referring to the first entry in Table 41–1, this network configuration is a computationally efficient method for computing the *N*-point moving average of $x(n)$. Also called a *recursive moving averager* or *boxcar averager*, this structure is equivalent to an *N*-tap direct convolution FIR filter with all the coefficients equal to $1/N$. Thus the moving averager's impulse response samples are $1/N$ in amplitude. This moving averager is efficient because it performs only one add and one subtract per output sample regardless of the value of *N*. (An *N*-tap direct convolution FIR filter must perform $N-1$ additions per output sample.) The moving averager's transfer function is

$$H_{\mathrm{ma}}(z) = (1/N)\frac{1-z^{-N}}{1-z^{-1}}. \qquad (41\text{–}2)$$

$H_{\mathrm{ma}}(z)$'s numerator produces *N* zeros equally spaced around the *z*-plane's unit circle located at $z(k) = e^{j2\pi k/N}$, where integer *k* is $0 \le k < N$. $H_{\mathrm{ma}}(z)$'s denominator places a single pole (infinite gain) at $z = 1$ on the unit circle, canceling the zero at that location.

Differentiator (First-Difference)

This is a discrete version of a first-order differentiator. (Purists call this network a *digital differencer*.) An ideal differentiator has a frequency magnitude response that's a linear function of frequency, and this network only approaches that ideal at low frequencies relative to f_s. This configuration of the network amplifies high-frequency spectral components, and this may be detrimental because noise is oftentimes high frequency in real-world signals. The network has a group delay of 0.5 samples.

Table 41–1 General Functions

Moving Averager (*N*-point):
$a_0 = 1, a_1 = 1, a_2 = 0, b_0 = 1/N, b_1 = 0, b_2 = 0, c_1 = 1, N = 8$

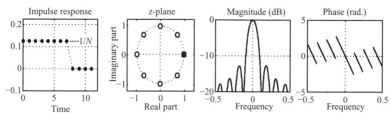

Differentiator (First-difference):
$a_0 = 1, a_1 = 0, a_2 = 0, b_0 = 1, b_1 = -1, b_2 = 0, c_1 = 0$

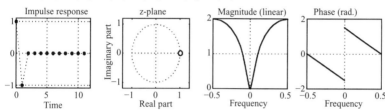

Differentiator (Central-difference):
$a_0 = 1, a_1 = 0, a_2 = 0, b_0 = 0.5, b_1 = 0, b_2 = -0.5, c_1 = 0$

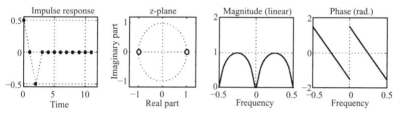

Integrator (Running sum):
$a_0 = 1, a_1 = 1, a_2 = 0, b_0 = 1, b_1 = 0, b_2 = 0, c_1 = 0$

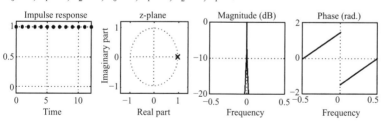

Leaky Integrator:
$a_0 = 1, a_1 = 1-\alpha, a_2 = 0, b_0 = \alpha, b_1 = 0, b_2 = 0, c_1 = 0, \alpha = 0.1$

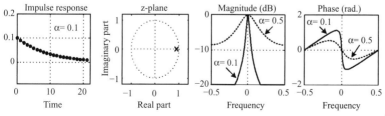

(*Continued*)

Table 41–1   *(Continued)*

1st-order Delay Network:
$a_0 = 1$, $a_1 = -R$, $a_2 = 0$, $b_0 = R$, $b_1 = 1$, $b_2 = 0$, $c_1 = 0$

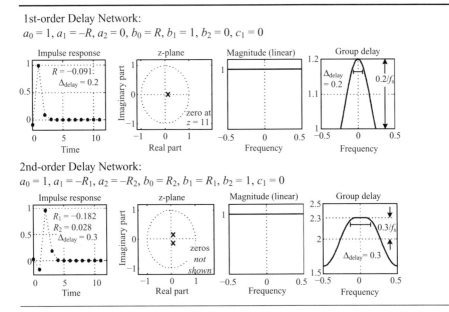

2nd-order Delay Network:
$a_0 = 1$, $a_1 = -R_1$, $a_2 = -R_2$, $b_0 = R_2$, $b_1 = R_1$, $b_2 = 1$, $c_1 = 0$

Differentiator (Central-Difference)

This second-order differentiator attenuates high-frequency (noise) spectral components, and has a group delay of one sample, which is convenient when the $y(n)$ output must be synchronized in time to some other signal sequence. Its frequency range of linear operation (starting at zero Hz), however, is smaller than the above first-difference differentiator.

Integrator (Running Sum)

This structure performs the *running summation* of the $x(n)$ inputs samples, making it a discrete-time approximation of continuous-time integration. This discrete integrator has a pole at $z = 1$, corresponding to a cyclic frequency of zero Hz, and this has an important consequence when we implement integrators. The pole forces us to ensure that the numerical format of our integrator hardware can accommodate summation results when the $x(n)$ input sequence has a nonzero average value (a constant amplitude bias). Stated in different words, the widths of our binary data registers must be large enough to guarantee that any nonzero-amplitude bias on $x(n)$ will not cause numerical overflow and corrupt the data within an integrator's accumulator register.

Leaky Integrator

This network configuration, also called an *exponential averager*, is a venerable structure used in many applications for reducing random noise that contaminates low-frequency, or constant-amplitude, signals of interest. It is a first-order infinite impulse response (IIR) filter where, for stable lowpass operation, the constant α lies in the range $0 < \alpha < 1$.

This nonlinear-phase lowpass filter has a single pole at $z = 1 - \alpha$ on the z-plane, and a transfer function of

$$H_{li}(z) = \frac{\alpha}{1 - (1-\alpha)z^{-1}}. \qquad (41\text{--}3)$$

Small values for α yield narrow passbands at the expense of increased filter response time. Table 41–1 shows the filter's behavior for $\alpha = 0.1$ as solid curves. For comparison, the frequency domain performance for $\alpha = 0.5$ is indicated by the dashed curves.

The DC (zero Hz) gain of the leaky integrator is unity. If a DC gain of $1/\alpha$ can be tolerated, then we can set b_0 to unity to eliminate one of the integrator's multipliers.

First-Order Delay Network

A subclass of a first-order IIR Filter, the coefficients in Table 41–1 yield an *all-pass* network having a relatively constant group delay at low frequencies. The network's delay is $D_{total} = 1 + \Delta_{delay}$ samples where Δ_{delay}, typically in the range of -0.5 to 0.5, is a fraction of the $1/f_s$ sample period. For example, when Δ_{delay} is 0.2, the network delay (at low frequencies) is 1.2 samples. The real-valued R coefficient is

$$R = \frac{-\Delta_{delay}}{\Delta_{delay} + 2} \qquad (41\text{--}4)$$

producing a z-plane transfer function of

$$H_{1,del}(z) = \frac{R + z^{-1}}{1 + Rz^{-1}}. \qquad (41\text{--}5)$$

with a pole at $z = -R$ and a zero at $z = -1/R$.

Performance for $\Delta_{delay} = 0.2$ ($R = -0.091$) is shown in Table 41–1, where we see the magnitude response being constant. The band, centered at DC, over which the group delay varies no more than $|\Delta_{delay}|/10$ from the specified D_{total} value, the bar in the group delay plot, ranges roughly from $0.1f_s$ to $0.2f_s$ for first-order networks. So if your signal is oversampled, making it low in frequency relative to f_s, this first-order all-pass delay network may be of some use. If you propose its use in a new design, you can impress your colleagues by saying this network is based on the *Thiran approximation* [1].

Second-Order Delay Network

A subclass of a second-order IIR filter, the coefficients in Table 41–1 yield an all-pass network having a relatively constant group at low frequencies. (Over a wider frequency range, by the way, than the first-order delay network.) This network's delay is $D_{total} = 2 + \Delta_{delay}$ samples where Δ_{delay} is typically in the range of −0.5 to 0.5. For example, when Δ_{delay} is 0.3, the network delay (at low frequencies) is 2.3 samples. The real-valued coefficients are

$$R_1 = \frac{-2\Delta_{delay}}{\Delta_{delay}+3} \quad and \quad R_2 = \frac{(\Delta_{delay})(\Delta_{delay}+1)}{(\Delta_{delay}+3)(\Delta_{delay}+4)}. \tag{41–6}$$

The band, centered at DC, over which the group delay varies no more than $|\Delta_{delay}|/10$ from the specified D_{total} value, the bar in the group delay plot, ranges roughly from $0.26f_s$ to $0.38f_s$ for this second-order network. Performance for $\Delta_{delay} = 0.3$ ($R_1 = -0.182$ and $R_2 = 0.028$) is shown in Table 41–1, where we see the magnitude response being constant.

The *flat* group delay band is wider for negative Δ_{delay} than when Δ_{delay} is positive. This means if you desire, for example, a group delay of $D_{total} = 2.5$ samples it's better to use an external unit delay and set Δ_{delay} to −0.5 rather than letting Δ_{delay} be 0.5. To ensure stability, Δ_{delay} must be greater than −1. Reference [1] provides methods for designing higher-order all-pass delay networks.

41.2 ANALYSIS AND SYNTHESIS FUNCTIONS

Goertzel Network

Referring to the first entry in Table 41–2, this traditional Goertzel network is used for single-tone detection because it computes a single-bin N-point discrete Fourier transform (DFT) centered at an angle of $\theta = 2\pi k/N$ radians on the unit circle, corresponding to a cyclic frequency of kf_s/N Hz. Frequency variable k, in the range $0 \le k < N$, need not be an integer. The behavior of the network is shown by the solid curves in Table 41–2. However, the frequency magnitude response of the Goertzel algorithm, for $N = 8$ and $k = 1$, is shown as the dashed curve.

After $N + 1$ input samples are applied, $y(n)$ is a single-bin DFT result. The DFT computational workload is $N + 2$ real multiplies and $2N + 1$ real adds. The network is typically stable because N is kept fairly low (in the hundreds) in practice before the network is reinitialized [2], [3].

Sliding DFT Network

This structure computes a single-bin N-point DFT centered at an angle of $\theta = 2\pi k/N$ radians on the unit circle, corresponding to a cyclic frequency of kf_s/N Hz. N is the DFT size and integer k is $0 \le k < N$. Real damping factor r is kept as close to, but less than, unity as possible to maintain network stability. After N input samples have been applied, this network will compute a new follow-on DFT result based on each new $x(n)$ sample (thus the term *sliding*) at a computational workload of only four real multiplies and four real adds per input sample [2], [3]. Setting coefficient $c_1 = -r^N$ allows the analysis band to be centered at an angle of $\theta = 2\pi(k + 1/2)/N$ radians, corresponding to a cyclic frequency of $(k + 1/2)f_s/N$ Hz.

Real Oscillator

There are many possible digital oscillator structures, but this network generates a real-valued sinusoidal $y(n)$ sequence whose amplitude is not a function of the output frequency. The argument for coefficient a_1 in Table 41–2 is $\theta = 2\pi f_t/f_s$ radians, where f_t is the oscillator's tuned frequency in hertz. To start the oscillator we set the $y(n-1)$ sample driving the a_1 multiplier equal to 1 and compute new output samples as the time index n advances. For fixed-point implementations, filter coefficients may need to be scaled so that all intermediate results are in the proper numerical range [4].

Quadrature Oscillator

Called the *coupled quadrature oscillator*, this structure provides $y(n) = \cos(n\theta) + j\sin(n\theta)$ outputs for a complex exponential sequence whose tuned frequency is f_t Hz. The exponent for a_1 in Table 41–2 is $\theta = 2\pi f_t/f_s$ radians. To start the oscillator, we set the complex $y(n-1)$ sample, driving the a_1 multiplier, equal to $1 + j0$ and begin computing output samples as the time index n advances. To ensure oscillator output stability in fixed-point arithmetic implementations, instantaneous gain correction $G(n)$ must be computed for each output sample. The $G(n)$ sample values will be very close to unity [5], [6].

Audio Comb

This structure is a second-order (the simplest) version of an infinite impulse response (IIR) comb filter used by audio folk to synthesize the sound of a plucked-string instrument. The input to the filter is random noise samples. The filter has frequency response peaks at DC and $\pm f_s/2$, with dips in the response located at $\pm f_s/4$. The filter's transfer function is

$$H_{ac}(z) = \frac{1}{1 - \alpha z^{-2}}. \tag{41–7}$$

Table 41–2   Analysis and Synthesis Functions

Goertzel Network:
$a_0 = 1$, $a_1 = 2\cos(\theta)$, $a_2 = -1$, $b_0 = 1$, $b_1 = -e^{-j\theta}$, $b_2 = 0$, $c_1 = 0$, $\theta = 2\pi k/N$ (rad.)

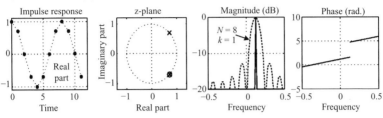

Sliding DFT Network:
$a_0 = re^{j\theta}$, $a_1 = 1$, $a_2 = 0$, $b_0 = 1$, $b_1 = 0$, $b_2 = 0$, $c_1 = r^N$, $\theta = 2\pi k/N$ (rad.)

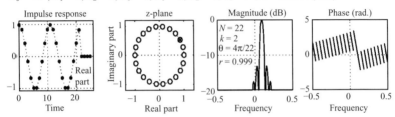

Real Oscillator:
$a_0 = 1$, $a_1 = 2\cos(\theta)$, $a_2 = -1$, $b_0 = 1$, $b_1 = 0$, $b_2 = 0$, $c_1 = 0$

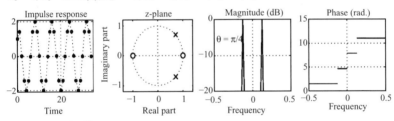

Quadrature Oscillator:
$a_0 = G(n)$, $a_1 = e^{j\theta}$, $a_2 = 0$, $b_0 = 1$, $b_1 = 0$, $b_2 = 0$, $c_1 = 0$, $\theta = 2\pi f_t/f_s$ (rad.)

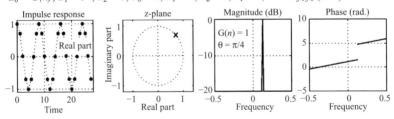

Audio Comb:
$a_0 = 1$, $a_1 = 0$, $a_2 = \alpha$, $b_0 = 1$, $b_1 = 0$, $b_2 = 0$, $c_1 = 0$

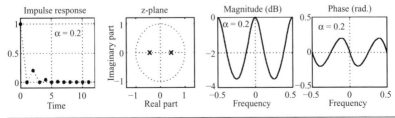

resulting in two poles located at $z = \pm\text{sqrt}(\alpha)$ on the z-plane. To maintain stability the real-valued α must be less than unity, and the closer α is to unity the more narrow the frequency response peaks.

For a more realistic-sounding synthesis, we can set $a_1 = \alpha$ and the top delay element of the biquad in Figure 41–1 may have its delay increased to, say, eight instead of one, yielding more frequency response peaks between 0 and $f_s/2$ Hz. In this music application, the filter's input is Gaussian white noise samples. Other plucked-string instrument synthesis networks have been used with success [7], [8].

41.3 FILTER FUNCTIONS

Comb Filter

Referring to the first entry in Table 41–3, this finite impulse response (FIR) comb filter is a key component on many filtering applications, as we shall see. Its transfer function, $H_{comb}(z) = 1 - z^{-N}$, results in N zeros equally spaced around the z-plane's unit circle located at $z(k) = e^{j2\pi k/N}$, where integer k is $0 \leq k < N$. Those $z(k)$ values are the N roots of unity when we set $H_{comb}(z)$ equal to zero yielding $z(k)^N = (e^{j2\pi k/N})^N = 1$. The N zeros on the unit circle result in frequency response nulls (infinite attenuation) located at cyclic frequencies of mf_s/N where integer m is $0 \leq m \leq N/2$. The peak gain of this linear-phase filter is 2.

If we set coefficient c_1 to -1 in the comb filter, making its transfer function $H_{alt,comb}(z) = 1 + z^{-N}$, we obtain an alternate linear-phase comb filter having zeros rotated counterclockwise around the unit circle by an angle of π/N radians, positioning the zeros at angles of $2\pi(k + 1/2)/N$ radians on the z-plane's unit circle. The rotated zeros result in frequency response nulls located at cyclic frequencies of $(m + 1/2)f_s/N$, where integer m is $0 \leq m \leq (N/2) - 1$. With this filter a frequency magnitude peak is located at 0 Hz (DC).

Bandpass Filter at $f_s/4$

This network is a bandpass filter centered at $f_s/4$ having a sin(x)/x-like frequency response and linear-phase over the passband. It has poles at $z = \pm j$, so for pole/zero cancellation the comb filter's delay (N) must be an integer multiple of four. This guaranteed-stable, multiplierless, bandpass filter's transfer function is

$$H_{bp}(z) = \frac{1 - z^{-N}}{1 + z^{-2}}. \tag{41–8}$$

First-Order IIR Filter

This is the Direct Form II version of a simple first-order IIR filter having a single pole located at a radius of R_p from the z-plane's origin at an angle of θ_p radians, and a single zero at a radius of R_z at an angle of $\pi + \theta_z$. For real-valued coefficients

Table 41–3 Filter Functions

Comb Filter:

$a_0 = 1$, $a_1 = 0$, $a_2 = 0$, $b_0 = 1$, $b_1 = 0$, $b_2 = 0$, $c_1 = 1$

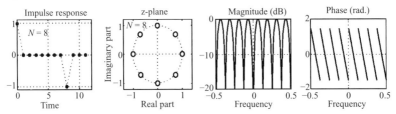

Bandpass Filter at $f_s/4$:

$a_0 = 1$, $a_1 = 0$, $a_2 = -1$, $b_0 = 1$, $b_1 = 0$, $b_2 = 0$, $c_1 = 1$

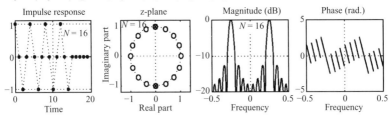

1st-order IIR Filter:

$a_0 = 0$, $a_1 = R_p e^{j\theta_p}$, $a_2 = 0$, $b_0 = 1$, $b_1 = R_z e^{j\theta_z}$, $b_2 = 0$, $c_1 = 0$

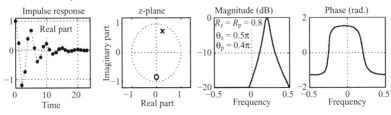

1st-order Equalizer:

$a_0 = 1$, $a_1 = R$, $a_2 = 0$, $b_0 = -R^*$, $b_1 = 1$, $b_2 = 0$, $c_1 = 0$

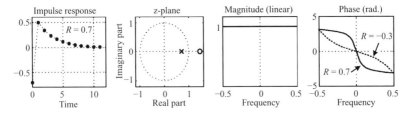

Table 41–3 (*Continued*)

2nd-order IIR Filter:
$a_0 = 1$, $a_1 = 1.194$, $a_2 = -0.436$, $b_0 = b_2 = 0.0605$, $b_1 = 0.121$, $c_1 = 0$

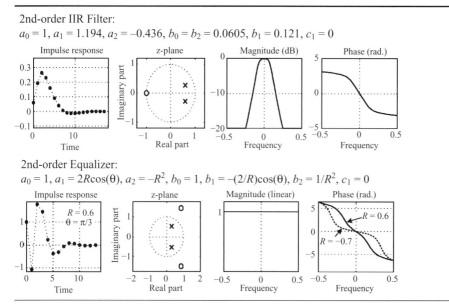

2nd-order Equalizer:
$a_0 = 1$, $a_1 = 2R\cos(\theta)$, $a_2 = -R^2$, $b_0 = 1$, $b_1 = -(2/R)\cos(\theta)$, $b_2 = 1/R^2$, $c_1 = 0$

($\theta_p = \theta_z = 0$) the filter can only exhibit either a lowpass or a highpass frequency response; no bandpass or bandstop filters are possible. The filter's transfer function is

$$H_{1,\text{iir}}(z) = \frac{1 + R_z e^{j\theta_z} z^{-1}}{1 - R_p e^{j\theta_p} z^{-1}}. \qquad (41\text{–}9)$$

The shape of the filter's frequency magnitude responses are nothing to write home about; its transition regions are so wide that they don't actually have distinct passbands and stopbands. Of course to ensure stability, R_p must be between zero and 1 to keep the pole inside the z-plane's unit circle, and the closer R_p is to unity the more narrowband is the filter.

First-Order Equalizer

This structure has a frequency magnitude response that's constant across the entire frequency band (an all-pass filter). It has a pole at $z = R$ on the z-plane, and a zero located at $1/R^*$, where * means conjugate. The value of R, which can be real or complex but whose magnitude must be less than unity to ensure stability, controls the nonlinear-phase response. The equalizer has a transfer function of

$$H_{1,\text{eq}}(z) = \frac{-R^* + z^{-1}}{1 - R z^{-1}}. \qquad (41\text{–}10)$$

These networks can be used as *phase equalizers* by cascading them after a filter, or network, whose nonlinear phase response requires crude linearization. The goal is to make the cascaded filters' combined phase as linear as possible. Table 41–3 shows the filter's behavior for $R = 0.7$ as solid curves. For comparison, the phase response for $R = -0.3$ is indicated by the dashed curve. These first-order all-pass filters can also be used for interpolation and audio reverberation for low-frequency signals.

Second-Order IIR Filter

This is the Direct Form II version of a second-order IIR filter, the *workhorse* of IIR filter implementations. Conjugate pole and zero pairs may be positioned anywhere on the z-plane to control the filter's frequency response. Because high-order IIR filters are so susceptible to coefficient quantization and potential data overflow problems, practitioners typically implement their IIR filters by cascading multiple copies of this second-order IIR structure to ensure filter stability and avoid *limit cycles*. The filters have a transfer function of (41–1) with the $c_1 = 0$. Lowpass, highpass, bandpass, and bandstop filters are possible. No single example shows all the possibilities of this structure, so Table 41–3 merely gives a simple lowpass filter example.

If an IIR filter design requires high performance, called "high Q", it turns out the Direct Form I version of a second-order IIR filter is less susceptible to coefficient quantization and overflow errors than the Direct Form II structure given here.

Second-Order Equalizer

This structure has a frequency magnitude response that's constant across the entire frequency band, making it an all-pass filter. It has two conjugate poles located at a radius of R from the z-plane's origin at angles of $\pm\theta$ radians, and two conjugate zeros at a reciprocal radius of $1/R$ at angles of $\pm\theta$. The positioning of the poles and zeros, using real-valued R, controls the nonlinear-phase response. Table 41–3 shows the equalizer's behavior for $R = 0.6$ and $\theta = \pi/3$ as solid curves. For comparison, the phase response for $R = -0.7$ and $\theta = \pi/3$ is indicated by the dashed curve.

These networks are primarily used for phase equalization by cascading them after a filter, or network, whose nonlinear-phase response requires linearization. However, it may take multiple cascaded biquad networks to achieve acceptable equalization.

41.4   ADDITIONAL FILTER FUNCTIONS

CIC Interpolation Filter

Referring to the first entry in Table 41–4, this network is a single-stage cascaded integrator-comb (CIC) interpolation filter used for time-domain interpolation. If a

Table 41–4 Additional Filter Functions

CIC Interpolation Filter:
$a_0 = 1$, $a_1 = 1$, $a_2 = 0$, $b_0 = 1$, $b_1 = 0$, $b_2 = 0$, $c_1 = 1$

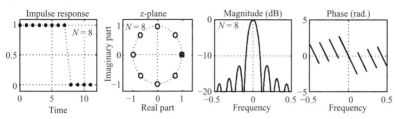

Complex Frequency Sampling Filter:
$a_0 = 1$, $a_1 = e^{j\theta_k}$, $a_2 = 0$, $b_0 = (-1)^k$, $b_1 = 0$, $b_2 = 0$, $c_1 = 1$, $\theta_k = 2\pi k/N$ (rad.)

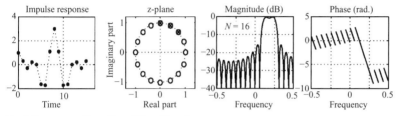

Real Frequency Sampling Filter, Type I:
$a_0 = 1$, $a_1 = 2\cos(\theta_k)$, $a_2 = -1$, $b_0 = |H_k|\cos(\phi_k)$, $b_1 = -|H_k|\cos(\phi_k - \theta_k)$, $b_2 = 0$,
$c_1 = 1$, $\theta_k = 2\pi k/N$ (rad.)

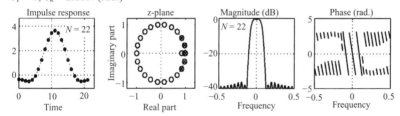

Real Frequency Sampling Filter, Type IV:
$a_0 = 1$, $a_1 = 2\cos(\theta_k)$, $a_2 = -1$, $b_0 = (-1)^k M_k$, $b_1 = 0$, $b_2 = (-1)^k(-M_k)$,
$c_1 = 1$, $\theta_k = 2\pi k/N$ (rad.)

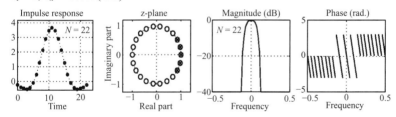

DC Bias removal Filter:
$a_0 = 1$, $a_1 = \alpha$, $a_2 = 0$, $b_0 = 1$, $b_1 = -1$, $b_2 = 0$, $c_1 = 0$

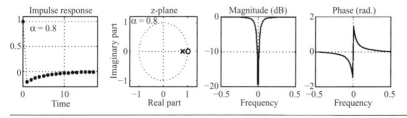

time-domain signal sequence is upsampled by N (by inserting $N - 1$ zero-valued samples in between each original sample) and applied to this lowpass filter, the filter's output is an interpolated-by-N version of the original signal. This lowpass filter's transfer function is

$$H_{\text{cic}}(z) = \frac{1 - z^{-K}}{1 - z^{-1}}. \tag{41-11}$$

To improve the attenuation of spectral images, we can cascade M copies of the comb filter followed by M cascaded biquad sections. Such cascaded filters will also have narrower passband widths at zero Hz.

In practice, the upsampling operation (zero stuffing) is performed after the comb filter and before the biquad network. This has the sweet advantage that the comb filter's delay length becomes $N = 1$, reducing the necessary comb delay storage requirement to one. CIC filters are often used as the first stage of multistage lowpass filtering in hardware interpolation-by-N applications because CIC filters require no multipliers [6]. When used as a lowpass filter in sample rate change applications, the filter is typically implemented with two's complement arithmetic.

Complex Frequency Sampling Filter (FSF)

This structure is a single section of a complex *frequency sampling filter* having a $\sin(x)/x$-like frequency magnitude response centered at an angle of $\theta_k = 2\pi k/N$ radians on the unit circle, corresponding to a cyclic frequency of kf_s/N Hz. N and k are integers where k is $0 \le k < N$. The larger is N, the more narrow is the filter's mainlobe width [6].

If multiple biquads are implemented in parallel (all driven by the single comb filter), with adjacent center frequencies, complex, almost linear-phase bandpass filters can be built. Table 41–4 shows the behavior of an $N = 16$, three-biquad, complex bandpass filter each centered at $k = 2$, 3, and 4, respectively.

Real Frequency Sampling Filter, Type I

This structure is a single section of a real-coefficient *frequency sampling filter* having a $\sin(x)/x$-like frequency magnitude responses centered at both $\pm\theta_k = \pm 2\pi k/N$ radians, where N is an integer. The larger is N, the more narrow is the filter's mainlobe width. Integer k is $0 \le k < N$.

If multiple biquads are implemented in parallel (all driven by the single comb filter), with adjacent center frequencies, almost linear-phase lowpass filters can be built. In this case, complex gain factors H_k are the desired peak frequency response of the kth biquad. Parameter ϕ_k is the desired relative phase shift, in radians, of H_k. Table 41–4 shows the behavior of an $N = 22$, three-biquad, lowpass filter each centered at $k = 0, 1$, and 2, respectively. In this example $|H_0| = 1$, $|H_1| = 2$, and $|H_2| = 0.74$.

These bandpass filters can have group delay fluctuations as large as $2/f_s$ in the pass-band. This recursive finite impulse response (FIR) filter is the most common frequency sampling filter discussed in the traditional DSP textbooks [6], [9], [10].

Real Frequency Sampling Filter, Type IV

This structure is similar in behavior to the type I frequency sampling filter, with important exceptions. First, in a multi-biquad lowpass filter implementation this filter yields an exactly linear phase response. Also, this filter provides deeper stopband attenuation than the type I filter.

The real-valued gain factors M_k are the desired peak frequency magnitude response of the kth biquad. Table 41–4 shows the behavior of an $N = 22$, three-biquad, lowpass filter with the biquads centered at $k = 0$, 1, and 2, respectively. In this example $M_0 = 1$, $M_1 = 1$, and $M_2 = 0.37$. Here's why you need to know about these filters: With judicious choice of the M_k gain factors, narrowband lowpass linear-phase FIR filters can be built, in some cases, whose computational workload is less than Parks-McClellan–designed FIR filters [6].

DC Bias Removal

This network, used to remove any DC bias from the $x(n)$ input, has a transfer function having a pole located at $z = \alpha$ and a zero at $z = 1$. Having a frequency response notch (null) at 0 Hz (DC, hence the name), the sharpness of the notch is determined by α, where for stable operation α lies in the range $0 < \alpha < 1$. The closer α is to unity the more narrow the notch at DC. This nonlinear-phase filter has a transfer function of

$$H_{dc}(z) = \frac{1 - z^{-1}}{1 - \alpha z^{-1}}. \qquad (41\text{–}12)$$

Table 41–4 shows the filter's behavior for $\alpha = 0.8$.

In those fixed-point implementations where the output $y(n)$ sequence must be truncated to avoid data overflow (i.e., $y(n)$ must have fewer bits than input $x(n)$), feedback noise shaping can be used to reduce the quantization noise induced by truncation [6], [11].

An alternative to truncation, to avoid overflow, is to limit the gain of the filter. On one hand, we could precede the network with a positive gain element whose gain is less than unity. On the other hand, we could use $b_0 = G$ and $b_1 = -G$, where $G = (1 + \alpha)/2$, in our implementation for this purpose, yielding a reduced-gain transfer function of

$$H_{alt,dc}(z) = \frac{G - Gz^{-1}}{1 - \alpha z^{-1}}. \qquad (41\text{–}13)$$

41.5 REFERENCES

[1] T. LAAKSO et al., "Splitting the unit delay," *IEEE Signal Proc. Magazine*, January 1996, pp. 30–60.

[2] E. JACOBSEN and R. LYONS, "The sliding DFT," *IEEE Signal Proc. Magazine*, DSP Tips & Tricks Column, vol. 20, no. 2, March 2003, pp. 74–80.

[3] E. JACOBSEN and R. LYONS, "The Sliding DFT, an Update," *IEEE Signal Proc. Magazine*, DSP Tips & Tricks Column, vol. 21, no. 1, January 2004.

[4] D. GROVER and J. DELLER, *Digital Signal Processing and the Microcontroller*. Prentice Hall, Upper Saddle River, NJ, 1999.

[5] C. TURNER, "Recursive Discrete-Time Sinusoidal Oscillators," *IEEE Signal Proc. Magazine*, vol. 20, no. 3, May 2003, pp. 103–111.

[6] R. LYONS, *Understanding Digital Signal Processing*, 2nd ed. Prentice Hall, Upper Saddle River, NJ, 2004.

[7] J. SMITH, "Physical Audio Signal Processing," [Online: http://ccrma-www.stanford.edu/~jos/waveguide/Comb_Filters.html.]

[8] Texas Instruments, "How Can Comb Filters Be Used to Synthesize Musical Instruments on a TMS320 DSP?" *TMS320 DSP Designers Notebook*, no. 56, 1995.

[9] V. INGLE and J. PROAKIS, *Digital Signal Processing Using MATLAB*, Brookes/Cole Publishing, Pacific Grove, CA, 2000, pp. 202–208.

[10] J. PROAKIS and D. MANOLAKIS, *Digital Signal Processing: Principles, Algorithms, and Applications*, 3rd ed. Prentice Hall, Upper Saddle River, NJ, 1996, pp. 630–637.

[11] C. DICK and F. HARRIS, "FPGA Signal Processing Using Sigma-Delta Modulation," *IEEE Signal Proc. Magazine*, vol. 17, no. 1, January 2000.

EDITOR COMMENTS

In what follows we expand upon the Figure 41–1 network's ability to implement integration. Other integrator implementations exist beyond the simple running sum integrator described in Table 41–1. Those alternate integrators, used to approximate continuous-time integration, go by the names *trapezoidal rule*, *Simpson's rule*, and *Tick's rule*. We mention these integrators because they provide a more accurate approximation to continuous-time integration than does the simple running sum integrator. The coefficients and transfer function pole/zero locations on the unit circle of the alternate integrators are provided in Figure 41–2.

The frequency magnitude response of an ideal integrator is $1/\omega$ and at low frequencies the three alternate integrators in Figure 41–2 approach that ideal more closely than does the running sum integrator. From zero Hz to roughly $0.3f_s$ Hz, the Tick's rule integrator provides the highest integration accuracy, with the Simpson's rule integrator being a very close second. However, if the integrators' input signals have appreciable noise spectral components near $f_s/2$ Hz, the Tick's rule and Simpson's rule integrators will amplify that noise. In a high-frequency noise scenario the Tick's rule and Simpson's rule integrators should be avoided; the running sum integrator should be used instead because it provides the most accurate integration in the presence of wideband noise and is simpler to implement than the trapezoidal rule integrator.

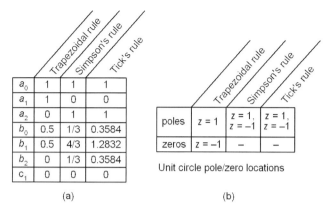

Figure 41–2   Alternate integrator coefficients and pole/zero locations.

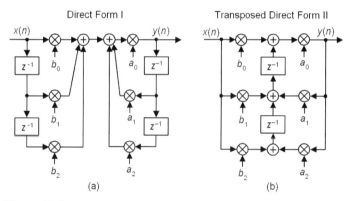

Figure 41–3   Alternate biquad structures: (a) Direct Form I; (b) Transposed Direct Form II.

As with the running sum integrator, the three Figure 41–2 integrators have z-domain transfer function poles (infinite gain) at $z = 1$ and, again, this is an important issue when we implement integrators. Those poles force us to ensure that the numerical format of our integrator hardware can accommodate (with no numerical overflow of) summation results when the $x(n)$ input sequence has a nonzero average value.

From an implementation standpoint, we remind the reader that the block diagram of the recursive biquad network in Figure 41–1 is called a Direct Form II structure. As such, the biquad network can be implemented in what are called the Direct Form I and Transposed Direct Form II structures, as shown in Figure 41–3. Those alternate biquad structures can be useful because they are less susceptible to coefficient quantization and stability problems, than is the Direct Form II structure, when fixed-point arithmetic is used. In closing, we also remind the reader that the general network's comb filter may precede the biquad network, as it does in Figure 41–1, or it may follow the biquad network.

Chapter 42

JPEG2000–Choices and Trade-offs for Encoders

Amy Bell
Institute for Defense Analysis

Krishnaraj Varma
Hughes Network Systems

A new, and improved, image coding standard has been developed, and it's called JPEG2000. In this chapter we describe the most important parameters of this new standard and present several "tips and tricks" to help resolve design trade-offs that JPEG2000 application developers are likely to encounter in practice.

42.1 JPEG2000 STANDARD

JPEG2000 is the state-of-the-art image coding standard that resulted from the joint efforts of the International Standards Organization (ISO) and the International Telecommunications Union (ITU) [1]; "JPEG" in JPEG2000 is an acronym for Joint Picture Experts Group. The new standard outperforms the older JPEG standard by approximately 2 decibels (dB) of peak signal-to-noise ratio (PSNR) for several images across all compression ratios [2]. Two primary reasons for JPEG2000's superior performance are the wavelet transform and *Embedded Block Coding with Optimal Truncation* (EBCOT) [3]. The standard is organized in 12 parts [4]. Part 1 specifies the core coding system while Part 2 adds some features and more sophistication to the core. Part 3 describes motion JPEG—a rudimentary form of video coding where each JPEG2000 image is a frame. Other important parts of the standard include security aspects, interactive protocols and application program interfaces for network access, and wireless transmission of JPEG2000 images.

The JPEG2000 standard is effectively a decoder standard. The parameters that were selected during the encoding process are explicitly written (or signaled) into the compressed data. The JPEG2000 standard describes how these values have to be read, interpreted, and used during the decoding process. The standard does not explicitly state on what basis these parameters are chosen on the encoding side. The

Streamlining Digital Signal Processing: A Tricks of the Trade Guidebook, Second Edition. Edited by Richard G. Lyons.
© 2012 the Institute of Electrical and Electronics Engineers. Published 2012 by John Wiley & Sons, Inc.

choices made on these parameters are important and affect the quality and available features of a compressed image. Here we provide useful pointers and hints that will help JPEG2000 application developers in making appropriate choices for encoding parameters. Throughout the remainder of this discussion, we use the term *encoder* to mean a piece of software or hardware that performs JPEG2000 encoding. For an encoder targeting a particular application, the parameter choices are made by the application developer and hard-coded into the encoder. For an encoder targeting a more generic application, the task of choosing some parameters might be handed down to the encoder that performs some rudimentary decision making based upon the application and the image to be encoded. In either case we refer to the act of making parameter choices as being performed by the encoder.

We limit our discussion to those parameters specified in the core processing system, Part 1 of the JPEG2000 standard. A comprehensive list of the parameters is depicted in Table 42–1; they are given in the order that they are encountered in the encoder.

The chosen values for some of these parameters are dictated by the target application. For example, most applications require the compressed image to be reconstructed at the original bit depth. The progression order and the number of quality layers are also determined by the requirements of the application. Other parameters like the magnitude refinement coding method or the MQ code termination method minimally impact the quality of the compressed image, the size of the compressed data, or the complexity of the encoder. For each parameter, JPEG2000 provides either a recommendation or a default; this provides a good, initial choice for the parameter.

We now elaborate on six parameters for which there exists a wide range of acceptable values and those chosen values significantly impact compressed image

Table 42–1   Parameters in Part 1 of the JPEG2000 Standard

---

1. Reconstructed image bit depth
2. Tile size
3. Color space
4. Reversible or irreversible transform
5. Number of wavelet transform levels
6. Precinct size
7. Code-block size
8. Coefficient quantization step size
9. Perceptual weights
10. Block coding parameters:
    (a) Magnitude refinement coding method
    (b) MQ code termination method
11. Progression order
12. Number of quality layers
13. Region of interest coding method

---

quality and codec efficiency. The six parameters are 2, 3, 5, 7, 8, and 13 in Table 42–1. We discuss the merits of the choices for these parameters based on the following performance measures: compressed data size, compressed image quality, computation time, and memory requirements.

42.2 TILE SIZE

JPEG2000 allows an image to be divided into rectangular blocks of the same size called *tiles*—each tile is encoded independently. Tile size is a coding parameter that is explicitly signaled in the compressed data. By tiling an image, the distinct features in the image can be separated into different tiles; this enables a more efficient encoding process. For example, a composite image comprised of a photograph and text can be divided into tiles that separate the two; then two very different approaches (e.g., original bit depth and 5-level transform for the photograph, and bit depth of 1 and 0-level transform for the text) are used to obtain significantly better overall coding efficiency. Dividing an image into tiles also allows the use of customized perceptual weights tuned to the feature in each tile. Furthermore, tiling provides a degree of random spatial access to the reconstructed image.

Choosing the tile size for an image is an important trade-off for the encoder. Figure 42–1 shows the *Woman* image compressed at $100:1$ using two different tile sizes: (a) 64×64 and (b) 256×256. Blocking artifacts are readily observed in Figure 42–1(a) (the smaller tile size). This is a common observation at moderate to high compression ratios; however, at low compression ratios ($<32:1$) a small tile size introduces minimal blocking artifacts. Alternatively, a large tile size presents two challenges. First, if the encoder/decoder processes an entire tile at once, this may require prohibitively large memory. Second, features may not be isolated into separate tiles, and the encoding efficiency suffers.

Recommendation: Do not tile small images ($\leq 512 \times 512$). Tile large images with a tile size that separates the features, but at high compression ratios, use a tile size greater than or equal to 256×256 to avoid blocking artifacts.

42.3 COLOR SPACE

We humans view color images in the red, green, and blue (RGB) color space. The encoder can choose to convert a color image into the luminance, chrominance blue, and chrominance red (YCbCr) color space. For most color images, YCbCr concentrates image energy as well or better than RGB. In RGB, energy is more evenly distributed across the three components; however, in YCbCr, most of the energy resides in the luminance (Y) component. For example, the two chrominance components typically account for only 20% of the bits in the compressed JPEG2000 image. However, if the RGB image is comprised of mostly one color, then the YCbCr representation cannot improve upon the efficient energy compaction in RGB. For these color images, RGB compression quality is superior to YCbCr compression quality.

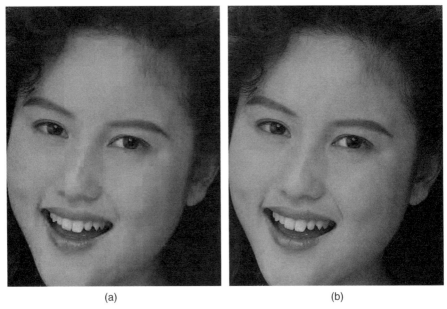

(a)　　　　　　　　　　　　　　　　　(b)

Figure 42–1 The *Woman* image compressed at 100:1 with tile size (a) 64 × 64 and (b) 256 × 256.

(a)　　　　　　　　　　　　　　　　　(b)

Figure 42–2 A grayscale version of the *Lighthouse* image compressed at 32:1 in (a) RGB and (b) YCbCr.

Figure 42–2 depicts a grayscale version of the *Lighthouse* image compressed at 32:1 in the (a) RGB color space and in the (b) YCbCr color space. Compression in YCbCr shows a higher quality compressed image in (b): the roof-edge, grass, and cloud texture, and other details are closer to the original, uncompressed image than in (a).

Recommendation: Convert the original, uncompressed RGB color image to the YCbCr color space except when the RGB image primarily consists of one color component.

42.4 NUMBER OF WAVELET TRANSFORM LEVELS

Each image color component is transformed into the wavelet domain using the two-dimensional discrete wavelet transform (DWT). The number of levels of the DWT is a parameter that is chosen by the encoder. By increasing the number of DWT levels, we examine the lower frequencies at increasingly finer resolution—thereby packing more energy into fewer wavelet coefficients. Thus, we expect compression performance to improve as the number of levels increases.

Figure 42–3 shows the *Goldhill* image compressed at 16 : 1 using (a) a one-level DWT and (b) a two-level DWT. The difference in quality between the two is minimal. At low compression ratios, quality improvement diminishes beyond two to three DWT levels. On the other hand, Figure 42–4 shows the same image compressed at 64 : 1 using (a) a one-level, and (b) a four-level DWT. In this case, the superior quality of the four-level DWT is clearly evident. (At high compression ratios, quality improvement diminishes beyond four to five DWT levels.) Wavelet-transform-based image compression tends to capture low-frequency coefficients before capturing higher frequencies. For a given compression ratio, assume that all of the lower frequencies in an image are completely captured after three levels. Increasing the number of levels beyond three resolves these lower frequencies more finely, but this does not improve compression quality because these frequencies have already been captured.

Recommendation: Use two to three DWT levels at low compression ratios and four to five DWT levels at high compression ratios.

42.5 CODE-BLOCK SIZE

The DWT coefficients are separated into nonoverlapping, square regions called code-blocks. Each code-block is independently coded using JPEG2000's MQ coder—a type of arithmetic encoding algorithm. Code-block size is a parameter dictated by the encoder and explicitly signaled in the compressed data.

As the code-block size increases, the memory required for the encoder/decoder increases. Therefore the size of the code-block may be limited by the available memory—particularly in hardware implementations. Moreover, if the simple scaling method is used to perform region of interest (ROI) coding (see below), then a large code-block size limits the precision of the ROI's boundary locations. Alternatively, a smaller code-block size allows a more precise definition of the ROI boundaries and, consequently, a higher-quality ROI in the compressed image. In the absence of ROI coding (and all other parameters being equal) the quality of the compressed image improves with increasing code-block size. A small code-block mitigates the efficiency of the MQ coder that, in turn, decreases compressed image quality. Finally,

(a)   (b)

Figure 42–3   The *Goldhill* image compressed at 16:1 using (a) one-level DWT and (b) two-level DWT.

(a)   (b)

Figure 42–4   The *Goldhill* image compressed at 64:1 using (a) one-level DWT and (b) four-level DWT.

encoding/decoding is faster for a larger code-block size since the overall overhead associated with processing all of the code-blocks is minimized.

Recommendation: In general, if there are memory limitations or if the scaling method of ROI coding is employed, then use a small code-block size ($<64 \times 64$). Otherwise, use the largest possible code-block size: 64×64. (JPEG2000 allows code-blocks to be of size $2^n \times 2^n$ where $n = 2, 3, 4, 5,$ or 6.)

42.6   COEFFICIENT QUANTIZATION STEP SIZE

A quantizer divides the real number line into discrete bins; the value of an unquantized wavelet coefficient determines which bin it ends up in. The quantized wavelet coefficient value is represented by its bin index (a signed integer). JPEG2000 employs a uniform deadzone quantizer with equal-sized bins—except for the zero bin, which is twice as large. The size of the nonzero bins is equal to the quantization step size. Quantization step size is specified by the encoder; this choice represents a trade-off between compressed image quality and encoding efficiency. It is worth noting that this trade-off does not exist for the reversible wavelet transform since the unquantized wavelet coefficients are already signed integers (consequently, the default quantization step size is 1).

JPEG2000's uniform deadzone quantizer is an embedded quantizer. This means that if the signed integers are truncated such that the n least significant bits are thrown away, then this is equivalent to an increase in the quantization step size by 2^n [5, 6]. Therefore quantization in JPEG2000 can be regarded as a two-step process. In the first step, a quantization step size is specified for each subband: the subband coefficients are represented by signed integers. In the second step, the signed integers within each code-block of each subband are optimally truncated. This is equivalent to optimally modifying the quantization step size of each code-block to achieve the desired compression ratio. Thus the resulting quantization only depends on the optimal truncation algorithm—so long as the quantization step size was chosen small enough. In summary, the quantization step size trade-off is: Choose too large and compression quality may be jeopardized; choose too small and achieve the desired quality, but compromise codec efficiency.

Figure 42–5 depicts how compressed image quality varies as a function of quantization step size. We define peak signal-to-noise ratio (PSNR) as the ratio of signal power at full dynamic range (255^2 for a bit-depth of 8) to the mean squared error between original and compressed images expressed in dB. Average PSNR was computed over 23 images (from the standard image set [7]), at four compression ratios, as quantization step size changed. In JPEG2000, quantization step size can

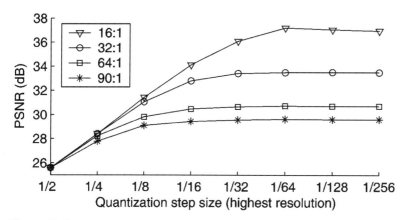

Figure 42–5   Compressed image quality (PSNR) as a function of quantization step size at four compression ratios.

be specified for the highest resolution subband and halved for each subsequent (lower resolution) subband. Figure 42–5 shows that there is a point of diminishing returns (the "knee" in the curve) for decreasing quantization step size—each compression ratio curve flattens out at a given step size. As expected, the higher the compression ratio, the faster the curve levels off (i.e., higher compression ratios cannot take advantage of smaller step sizes). The knee of each curve represents the largest step size for which quantization due to optimal truncation is the dominant factor affecting compressed image quality. In general, if B is the bit-depth of the original image, then $1/2^B$ is a conservative (i.e., to the right of the knee) quantization step size for the highest resolution subband.

Recommendation: In general, for fixed-point codecs, the available bit-width determines quantization step size. The design of such a system must ensure that the bit-width of the highest resolution subband corresponds to a quantization step size that is in the flat region of the curve for the desired compression ratio. For floating point and software codecs, $1/2^B$ is a good, conservative value to be used for quantization step size of the highest resolution subband.

42.7   REGION OF INTEREST CODING METHOD

Region of interest (ROI) coding is the JPEG2000 feature that allows a specified region of the image to be compressed at a higher quality than the remainder of the image. There are two methods for ROI coding: the scaling method and the maxshift method.

In the scaling method, the coefficients in each code-block of the ROI are multiplied by a weight that increases their value. In this way, the optimal truncation algorithm allocates more bits to these code-blocks and they are reconstructed at a higher quality. A conceptually simple method, it has two disadvantages: (1) the ROI coordinates and the scaling factor must be explicitly signaled in the compressed data; and (2) the ability to capture a ROI of a particular size is dictated by the code-block size. For example, consider a ROI of size 256×256. In a five-level DWT, this ROI corresponds to an 8×8 area in the lowest resolution subband. Thus the code-block size must be less than or equal to 8×8. Otherwise the region will extend over the intended boundary (at the lower resolutions) and the reconstructed image will depict a progressive deterioration in quality around the ROI. Figure 42–6 depicts the impact of the code-block size on quality in ROI coding. The image was compressed at $200:1$ with five levels of decomposition and an ROI scale factor of 2048. Two different code-block sizes were employed: the 8×8 code-block defined the ROI better than the 64×64. A close-up view of Figure 42–6(a) in (c) and Figure 42–6(b) in (d) shows the smaller code-block size's higher quality. The disadvantage with the larger code-block is that some of the bits that should have been used to preserve the quality of the ROI are diverted to the surrounding area. Consequently, the 8×8 code-block results in better subjective quality and objective performance (PSNR is 34.97 dB for 8×8 and 31.52 dB for 64×64).

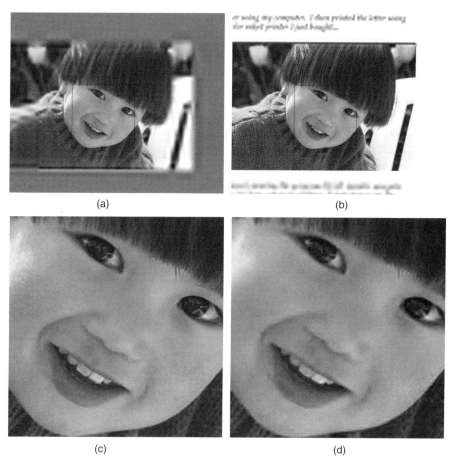

(a)

(b)

(c)

(d)

Figure 42–6 ROI simple scaling performed on the boy's face in the standard image *CMPND2*. The ROI scale factor is 2048 for two code-block sizes (a) 8 × 8 and (b) 64 × 64.

In the maxshift method, an arbitrarily shaped mask specifies the region of interest [8]. All coefficients—at all resolutions—that fall within the mask are shifted up in value by a factor called the Maxshift factor. This shifting ensures that the least significant bit of all of the ROI coefficients is higher than the highest encoded bit-plane. As a result, the ROI is completely encoded before the remainder of the image. This method permits regions of arbitrary shape and size. Furthermore, the ROI does not extend beyond the specified area, nor does it depend on the code-block size and the number of wavelet transform levels. However, unlike the scaling method, this method reconstructs the entire region of interest before the rest of the image; therefore, there may be a significant quality difference between the ROI and non-ROI areas (particularly at high compression ratios).

Recommendation: As discussed in a previous section, larger code-block sizes correspond to higher compressed image quality; however, smaller code-block sizes are required for the ROI scaling method. So use the ROI scaling method if rectangular regions are of interest, but take care about how the small code-block size affects overall quality and codec efficiency. Code-block size is not an issue with the ROI maxshift method. Use the ROI maxshift method when large code-block size and/or arbitrary (nonrectangular) regions are desired. One final consideration is the compressed image quality outside the ROI. The scaling method permits a more flexible distribution of ROI and non-ROI quality; degradation in the non-ROI is more severe with the maxshift method—particularly at high compression ratios.

42.8 REFERENCES

[1] ITU *T.800: JPEG2000 Image Coding System Part 1*, ITU Standard, July 2002. [Online: www.itu.org.]

[2] A. SKODRAS, C. CHRISTOPOULOS, and T. EBRAHIMI, "The JPEG 2000 Still Image Compression Standard," *IEEE Signal Processing Magazine*, September, 2001, pp. 36–58.

[3] D. TAUBMAN, "High Performance Scalable Image Compression with EBCOT," *IEEE Trans. Image Processing*, vol. 9, no. 7, July 2000, pp. 1158–1170.

[4] JPEG2000 Standard, [Online: http://www.jpeg.org/jpeg2000/index.html.]

[5] D.S. TAUBMAN and M.W. MARCELLIN, *JPEG2000–Image Compression Fundamentals, Standards and Practice*, Kluwer Academic Publishers, Norwell, MA, 2002.

[6] M. MARCELLIN, M. LEPLEY, A BILGIN, T. FLOHR, T. CHINEN, and J. KASNER, "An Overview of Quantization in JPEG2000," *Signal Processing: Image Communications (Special Issue on JPEG2000)*, vol 17, no. 1, 2002, pp. 73–84.

[7] *ITU T.24: Standardized Digitized Image Set*, ITU Standard, June 1998. [Online: www.itu.org.]

[8] J. ASKELÖF, M.L. CARLANDER, and C. CHRISTOPOULOS, "Region of Interest Coding in JPEG2000," *Signal Processing: Image Communications (Special Issue on JPEG2000)*, Vol 17, no. 1, 2002, pp. 105–111.

Chapter 43

Using Shift Register Sequences

Charles Rader
Retired, formerly with MIT Lincoln Laboratory

$\mathbf{T}$his chapter discusses the time and frequency domain behavior of simple linear feedback shift registers to show how they can be used to generate random numbers, conveniently test high-speed digital logic, and efficiently produce useful signaling waveform sequences.

Suppose we have a clocked shift register that holds N bits. On each clock pulse, the bits are shifted to the right, the rightmost bit is lost, and a new bit enters from the left. Some of the shift-register stages are tapped and we compute the new bit as some logical function of the bits appearing at those taps, so the contents of the shift register change after each clock pulse.

We illustrate this in Figure 43–1 and Table 43–1 below, using a four-stage register containing a_1, a_2, a_3, a_4 and the logical function is the exclusive-or, $\oplus$, of the bits in the last two flip-flops.

The logic function output, hereafter called the shift register output, will be the sequence of bits . . . 100110101111000 . . . repeated endlessly. The repetition period is 15. This generated sequence has a number of interesting and useful properties and we can generalize these properties to shift registers with any number of bits N.

43.1 COMPLETE SET OF POSSIBLE STATES

Note that an N stage shift register can have exactly 2^N possible states and when the shift register contains any given state its future history is determined. Therefore any shift register sequence must repeat its state after a period that cannot be greater than 2^N. But we are going to limit our consideration to shift registers for which the logic function to create the input is made up of the exclusive-or of some of the shift-register taps. With that limitation, the all-zero state would repeat itself. So if the shift

Streamlining Digital Signal Processing: A Tricks of the Trade Guidebook, Second Edition. Edited by Richard G. Lyons.
© 2012 the Institute of Electrical and Electronics Engineers. Published 2012 by John Wiley & Sons, Inc.

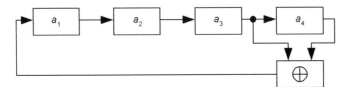

Figure 43–1   Four-stage shift register.

Table 43–1   Four-Stage Shift Register Sequences

$a_3 \oplus a_4 \rightarrow$	$a_1 \rightarrow$	$a_2 \rightarrow$	$a_3 \rightarrow$	a_4
1	0	0	0	1
0	1	0	0	0
0	0	1	0	0
1	0	0	1	0
1	1	0	0	1
0	1	1	0	0
1	0	1	1	0
0	1	0	1	1
1	0	1	0	1
1	1	0	1	0
1	1	1	0	1
1	1	1	1	0
0	1	1	1	1
0	0	1	1	1
0	0	0	1	1
1	0	0	0	1
:	:	:	:	:

register is initialized with anything other than all zeros, the longest repeat period it can have is $2^N - 1$. Our four-stage example achieves that bound.

Treat the contents of the shift register as a binary integer. The sequence of values taken by the register are 1,8,4,2,9,12,6,11,5,10,13,14,15,7,3, . . . These are all the integers from 1 to 15, in a permuted order, each occurring once before any is repeated.

To generalize for a shift register of length N we must be careful in how we choose which taps to feed into the exclusive-or function. If we choose the taps correctly the shift register's successive states will be the integers from 1 to $2^N - 1$ periodically in a permuted order, each appearing once before any is repeated. The sequence of bits entering the input of such a shift register is called a *maximal-length-shift-register-sequence* or an *m-sequence* for short.

In the general case of length N, the logical function for the new bit that we shift into stage 1 has the form $a_{N-i} \oplus a_{N-j} \oplus a_{N-k} \oplus \ldots$, where i, j, and k, $\ldots$ are in the set $[0, 1, \ldots, N-1]$. Thus we could have $N = 32$ and compute the shift register input as $a_1 \oplus a_2 \oplus a_{22} \oplus a_{32}$. It is easy to see that we must include the last tap a_N ($i = 0$) because otherwise the shift register sequence would be a delayed version of a sequence generated by a shorter shift register. It is less obvious how we should choose the other taps to produce a shift register sequence with the maximum length property. A random choice won't always work. But there is an elegant mathematical theory [1] that lets a mathematician predict which sets of taps will produce a maximum length (length $2^N - 1$) sequence. We don't even need to understand the theory because there are published tables that tell us which set of taps will work for shift register lengths N as large as we might care about. Usually those tables give only one allowable set of taps for each shift register length, although there are many sets of taps that would work equally well.

It is important, however, to know that some of the literature on shift register sequences gives the proper tap weights using a polynomial notation. For our four-stage shift register with the update rule $a_n = a_{n-3} \oplus a_{n-4}$ the tap weights would be given as the polynomial $1 + x^3 + x^4$ and in the case of the 32-stage shift register with the update rule $a_n = a_{n-1} \oplus a_{n-2} \oplus a_{n-22} \oplus a_{n-32}$ the polynomial would be $1 + x^1 + x^2 + x^{22} + x^{32}$. Hopefully the pattern is clear.

Table 43–2 provides an abbreviated list giving an allowed set of taps, in the polynomial notation, for some small values of N and for a few selected larger values.

For many values of N, it is possible to achieve the maximum length property with only two taps. For example, with $N = 36$ we can use $a_{25} \oplus a_{36}$ and produce a sequence of length $2^{36} - 1$ which is more than 10^{10}. For other values of N we need at least four taps. The published tables usually give a tap set using the smallest allowable number of taps.

Table 43–2   Maximum Length Polynomials

N	Good taps
3	$1 + x^2 + x^3$
4	$1 + x^3 + x^4$
5	$1 + x^3 + x^5$
6	$1 + x^5 + x^6$
7	$1 + x^6 + x^7$
8	$1 + x^4 + x^5 + x^6 + x^8$
9	$1 + x^5 + x^9$
10	$1 + x^7 + x^{10}$
11	$1 + x^9 + x^{11}$
12	$1 + x^1 + x^4 + x^6 + x^{12}$
32	$1 + x^1 + x^2 + x^{22} + x^{32}$
36	$1 + x^{25} + x^{36}$
250	$1 + x^{103} + x^{250}$

We can easily choose N large enough that the sequence, for all practical purposes, never repeats.

Often the bit repeats its value for several clocks in a row before it changes. If a bit keeps its value constant for m consecutive clocks we will say that it is within a run of length m. We will find in the periodic repetition of the sequence that half of the runs are one bit long, a quarter of the runs are two bits long, an eighth of the runs are three bits long, and so on. The longest is N bits long and consists of N ones.

Suppose we look at the shift register output only after every Kth clock. In other words, suppose we form a new sequence by looking at every Kth state of the original sequence. If K does not share any factors in common with $2^N - 1$ the new sequence is also maximum length. In particular, $2^N - 1$ is always odd, so looking at every other state, or every fourth state, and so on, gives us different maximum length sequences. For example, with the four-stage shift register we used as an example, here is every second bit:

$$\ldots 0\,1\,0\,0\,1\,1\,0\,1\,0\,1\,1\,1\,1\,0 \ldots$$

The original infinite length sequence has been recovered from itself, with only a time shift, by subsampling!

This is not what happens when we look at the state of the entire shift register on every Kth clock. The register still goes through all 15 states before repeating itself, but the sequence of states is a different permutation of 1 to 15.

$$\ldots 8, 2, 12, 11, 10, 14, 7, 1, 4, 9, 6, 5, 13, 15, 3, \ldots$$

Using different tap sets also gives different maximum length sequences.

We can also run the shift register backward in time, which is the same as running the shift register with a reversed set of taps. For our four-stage example we had $a[n] = a[n-3] \oplus a[n-4]$ using taps 3 and 4. If, on both sides of the equation, we exclusive-or with $a[n] \oplus a[n-4]$ we get $a[n-4] = a[n-3] \oplus a[n]$ and hence $a[n] = a[n+N-3] \oplus a[n+N]$. If we view this as the update rule for a shift register run backwards in time we see that this is equivalent to a shift register using taps 1 and 4 instead of taps 3 and 4.

43.2 USE FOR TESTING LOGIC CIRCUITS

It is a pleasant surprise to find that such complex binary sequences can be obtained with such simple logic. For $N-1$ of the N flip-flops the input is simply taken from the output of the adjacent flip-flop and for the remaining flip-flop the logic function is usually a two-input exclusive-or (but for some N it is a four-input exclusive-or).

Suppose you want to exhaustively test a combinational logic circuit with N logic inputs, appearing in all 2^N combinations. Of course you could produce those test inputs with a binary counter. But at the highest circuit speed, can the counter update itself fast enough? The maximum length shift register can update in the time required for a single $\oplus$ operation. Do not forget, however, that the all-zero state will need to be tested also—it will not be one of the shift register outputs.

A long time ago, I needed to design a tester for a computer that was to be built using the fastest available family of integrated logic. Of course, the logic circuits making up the tester had to be able to run at least as fast as the logic circuits making up the computer being tested. My tester had to be able to present the computer's processor with any arbitrary sequence of at least 15 consecutive computer instructions. These instructions were stored in 15 locations in a random access memory so I needed to generate 15 different addresses, one right after the other, to access the memory. A naive approach would have been to use a binary counter and generate the addresses in the sequence 0,1,2,3, . . . but the logic function of a binary counter would have had a propagation delay that would have limited the clock rate. Instead, I used a four-stage shift register with the taps as in the example above. The contents of the shift register after each clock were used as the address lines of the (16-bit) memory chips. My tester's worst logic propagation delay was that of an exclusive-or circuit. It made no difference that the memory cells were accessed in the wacky order 1,8,4,2,9,12,6,11,5,10,13,14,15,7,3, . . . instead of the more familiar order 1,2,3,4, . . . because I could store the instructions in the memory cells using that same order.

Today we have much larger memories, but the same trick could be used to store an arbitrary digital sequence in a very large memory and to generate addresses for that memory at a very high speed.

Many special-purpose computers today are designed with separate memories for storing programs and data. We would not want to store and address data with a shift register state but there is no reason we could not access the program memory that way. The conventional "program counter" could be replaced by a shift register that, after most instructions, is clocked to point to the next instruction. But for jump-type instructions, the address of the next instruction, stored in the current instruction, simply replaces the shift register contents.

43.3 CIRCULAR BUFFER

We often want to use a data sample delayed by P clocks. Of course, we could do this by putting the sample into a shift register with P stages, where each stage holds an entire sample. But that is not the best way to accomplish the delay because on every clock all the stages are changed, costing power and chip real estate. Instead we can use P words of memory organized as a circular buffer. On each clock we read one word of the memory that gives us the delayed sample, and then we write the newest sample into the same memory word. On the next clock the same thing is done but with a different word. We revisit each memory word only after P clocks. Most engineers would accomplish this by addressing the memory from a counter, incrementing the counter on each clock, starting with 0 and resetting the counter to 0 when it would otherwise have incremented to P.

But there is no good reason why the P memory words need to be consecutive. The trick I used for generating memory addresses in the computer tester described above would work equally well for indexing a circular buffer for $P = 15$.

Suppose we want a period P somewhere between 2^{N-1} and $2^N - 2$. We add some extra logic. Starting with register contents $0 \ldots 01$, we figure out in advance what it will become after P clocks and we detect that case and use it as a cue to reset the shift register to its starting state. Alternatively, we can decide that we want to reset the shift register when we detect $11 \ldots 1$, and pick the starting state which will become $11 \ldots 1$ after P clocks. So we can easily make a circular buffer of any length, and we can update the circular buffer faster than we can increment a binary counter.

43.4 SPECTRAL PROPERTIES

The output of the shift register also has interesting spectral properties. In what follows, we'll assume that we've chosen taps to produce the maximum length period, $M = 2^N - 1$. Not surprisingly, there are $2^{N-1} - 1$ zeros and 2^{N-1} ones in the first period of the sequence. We can hardly say that they are in a random order, but they have an interesting random-like property.

Suppose we computed the M-point DFT of one period of the sequence of bits x_n coming out of the shift register. This is like expanding the infinite length bit sequence in a Fourier series. We will find that the DC component X_0 is 2^{N-1} and all the remaining $M - 1$ DFT coefficients X_k have equal magnitudes! By the Parseval theorem we have

$$2^{2N-2} + (2^N - 2)|X_k|^2 = (2^N - 1)(2^{N-1}), \text{ and } |X_k| = \text{sqrt}(2^{N-2}).$$

So the sequence is almost *white* but its DC component is dominant.

It is more useful to treat the output of the shift register as ± 1. That is, when the shift register outputs x_n we use the value $y_n = (-1)^{x_n}$. The M-point DFT of that sequence has a DC coefficient $Y_0 = -1$ and all other coefficients Y_k have equal magnitudes of $|Y_k| = \text{sqrt}(2^N)$. Again, the sequence is almost white but now its DC component is very small.

The nearly flat Fourier series implies a nearly peaky circular autocorrelation function. The circular autocorrelation function of y_n has a peak of $2^N - 1$ for lag zero, and takes the value -1 for all the other lags.

More important for most applications is that the linear autocorrelation function for a complete period of a moderately long period shift register sequence has a fairly good (impulse-like) autocorrelation function. For example, an 11-stage shift register sequence of ± 1, length 2047, has an autocorrelation function consisting of a peak of 2047 for lag 0, and an off-peak range of -48 to 47. Two thirds of the energy of that autocorrelation function is in the zeroth lag.

A waveform with a sharp peak in its autocorrelation function is easy to extract from white noise by matched filtering, and the shift register sequences are so easy to generate that they are often used as training waveforms for synchronization between a transmitter and a receiver.

43.5 RANDOM NUMBER GENERATORS

Shift register sequences have been used as random number generators for use as noise sources in simulations of signal processing systems. But we need to be careful. The single-bit output of a shift register sequence is just that, a single bit, and hence not very good as a random signal. It is tempting to use, instead, the full contents of the shift register as a pseudorandom integer s_n in the range $1 \leq s_n \leq 2^N - 1$, generated in one clock. We have seen that s_n is, by definition, uniformly distributed, but it is not even close to uncorrelated. Consider that if its output now is s_n its succeeding output can only be either $s_n/2$ or $s_n/2 + 2^{N-1}$.

To use a shift register as a reasonably good random number generator we use a large number, L, of shift registers operating in parallel, which gives us an L-bit pseudorandom number on every clock. The L shift registers can be identical to one another as long as their contents are initialized differently so that each shift register is in a different part of its repetition period and will not reach any other shift register's state for thousands of clocks. Thus, if we set up N whole L-bit words X[n], n = 1, . . . , N we would compute (for $N = 32$)

$$X[n] = X[n-1]\langle\oplus\rangle X[n-2]\langle\oplus\rangle X[n-22]\langle\oplus\rangle X[n-32]$$

where $\langle\oplus\rangle$ means $\oplus$ for each bit position in a whole N-bit word. But we should only use this as a quick-and-dirty random number generator, and we should understand that a random initialization could, by bad luck, produce long strings of words with several of their bits following identical patterns.

It is helpful if the choice of N permits a two-tap maximum length shift register. Here's a particularly nice choice [2]:

$$X[n] = X[n-103]\langle\oplus\rangle X[n-250]$$

which gives pseudo random numbers with a period of more than 10^{75}. Since 250 words is a rather large memory, it is probably best to use a circular buffer to simulate the shift register. Then we can use an eight-stage single-bit-per-stage shift-register ($a[n] = a[n-8] \oplus a[n-6] \oplus a[n-5] \oplus a[n-4]$) that is reset to 00000001 on the clock after it reaches its 250*th* value (01011000).

If you want to use pseudorandom numbers in simulations to get rigorously correct results, the choice of a random number generator is way beyond the scope of this little note.

43.6 CONCLUSIONS

Linear feedback shift registers have two virtues. They can be implemented very easily in digital logic, and they produce a sequence of outputs with useful properties.

The shift register's N-bit content sequences through all possible states, except $0 \ldots 0$, in the least possible time. Therefore it can be used for exhaustively testing high-speed combinatorial and sequential logic. The spectral properties of the sequence of single bits make these sequences useful as signaling waveforms and for spread-spectrum applications. A collection of linear shift registers, one for each bit in a word, can be the basis of a very cheap random number generator.

Although the logical description of these shift registers is very simple, most digital computers do not simulate them efficiently. But for signal processing using either specially designed chips or field programmable gate arrays, their efficiency is very attractive.

43.7 REFERENCES

[1] S. GOLOMB, *Shift Register Sequences*, Aegean Park Press, Laguna Hills, CA, 1981, pp. 1–257.
[2] S. KIRKPATRICK and E. STOLL, "A Very Fast Shift-Register Sequence Random Number Generator," *Journal of Computational Physics*, vol. 40, 1981, pp. 517–526.

Chapter 44

Efficient Resampling
Implementations

Douglas W. Barker
ITT Corporation

The process of sample rate conversion (resampling) of a discrete sequence has many applications in digital signal processing, particularly in the field of digital communications. This chapter describes efficient implementations used to change the sample rate of a discrete signal by an arbitrary factor.

44.1 RESAMPLING USING POLYPHASE FILTERS

The fundamental process of resampling a discrete sequence by an integer factor involves either inserting a sequence of zero-valued samples between each input sample followed by lowpass filtering (interpolation), or lowpass filtering an input signal and discarding some of those filter output samples (decimation). Polyphase filters were developed to improve the computational efficiency of the lowpass filtering in interpolation by (a) eliminating the multiply-by-a-zero-sample operations, and (b) avoiding the computation of the lowpass filter output samples to be discarded in decimation [1], [2]. With these thoughts in mind, we assume the reader is familiar with the nature of polyphase resampling filters, and here we present practical and efficient polyphase filter implementations that the reader should consider for their next resampling design.

44.2 RESAMPLING BY AN INTEGER FACTOR

To reduce input signal data storage requirements, it's smart to implement resampling filtering using a single tapped-delay line filter. Our recommended design of a single delay line, multiple coefficient set interpolator is shown in Figure 44–1(a). In this

Streamlining Digital Signal Processing: A Tricks of the Trade Guidebook, Second Edition. Edited by Richard G. Lyons.
© 2012 the Institute of Electrical and Electronics Engineers. Published 2012 by John Wiley & Sons, Inc.

449

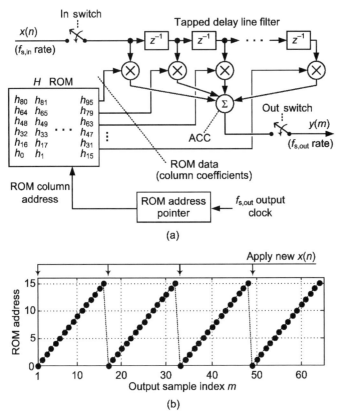

Figure 44–1 ROM-based interpolation by an integer factor $L = 16$: (a) structure; (b) ROM column addresses.

figure, for example, $L = 16$ sets of coefficients (each set having length $R = 6$) are precomputed and stored as the columns of a two-dimensional matrix, H, in read-only memory (ROM). The matrix of coefficients, H, has R rows and L columns.

We implement interpolation by integer $L = 16$, with $f_{s,in}$ and $f_{s,out} = 16 f_{s,in}$ being the input and output sample rates, respectively, as follows: the *in-switch* closes momentarily and applies a single $x(n)$ sample to the tapped-delay line filter. The *out-switch* remains closed. At the $f_{s,out}$ output data rate the ROM address pointer begins incrementing by one and scans through the 16 ROM column addresses causing the ROM to output R coefficients for each column address. For each set of R coefficients we compute an interpolated $y(m)$ output sample, and reset the tapped-delay line's ACC accumulator to zero after each computation.

The ROM column addresses, repeatedly incrementing from zero to $L - 1 = 15$, are shown in Figure 44–1(b) for four $x(n)$ input samples. We view the ROM address pointer operation as a kind of modulo-L numerically controlled oscillator (NCO).

Again, each dot in Figure 44–1(b) represents the computation of an interpolated $y(m)$ output sample. Each time the ROM address pointer is reset to zero, a time period of $1/f_{s,in}$ indicated by the down arrows in the figure, the *in-switch* closes momentarily shifting a new $x(n)$ input sample into the tapped-delay line.

Note that the *in-switch* in Figure 44–1(a) is for descriptive purposes only. In reality, upon experiencing an overflow condition, the ROM address pointer outputs what we will call an *NCO overflow* control signal that is applied to an external system causing that system to supply a new $x(n)$ input sample to our interpolator.

44.3 RESAMPLING BY AN ARBITRARY FACTOR

The interpolation and decimation processes, described earlier, can be extended to produce output sample rates that are arbitrary (noninteger) multiples of the input sample rate. Such interpolation and decimation lead to two slightly different design methods, and structures, and are therefore discussed separately.

44.4 INTERPOLATION BY AN ARBITRARY FACTOR

For the case of interpolation by an arbitrary factor, we use the network in Figure 44–1(a), but we change the ROM address pointer mechanism to the process shown in Figure 44–2(a), where the notation $\lfloor q \rfloor$ means the integer portion of q.

In Figure 44–2(a) we implement an NCO that operates as a modulo-C accumulator. In Figure 44–1(a) the interpolation factor L was equal to the number of columns in the ROM memory, but that is not the case when interpolating by an arbitrary

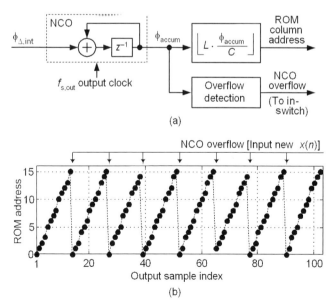

Figure 44–2   Interpolation by an arbitrary factor: (a) ROM address generation; (b) ROM addresses when $L = 16$ and $F_{int} = 12.6374$.

factor. For this discussion, we'll let L continue to represent the number of columns in the ROM memory. The value L, then, determines the desired interpolation accuracy, where larger L leads to more accurate interpolation. We now define F_{int} as the desired interpolation factor

$$F_{int} = \frac{f_{s,out}}{f_{s,in}} \qquad (44\text{--}1)$$

where frequency $f_{s,in}$ is our input signal's sample rate and $f_{s,out}$ is our desired upsampled output sample rate in Hz.

To the NCO in Figure 44–2(a) we apply a fixed phase increment value $\phi_{\Delta,int}$ defined by

$$\phi_{\Delta,int} = \frac{C}{F_{int}} = C \cdot \frac{f_{s,in}}{f_{s,out}} \qquad (44\text{--}2)$$

where the subscript "int" means interpolation. At the $f_{s,out}$ rate the value $\phi_{\Delta,int}$ is added to the NCO's accumulator producing the value ϕ_{accum}. The ratio ϕ_{accum}/C is a value between zero and one. In fixed-point implementations $\phi_{\Delta,int}$ is rounded to the nearest integer. This forces F_{int} in (44–1) to be a rational number, but we can approach any desired F_{int} value arbitrarily closely by increasing the value of C. Thus the constant C is made as large as possible in any given implementation.

The function $\lfloor L \cdot \phi_{accum}/C \rfloor$ produces the integer part of $L \cdot \phi_{accum}/C$ selecting the appropriate one out of L ROM column addresses, causing the ROM to output R coefficients. For each set of R coefficients, as before, we compute an interpolated $y(m)$ output sample and reset the tapped-delay line's ACC accumulator to zero after each computation. However, when interpolating by an arbitrary factor the ROM column address no longer increments by one. For example, when $L = 16$ and $F_{int} = 12.6374$ the ROM column addresses are those shown in Figure 44–2(b), where we see the irregular progression of the integer ROM addresses. When the modulo-C NCO accumulator overflows, the NCO overflow control line causes the *in-switch* to momentarily close and shift a new $x(n)$ input sample into the tapped-delay line.

For practicality, we let L and C be integer powers of two so that, in Figure 44–2(a), the $L \cdot \phi_{accum}/C$ computation is performed, multiplier-free, as

$$L \cdot \frac{\phi_{accum}}{C} = \frac{\phi_{accum}}{2^{\log_2(C)-\log_2(L)}}. \qquad (44\text{--}3)$$

This means we compute $\lfloor L \cdot \phi_{accum}/C \rfloor$ by merely shifting ϕ_{accum} to the right by $[\log_2(C) - \log_2(L)]$ bits. This process gives us an efficient way to extract the proper L bits of ϕ_{accum} to select the correct column of the ROM.

By shifting ϕ_{accum} to the right the fractional portion of the quotient in (44–3) is truncated causing a small error in the resulting interpolation. To keep the interpolation error less than the quantization error of the $x(n)$ input signal, represented by b-bit samples, the number of L columns in the coefficient matrix H should be chosen so that

$$L > \frac{2^{(b-1)} \cdot B}{f_{s,in}} \tag{44-4}$$

where B is the two-sided bandwidth (centered at zero Hz) of the lowpass $x(n)$ input signal measured in Hz [3].

44.5 COMPUTATION OF THE INTERPOLATION COEFFICIENTS

Each column of the H ROM in Figure 44–1(a) contains the R coefficients of a sub-filter, where each subfilter is obtained by standard polyphase decomposition of a prototype filter designed using commercial filter design software. The prototype filter is a linear phase tapped-delay line filter having $L \cdot R$ coefficients, and it is designed to eliminate all its input spectral components whose frequencies are above $f_{s,in}/2$. The prototype filter design is based entirely upon the factors L and R. This has a very practical, implication; a single filter design specification (the passband width is always $f_{s,in}/2$ Hz) can be used to design any filter needed to interpolate any signal of lesser sample rate than $f_{s,out}$ to an $f_{s,out}$ rate with at least $1/(L \cdot f_{s,in})$ accuracy in interpolation timing.

44.6 DECIMATION BY AN ARBITRARY FACTOR

The complementary operation to interpolation is decimation, whose ROM address generation is shown in Figure 44–3.

Our resampling network effectively interpolates $x(n)$ by L' and decimates by M, where $M > L'$, and L' is defined as

$$L' = \frac{(M \cdot f_{s,out})}{f_{s,in}}. \tag{44-5}$$

Decimation is implemented as follows: the fixed integer phase increment value $\phi_{\Delta,dec}$, given by

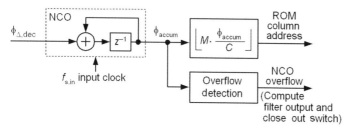

Figure 44–3 Decimation ROM address generation.

$$\phi_{\Delta,\text{dec}} = \left\lfloor C \cdot \frac{f_{s,\text{out}}}{f_{s,\text{in}}} \right\rfloor \qquad (44\text{-}6)$$

in Figure 44–3(a), is added to the NCO's accumulator at the $f_{s,\text{in}}$ rate and, again, the ratio ϕ_{accum}/C is a value between zero and one. As before, the NCO operates as a modulo-C accumulator and L represents the number of columns in the ROM memory. For decimation the integer value M determines the resampling accuracy, and M is chosen to be an integer power of 2 (for the same reason that we chose L to be an integer power of two in the above interpolation case). As such, M is determined using expression (44–4) with the L replaced by M.

The value L' in (44–5), then, is determined by M and our desired overall decimation resampling rate, and we use this value L' to compute the coefficients of the H ROM in Figure 44–1(a).

At the $f_{s,\text{in}}$ rate, we apply a new $x(n)$ input sample to the tapped-delay line and increment the NCO by $\phi_{\Delta,\text{dec}}$. When the NCO overflows we compute the function $\lfloor M \cdot \phi_{\text{accum}}/C \rfloor$ defining the appropriate one out of L ROM column addresses, causing the ROM to output R coefficients. (The integer value $\lfloor M \cdot \phi_{\text{accum}}/C \rfloor$ is computed, multiplier-free, as before in (44–3).) The R coefficients are then used to compute a $y(m)$ output sample, the out-switch in Figure 44–1(a) is momentarily closed, and the tapped-delay line's ACC accumulator is reset to zero. With this operation in mind, we see that the NCO is driven with a phase increment that causes an overflow at the desired $f_{s,\text{out}}$ output sample rate.

Once the value of M is chosen based on the interpolation accuracy requirements for the output signal, then we compute the value L' using (44–5). Next we compute the number of columns in the H coefficient ROM matrix, L, using

$$L = \lceil L' \rceil \qquad (44\text{-}7)$$

where $\lceil L' \rceil$ means take the next integer larger than L', if L' is not an integer.

44.7 COMPUTATION OF THE DECIMATION COEFFICIENTS

Calculating the filter coefficients for the decimation case is somewhat tricky because the rows and columns of the H coefficient matrix are not spaced regularly in time. We cannot simply use a commercial filter design routine because such software only produces prototype filter impulse response samples at regularly spaced time points. The method we chose to compute the coefficients is to evaluate the continuous-time impulse response of the desired prototype FIR filter at the time points of H using a time-point matrix T. Matrix T has the same rows and columns of H and is computed first. The time-points in T are specified in units based on the interpolated sample rate of $L' \cdot f_{s,\text{in}} = M \cdot f_{s,\text{out}}$. The elements of T are then given by

$$T(r, c) = rL' + c \qquad (44\text{-}8)$$

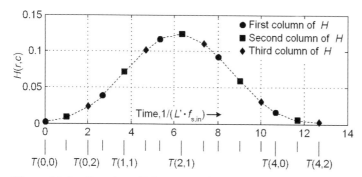

Figure 44–4 H matrix coefficients when $L' = 2.666$, $L = 3$, and $R = 5$.

where row index r ranges from zero to $R - 1$, and column index c ranges from zero to $L - 1$. The filter coefficients are calculated by evaluating the prototype filter's impulse response at the time-points in T. Notice that the spacing between the columns (phases) is equal to unity, while the spacing between the rows is L', which is equal to the time-spacing of the input data.

For example, if $L' = 2.666$, $L = 3$, and $R = 5$, the T matrix would be

$$T(r, c) = \begin{matrix} 0 & 1.000 & 2.000 \\ 2.666 & 3.666 & 4.666 \\ 5.333 & 6.333 & 7.333 \\ 8.000 & 9.000 & 10.000 \\ 10.666 & 11.666 & 12.666. \end{matrix}$$

If the prototype filter's impulse response is the dashed curve in Figure 44–4, the H decimation coefficients in this example are the dots. The tick marks at the bottom of Figure 44–4 show the values of time matrix T. The irregular spacing of the coefficients, in time, is due to the fractional nature of L'.

It is important to notice that the value of the NCO that occurs after rollover, $M \cdot \phi_{accum}/C$, represents the interval of time between the current state of the NCO and the point in time that must be interpolated. The larger the rollover value, the further back in time we must interpolate. For this reason, the columns of H must be reversed so that larger NCO overflow values produce interpolations that are further back in time.

44.8 ALTERNATE METHOD FOR COMPUTING DECIMATION COEFFICIENTS

Another method for calculating the filter coefficients for decimation involves oversampling the impulse response of the prototype filter by a factor of P, then selecting the coefficients closest to the time points of T in (44–8). This allows the use of any canned filter design program. Oversampling the impulse response involves increasing the filter sample rate by the factor P, while holding the cut-off frequency and all other specifications the same, while multiplying the filter order by P.

44.9   FPGA IMPLEMENTATION ISSUES

The structure of Figure 44–1(a) is easily implemented in hardware using a field-programmable gate array (FPGA). The FPGA clock rate can be fixed at the $f_{s,in}$ clock rate in the case of the downsampler, or the $f_{s,out}$ clock rate in the case of the upsampler. No flip-flops in the design are required to operate on any other clock; this leads to a very efficient and elegant solution. Both resampling operations can take advantage of the hardware multipliers now available in FPGA's and RAM blocks can be used for coefficient storage. The number of hardware multipliers available impose a practical limitation on R, while the number of RAM blocks impose a practical limitation on L. For the case of the upsampler, ROM blocks may be used to store the coefficients even if the input sample rate is variable, whereas for the downsampler, RAM blocks must be used if one desires to change the output sample rate. In this case, an external microprocessor must calculate a new H every time the nominal output sample rate is to be changed.

Another peculiarity of the implementation is that for interpolation, the complexity of the prototype FIR filter does not change when the ratio of input sample rate to output sample rate changes. This is a consequence of the fixed interpolation factor. On the contrary, for decimation, larger sample rate ratios lead to higher filter complexity, that is, one must increase R proportional to the sample rate ratio in order to achieve the same filter specifications. A simple implementation trick can be used here to alleviate this problem to a certain degree. It is noted that when the sample rate ratio, $f_{s,out}/f_{s,in}$, is less than 0.5, we can reuse all R multipliers again for each computed output sample. This allows us to double R for ratios less than 0.5, while not increasing the number of hardware multipliers at all. It should be noted that we must double the number of coefficients in H since R has doubled, but we should also note that L' drops in half so the net affect is that there is no change in coefficient storage requirements. Similarly, when the ratio is less than 0.25, we may quadruple R as we can at every octave change in sample rate ratio. There is, however, one side effect, which is the length of the tapped-delay line in Figure 44–1(a); this will double, quadruple, and so on, and hence poses practical limitation on the sample rate ratio.

44.10   CONCLUSIONS

This material has presented methods for efficiently resampling a discrete-time signal. We described techniques to compute the coefficients for both interpolation and decimation filters, and provided information with regard to minimizing the resamplers' timing jitter errors.

MATLAB code illustrating the operation of our efficient resampling techniques is made available at http://booksupport.wiley.com.

44.11 REFERENCES

[1] A. OPPENHEIM, R. SCHAFER, and J. BUCK, *Discrete-Time Signal Processing*, 2nd ed. Prentice-Hall, Upper Saddle River, NJ, 1989, pp. 179–184.

[2] R. CROCHIERE and L.R. RABINER, *Multirate Digital Signal Processing*, Prentice-Hall, Englewood Cliffs, NJ, 1983, Chapters 2–3.

[3] F. HARRIS, *Multirate Signal Processing for Communications Systems*, Prentice Hall, Upper Saddle River, NJ, 2004, pp. 172–175.

EDITOR COMMENT

A frequency-domain implementation of this resampling process is discussed in Chapter 45.

Chapter 45

Sampling Rate Conversion in the Frequency Domain

Guoan Bi
Nanyang Technological University

Sanjit K. Mitra
University of Southern California

Sampling rate conversion (SRC) is usually performed in the time domain by using the operations of up-sampling, filtering, and down-sampling. However, it is also possible to perform the SRC in the frequency domain by formulating the desired spectrum from the spectrum of an input signal. This chapter shows how to perform SRC for both integer- and fractional-rate conversion by manipulating the discrete Fourier transform (DFT), implemented using the fast Fourier transform (FFT), of a time-domain signal. The analysis on error performance and the required computational complexities shows that by using the FFT, for both short- and long-input sequences, improvements in conversion accuracy are achieved at reduced computational costs.

45.1 SRC BASICS

The basic time-domain operations used to perform SRC are up-sampling by an integer-valued interpolation factor I, lowpass filtering, and down-sampling by an integer-valued decimation factor D, as shown in Figure 45–1(a) [1], [2]. The corresponding effects of SRC in the frequency domain are also well explained in the literature. However, although some attempts have been made in the past, SRC based on frequency-domain processing has not received sufficient attention. Here we remedy that situation.

A DFT-based method for interpolation was first reported in [3]. Later it was shown in [4] that decimation was also achieved in the DFT domain. Interpolation

Streamlining Digital Signal Processing: A Tricks of the Trade Guidebook, Second Edition. Edited by Richard G. Lyons.
© 2012 the Institute of Electrical and Electronics Engineers. Published 2012 by John Wiley & Sons, Inc.

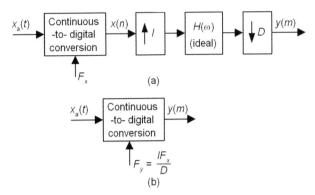

Figure 45–1   Time domain SRC: (a) computing $y(m)$ from $x(n)$; (b) generating $y(m)$ directly.

by an integer factor was further investigated with a focus on the issues of computational complexity and the method of formulating the spectrum of the interpolated signals [5], [6]. Finally, an experimental study on the error performance of interpolation was reported in [7]. These reported works did not consider SRC for fractional-rate conversion, nor did they consider how to accommodate long input signal sequences.

45.2 REQUIREMENTS IN THE FREQUENCY DOMAIN

Let $x_a(t)$ denote an analog signal and $x(n)$ the corresponding discrete-time sequence obtained with a sampling frequency F_x as shown in Figure 45–1(a). The objective of the SRC is to convert the sequence $x(n)$ into another sequence $y(m)$ that is ideally the one obtained by sampling $x_a(t)$ with another sampling frequency F_y, where $F_y/F_x = I/D$, as shown in Figure 45–1(b). Factors D and I are the decimation and interpolation factors, respectively. For this discussion we assume the lowpass filter $H(\omega)$ in Figure 45–1(a) is ideal in its operation and that the two $y(m)$ sequences in Figure 45–1 are identical.

The magnitude response of the digital filter is specified by

$$|H(\omega_v)| = \begin{cases} I, & |\omega_v| \leq \min\left(\dfrac{\pi}{I}, \dfrac{\pi}{D}\right), \\ 0, & \text{otherwise,} \end{cases} \qquad (45\text{–}1)$$

where ω_v is the post-interpolation sample rate $\omega_v = 2\pi f/IF_x$ radians/sample. Various design techniques to reduce the computational complexity for the filtering operation have been investigated to achieve the desired performance [1], [2]. The effects of the filter operation in the frequency domain have also been well explained.

The spectrum of the output signal $y(m)$ after the SRC is expressed as

$$Y(\omega_y) = \begin{cases} \dfrac{I}{D} \cdot X\left(\dfrac{\omega_y}{D}\right), & 0 \le \omega_y \le \min\left(\pi, \dfrac{D\pi}{I}\right), \\ 0, & \text{otherwise}, \end{cases} \qquad (45\text{--}2)$$

where $\omega_y = D\omega_x/I$ radians/sample, and $X(\omega_x)$ is the spectrum of the input signal $x(n)$. Equation (45–2) shows that $Y(\omega_y)$, the spectrum of the desired $y(m)$, can be obtained by scaling the magnitude of the spectrum $X(\omega_x)$ and the frequency in the horizontal direction with constant factors. Because $x(n)$ is assumed to be real, the spectrum of $X(\omega_x)$ in the frequency range from π and 2π is the conjugate mirror image of the spectrum in the range from 0 and π.

45.3 SAMPLE RATE DECREASE ($D > I$)

Let us define $X(k)$ to be the N-point DFT of the sequence $x(n)$ and $Y(k)$ to be the N_1-point DFT of the sequence $y(m)$, where $x(n)$ and $y(m)$ are obtained by sampling $x_a(t)$ at sampling frequencies F_x and F_y, respectively. For our SRC, $\hat{Y}(k)$, which is the N_1-point DFT of the output sequence $\hat{y}(m)$ where $N_1 = IN/D$, can be formulated by manipulating $X(k)$. Based on (45–2) $\hat{Y}(k)$, for $D > I$, is obtained by

$$\hat{Y}(k) = \begin{cases} \dfrac{I}{D} X(k), & 0 \le k < \dfrac{N_1}{2}, \\ X\left(\dfrac{N}{2}\right), & k = \dfrac{N_1}{2}, \\ \dfrac{I}{D} X(k + N - N_1), & \dfrac{N_1}{2} < k < N_1. \end{cases} \qquad (45\text{--}3)$$

According to (45–3), the values of $X(k)$, where $(N_1/2) \le k \le (N/2) - 1$ and $(N/2) + 1 \le k \le N - (N_1/2)$ are ignored. The other values of $X(k)$ are rearranged as shown in Figure 45–2 (a) and (b) to obtain the $\hat{Y}(k)$ spectrum. The desired $\hat{y}(m)$ time-domain sequence, ideally equal to $y(m)$, is then obtained by performing an inverse-DFT on $\hat{Y}(k)$.

Figure 45–3 shows the computed time-domain signal $\hat{y}(m)$ and its deviation from $y(m)$ whose DFT is given in Figure 45–2(c) when $D = 4$ and $I = 3$. We see in Figure 45–3 that the time domain, $\hat{y}(m)$ has small differences from the desired signal $y(m)$. However, these errors become larger towards both ends of the signal segments.

45.4 SAMPLE RATE INCREASE ($D < I$)

When $D < I$, the spectrum $X(k)$ is again used to form $\hat{Y}(k)$ by inserting selected numerical values into $\hat{Y}(k)$'s frequency region between $D\pi/I$ and $2\pi - D\pi/I$. In terms of the DFT $X(k)$, $\hat{Y}(k)$, the DFT of output signal $\hat{y}(m)$, is expressed by

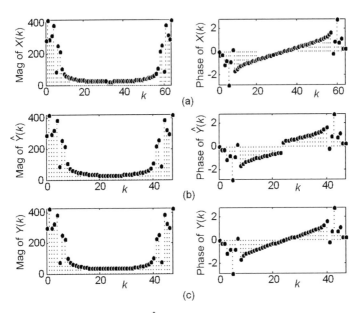

Figure 45–2   Formulating the $\hat{Y}(k)$ spectrum of the $\hat{y}(m)$ output signal when $D = 4$, $I = 3$ and $N = 64$.

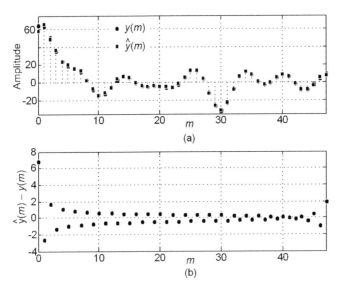

Figure 45–3   Figure 45–2 SRC errors: (a) $\hat{y}(m)$ and $y(m)$ sequences; (b) error sequence $\hat{y}(m) - y(m)$.

$$\hat{Y}(k) = \begin{cases} \dfrac{I}{D}X(k), & 0 \le k < \dfrac{N}{2}, \\[2ex] C_I, & \dfrac{N}{2} \le k \le N_1 - \dfrac{N}{2}, \\[2ex] \dfrac{I}{D}X(k - N_1 + N), & N_1 - \dfrac{N}{2} < k < N_1. \end{cases} \qquad (45\text{-}4)$$

In (45–4), the value of C_I should be carefully selected to minimize errors that cause $\hat{y}(m)$ to be different from the desired $y(m)$. Figure 45–4 shows the generation of $\hat{Y}(k)$ from $X(k)$ where $C_I = X(N/2)$. The value of C_I has a direct impact on the deviation of $\hat{Y}(k)$ from $Y(k)$. In general, we seek suitable values of C_I that result in as small errors as possible. To minimize any additional computation associated with C_I, we generally assume C_I to be a constant. For example, $C_I = 0$ is used in [5]. To minimize the errors due to the insertion of zeros, $\hat{Y}(N_1/2)$ was set equal to $X(N/2)$ in [6]. Other C_I values have been studied and the impact of their errors has been investigated by simulation [7].

For both the cases illustrated in Figures 45–2 and 45–4, the final output sequence $\hat{y}(m)$ is obtained from the inverse DFT of $\hat{Y}(k)$. For comparison purposes, Figures 45–2(c) and 45–4(c) show the ideal $Y(k)$ spectra, obtained from a Figure 45–1(a) sampling scenario, which generally are different from the formulated $\hat{Y}(k)$ spectra. Therefore, the computed $\hat{y}(m)$ has errors compared with the true $y(m)$ sequence in Figure 45–1(a).

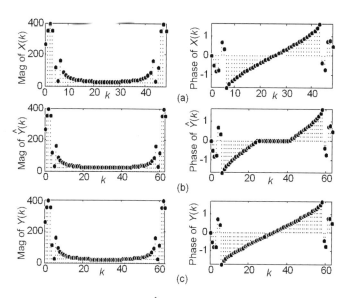

Figure 45–4   Formulating the $\hat{Y}(k)$ spectrum of the $\hat{y}(m)$ output signal when $D = 3$, $I = 4$, and $N = 48$.

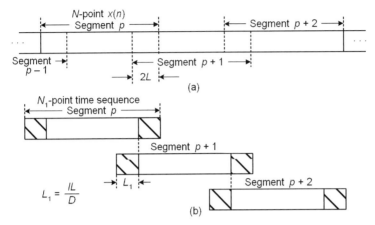

Figure 45–5   Processing long sequences: (a) $x(n)$ input segmentation and overlapping; (b) $\hat{y}(m)$ output overlap and cascading.

45.5 OVERLAP APPROACH FOR LONG SEQUENCES

The method described above is only suitable for short input sequences due to the limitation of the maximum DFT length for practical applications. For long input sequences, a widely used approach is to divide the long input sequence into many shorter segments that are processed individually. The outputs from these processed segments are then combined to form the required output. Such an arrangement is widely used in short-time Fourier transforms and other processing techniques.

If each $x(n)$ input time-domain segment has N points, Figure 45–5(a) shows that the adjacent input segments are overlapped by $2L$ points, and from each segment an N-point $X(k)$ DFT is computed. The previously described methods are then used to obtain an N_1-point $\hat{Y}(k)$ DFT segment for each $X(k)$ segment, where $N_1/N = I/D$. The value of L should be sufficiently large to effectively minimize the errors that are distributed near both ends of each output time-domain segment.

After performing an N_1-point inverse DFT on each $\hat{Y}(k)$ segment, the corresponding N_1-point time-domain segments are obtained. Finally, the cascading operation with L_1-point overlapping, where $L_1/L = I/D$ for each pair of adjacent time segments, is performed to recover the final $\hat{y}(m)$ time-domain sequence with the desired F_y sampling frequency, as shown in Figure 45–5(b), where the shaded time samples are discarded.

45.6 MEAN SQUARED ERRORS

In general, it is desirable to increase the length of the $2L$ overlap, and the length N of the DFT, so that the mean squared error (MSE) of $\hat{y}(m)$ can be effectively reduced. Let us now consider an example for the error performance affected by these factors. Figure 45–6 presents the MSEs, defined by $20\log\{[y(m) - \hat{y}(m)]^2 / N_1\}$, where N_1 is

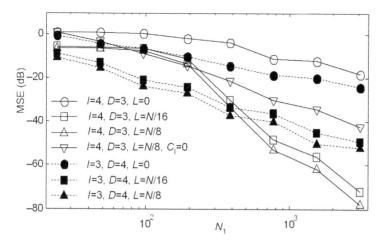

Figure 45–6   SRC errors versus overlapped time segment length, N_1.

the length of an overlapped output time segment. This experiment shows that increasing the length of overlap definitely helps to minimize the errors. However, the effectiveness of error minimization is decreased when the length of overlap is too large because the errors located far away from the ends of each segment are relatively smaller than those located near the ends. We found that for $I > D$, the reduction of the MSEs is effective only when the length of the $X(k)$ DFT is $N_1 > 192$. In addition, a better selection of the C_I value has substantial influence on error minimization. For example the errors obtained using $C_I = 0$ is much larger than those measured by using $C_I - X(N/2)$.

45.7   COMPUTATIONAL COMPLEXITY

Because factors I and D can be any integer value, we require flexible FFT algorithms that accommodate various input sequence lengths. For example, fast $q \cdot 2^n$-length FFT algorithms, where q is a small integer, were reported in [8], [9]. By using computationally efficient FFT algorithms for real input sequences, we have estimated the required computational complexity in terms of the number of real additions and real multiplications per data point.

Figure 45–7 shows the computational complexity for two SRC cases with zero overlap. The analysis shows that the required computational complexity increases with the selected FFT size. This means that to achieve better conversion accuracy, FFT sizes should be increased, which accordingly requires more computation. When the length of overlap is increased, the computational complexity is proportionally increased by a factor of $N/(N - 2L)$.

We compared the proposed SRC method with one using a polyphase FIR filter in the time domain. In this comparison, $I = 4$, $D = 3$, and the MSE $= -36$ dB. The $x(n)$ input signal had seven sinusoidal terms whose frequencies ranged from 200 to

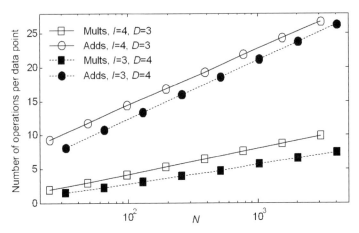

Figure 45–7   Number of frequency-domain SRC computations per data point versus $X(k)$ FFT size.

1930 Hz and an input sampling frequency of $F_x = 8000$ Hz. To achieve the desired level of MSE, the FIR filter has a transition bandwidth of 100 Hz (from 1950 and 2050 Hz), passband ripple = 0.1 dB, and a stop-band ripple = 60 dB. It was estimated that the FIR filter required at least 35 multiplications per data sample. For the proposed frequency-domain SRC method, we chose $N = 768$ and $L = N/16$ to achieve a MSE level of –45 dB. Our proposed approach required only eight multiplications per data sample.

45.8   CONCLUSIONS

This material describes a frequency-domain method for sample rate conversion (SRC). We showed that computationally simple SRC can be implemented by combining an efficient FFT procedure with a trivial deletion (sample rate decrease as in (45–3)), or insertion (sample rate increase as in (45–4)), of selected high-frequency spectral sample values. Using efficient FFTs, the computational complexity is substantially reduced compared with conventional time-domain SRC methods. Furthermore, with the overlapping technique the conversion accuracy can be also increased for long sequences.

45.9   REFERENCES

[1] P. VAIDYANATHAN, *Multirate Systems and Filter Banks*, Prentice Hall, Englewood Cliffs, NJ, 1993.

[2] S.K. MITRA, *Digital Signal Processing: A Computer-Based Approach*, 4th ed. McGraw Hill, New York, 2011.

[3] B. GOLD and C. RADER, *Digital Processing of Signals*, McGraw Hill, New York, 1969.

[4] M. YEH, J. MELSA, and D. COHN, "A Direct FFT Scheme for Interpolation, Decimation and Amplitude Modulation," *Proc. 16th Asilomar Conf. Circuits, Syst., Computers*, 1982, pp. 437–441.

[5] K. PRASAD and P. SATYANARAYANA, "Fast Interpolation Algorithm Using FFT," *Electr. Letters*, vol. 22, no. 4, February 1986, pp. 185–187.

[6] J. ADAMS, "A Subsequence Approach to Interpolation Using the FFT," *IEEE Trans. on Circuits and Systems*, vol. CAS-34, no. 5, May 1987, pp. 568–570.

[7] D. FRASER, "Interpolation by the FFT Revisited—An Experimental Investigation," *IEEE Trans. on Acoustics, Speech and Signal Proc.*, vol. 37, no. 5, May 1989, pp. 665–675.

[8] G. BI, Y. CHEN, and Y. ZHENG, "Fast Algorithms for Generalized Discrete Hartley Transform of Composite Sequence Lengths," *IEEE Trans. on Circuits and Systems II*, vol. 47, no. 9, September 2000, pp. 893–901.

[9] G. BI and Y. CHEN, "Fast DFT Algorithms for Length $N = q*2^m$," *IEEE Trans. on Circuits and Systems II*, vol. 45, no. 6, June 1998, pp. 685–690.

Chapter 46

Enhanced-Convergence Normalized LMS Algorithm

Maurice Givens
Gas Technology Institute

$\mathbf{L}$east mean square (LMS) algorithms have found great utility in many adaptive filtering applications. However, this chapter shows how the traditional constraints placed on the update gain of normalized LMS algorithms are overly restrictive. Here we present relaxed update gain constraints that significantly improve normalized LMS algorithm convergence speed.

46.1 BACKGROUND

The scalar form of the LMS algorithm is well known as

$$\mathbf{h}(k+1) = \mathbf{h}(k) + \mu e(k)\mathbf{x}(k) \qquad (46\text{–}1)$$

with μ = scalar update gain, $e(k)$ the error at time = k, and $\mathbf{x}(k) = [x(k), x(k-1), \ldots, x(k-N+1)]^{\mathrm{T}}$ is a N-dimensional column vector at time = k. The scalar value for μ results in a constant, uniform update gain. The update gain, μ, can be replaced with the matrix

$$\mathbf{M} = \begin{bmatrix} \mu & & 0 \\ & \ddots & \\ 0 & & \mu \end{bmatrix} \qquad (46\text{–}2)$$

resulting in an LMS algorithm defined as

$$\mathbf{h}(k+1) = \mathbf{h}(k) + \mathbf{M}e(k)\mathbf{x}(k). \qquad (46\text{–}3)$$

The update gain is constant and uniform across all coefficients of the vector $\mathbf{h}(k)$.

Streamlining Digital Signal Processing: A Tricks of the Trade Guidebook, Second Edition. Edited by Richard G. Lyons.
© 2012 the Institute of Electrical and Electronics Engineers. Published 2012 by John Wiley & Sons, Inc.

An example of a variable step size with an uniform update gain is the normalized LMS algorithm (NLMS) defined as

$$\mathbf{h}(k+1) = \mathbf{h}(k) + \frac{\mu e(k)\mathbf{x}(k)}{\mathbf{x}^{\mathrm{T}}(k)\mathbf{x}(k)} \tag{46–4}$$

where the effective update gain is $\mu/\mathbf{x}^{\mathrm{T}}(k)\mathbf{x}(k)$, and varies with σ_x^2, the variance of the input. Although the update gain is variable, it is still uniform across all coefficients of $\mathbf{h}$.

In some research [1]–[3], the update gain matrix $\mathbf{M},$ has been replaced by

$$\mathbf{M} = \begin{bmatrix} \mu_1 & & \mathbf{O} \\ & \ddots & \\ \mathbf{O} & & \mu_N \end{bmatrix} \tag{46–5}$$

with each μ_m ($1 \le m \le N$) not necessarily equal, to form a nonuniform gain matrix. Some performance characteristics for this form have been published [1]. The upper bound for μ_m of $\mathbf{M}$ has been stated, generally, as $\mu_m \le 2/\lambda_{\mathrm{max}}$ [4], or $\mu_m \le 2/3 \cdot \mathrm{tr}[\mathbf{R}]$ [5], [6]. While these are upper bounds, they are not, in general, the largest upper bounds that maintain stability of the algorithm.

Improvement in convergence speed for the NLMS algorithm has been sought by numerous researchers. Each has been limited, however, to the criterion that the maximum limit for the update gain is either $\mu_m \le 2/\lambda_{\mathrm{max}}$ or $\mu_m \le 2/3 \cdot \mathrm{tr}[\mathbf{R}]$. This chapter uses an update gain greater than the accepted maximum by limiting the sum of μ_m of matrix $\mathbf{M}$ to the number of coefficients in the adaptive filter. We express our limitation as

$$\sum_{m=1}^{N} \mu_m \le N. \tag{46–6}$$

That is, if the adaptive filter has $N = 256$ coefficients, then the sum of μ_m is restricted to being no greater than 256. Our restriction in (46–6) does two things:

- Increases the update gain, subject to this limitation, during convergence for those coefficients that are coincident with the nonzero (or relatively large) portion of the amplitude of the unknown impulse response, which we call active coefficients.

- Makes small, or zero, those coefficients of the adaptive filter that are inactive.

46.2 EXAMPLES

Let's use an example to illustrate our improved update gain constraint. Using the classic system identification configuration of Figure 46–1, the unknown system is

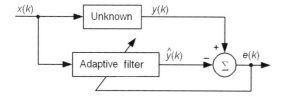

Figure 46–1 Classic system identification adaptive filter configuration.

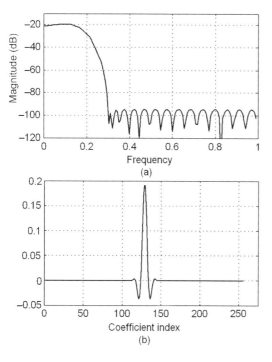

Figure 46–2 Unknown system: (a) power spectral density; (b) impulse response.

modeled as a lowpass function whose power spectral density and impulse response are shown in Figure 46–2. The purpose of the system is to adjust the adaptive filter's coefficients to reduce the $e(k)$ error signal to zero, in which case the adaptive filter's **h** coefficients identify (match) the impulse response of the unknown system.

Three versions of Figure 46–1's adaptive filter, having $N = 256$ coefficients, were implemented. One adaptive filter is allowed to converge using a constant, uniform update gain of 1, and the other two converged using nonuniform update gains. The nonuniform update gains were a constant value of 7 for the active coefficients, and zero for the inactive coefficients for the second adaptive filter. The values used for the third adaptive filter were 1 for the active coefficients, and zero for the inactive coefficients as shown in Figure 46–3. The comparative convergence of the three adaptive filters is shown in Figure 46–4.

For the second filter's nonuniform update gain of 7, very much higher than the traditionally accepted update gain upper bound, the algorithm remained stable and

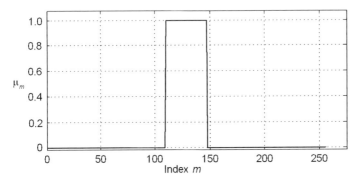

Figure 46–3   Nonuniform update gain = 1 matrix elements.

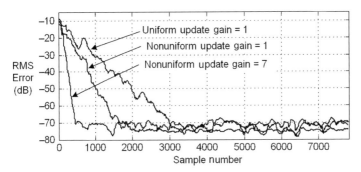

Figure 46–4   Performance comparison of filter update gain configurations.

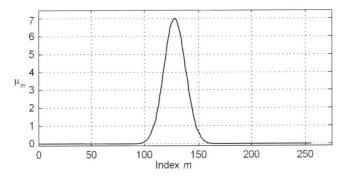

Figure 46–5   Gaussian nonuniform update gain matrix elements.

the convergence speed is very much improved. Improved convergence performance can be realized if the update gain of the adaptive filter is increased in accordance with our specified limit in (46–6), instead of the traditional limits.

An additional example of nonuniform update gain matrix elements, μ_1 to μ_N, is shown in Figure 46–5. Those update gain elements are the result of the Gaussian equation

$$\mu_m = \frac{1}{\sigma\sqrt{2\pi}} \varepsilon^{\frac{-(m-\bar{m})^2}{2\sigma^2}} \tag{46-7}$$

with $\bar{m}$ set such that it is coincident with the middle of the nonzero (active) coefficients of the unknown impulse response in Figure 46–2(b). In (46–7), $\sigma = 10$ and, again, index m ranges from 1 to N, the number of coefficients in the adaptive filter.

The long-term performance of the Gaussian update gain configuration indicates that it should be used for initial convergence with reversion to a more traditional gain configuration after initial convergence.

46.3 REFERENCES

[1] R. HARRIS, D. CHABRIES, and P. BISHOP, "A Variable Step (VS) Adaptive Filter Algorithm," *IEEE Trans. Acoust., Speech, Signal Processing*, vol. ASSP-34, April 1986, pp. 309–316.

[2] J. EVANS, P. XUE, and B. LIU, "Analysis and Implementation of Variable Step Size Adaptive Algorithms," *IEEE Trans. Signal Processing*, vol. 41, August 1993, pp. 2517– 2535.

[3] T. HAWEEL and P. CLARKSON, "A Class of Order Statistic LMS Algorithm," *IEEE Trans. Signal Processing*, vol. 40, January 1992, pp. 44–53.

[4] T. ABOULNASR and K. MAYYAS, "A Robust Variable Step-Size LMS-Type Algorithm: Analysis and Simulations," *IEEE Trans. Signal Processing*, vol. 45, March 1997, pp. 631–639.

[5] D. PAZAITIS and A. CONSTANTINIDES, "A Novel Kurtosis Driven Variable Step-Size Adaptive Algorithm," *IEEE Trans. Signal Processing*, vol. 47, March 1999, pp. 864–872.

[6] R. KWONG and E. JOHNSTON, "A Variable Step Size LMS Algorithm," *IEEE Trans. Signal Processing*, vol. 40, July 1992, pp. 1633–1642.

Index

Streamlining Digital Signal Processing: A Tricks of the Trade Guidebook, Second Edition. Edited by
Richard G. Lyons.
© 2012 the Institute of Electrical and Electronics Engineers. Published 2012 by John Wiley & Sons, Inc.